Springer Textbooks in Earth Sciences, Geography and Environment

The Springer Textbooks series publishes a broad portfolio of textbooks on Earth Sciences, Geography and Environmental Science. Springer textbooks provide comprehensive introductions as well as in-depth knowledge for advanced studies. A clear, reader-friendly layout and features such as end-of-chapter summaries, work examples, exercises, and glossaries help the reader to access the subject. Springer textbooks are essential for students, researchers and applied scientists.

More information about this series at http://www.springer.com/series/15201

Emin Özsoy

Geophysical Fluid Dynamics II

Stratified / Rotating Fluid Dynamics of the Atmosphere—Ocean

 Springer

Emin Özsoy
Eurasia Institute of Earth Sciences
Istanbul Technical University
Istanbul, Turkey

ISSN 2510-1307   ISSN 2510-1315   (electronic)
Springer Textbooks in Earth Sciences, Geography and Environment
ISBN 978-3-030-74936-1   ISBN 978-3-030-74934-7   (eBook)
https://doi.org/10.1007/978-3-030-74934-7

This Springer imprint is published by the registered company Springer Nature Switzerland AG
The registered company address is: Gewerbestrasse 11, 6330 Cham, Switzerland

... to a global community of scientists and students of nature, continuing to develop and share visions in this age of great uncertainties* ...

(* climate change/pandemia)

This ASTER thermal infrared image of June 16, 2000 inspires the book cover, but also reveals common fates and climates of Black and Mediterranean Seas, connected by the Bosphorus Strait. Besides being a natural channel connecting the two seas, the Bosphorus exhibits a unique hydrodynamic regime of strongly stratified exchange flow with the greatest contrasts in seawater properties between any sea basins of comparable size on the entire earth. The Bosphorus Jet shown in dark blue is the main driver of circulation and mixing in the Marmara Sea and a critical element in this environment. Multi-scale hydrodynamical, geophysical and environmental processes determine the ocean climates through the gradient zone across the Turkish Straits System extending from the Black Sea, Bosphorus Strait, Marmara Sea, Dardanelles Strait, Aegean Sea and finally the rest of the ocean. The adjoining isthmus covers rich nature reserves of forests, farmlands and freshwater sources scorched by the megalopolis sprawling over two continents. Mega-projects (either more recent than the image date or planned but hopefully not to be realized) are destined to be ravenous disasters of climate in this rich heartland of nature. The forests are displayed in red color, city in grey and waters in shades of blue in this multi-spectral thermal infrared and visible image.(Credits and further description: https://www.jpl.nasa.gov/images/istanbul-turkey)

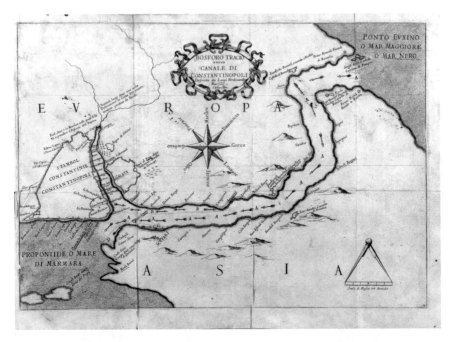

A map of Marsili (1681) displaying geography and surface currents in the Bosphorus Strait (referred to as *Bosforo Tracio, Canale di Constantinopoli, İstanbul Boğazi* in the maps). The surface currents from the Black Sea were noted to be balanced by an undercurrent in reverse direction transporting Mediterranean water with higher density. Also noted are the reversals of current near the European coast

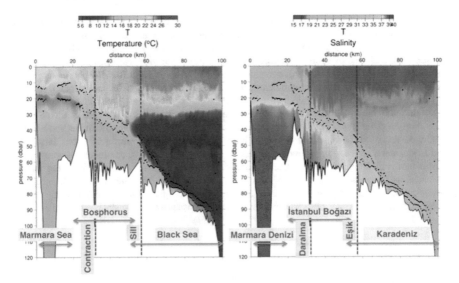

Temperature (left) and salinity (right) distributions varying from the Marmara Sea to the Black Sea, with rapid changes in properties both in horizontal and vertical directions, mixing across a sharp interface between layers (Özsoy *et al.*, 2001)

Preface

The present textbook is the second one of two volumes, aiming basic review of Geophysical Fluid Dynamics (GFD) for the student and researcher interested in basic, step-wise development from first principles to simple models that mimic complex behavior of ocean and atmosphere.

Fluid dynamics in general and GFD in particular aim to study properties and complex dynamical behavior of fluids, *i.e.* liquids, gases, plasma and mixtures based on the *continuum hypothesis*. GFD problems extend from the description of motion in earthly fluids of atmosphere and ocean in the present book to magneto-hydrodynamics and relativistic dynamics theories of matter spanning the universe.

GFD of the earth's atmosphere and ocean are described by essential physics and thermodynamics, including transport and diffusion of interacting chemical and biological species within. The fundamental laws of GFD are generally expressed by partial differential equations, extending to nonlinear, complex processes. By its very nature, GFD relies on analytical and numerical mathematics supported by observations and experiments, aiming to understand and predict processes in earth's climate at planetary, regional and local scales. Scientific methods and technologies using computers, digital communication, advanced instruments and satellites have enabled to integrate and enlarge borders of knowledge to interpret changes in the environment.

The quest to advance GFD theory for earthly fluids has brought into attention the two most distinctive effects at geophysical scales: (i) earth's rotation and (ii) buoyancy (density stratification) in non-homogeneous fluids. Despite only a few changes in similar sets of governing equations applied to rotating and stratified fluids, a multitude of new effects is observed in response to basic restoring forces of gravity, rotation and density gradient.[1] An expert review of the theory for non-homogeneous fluids has been given in a landmark reference.[2]

---

[1]Greenspan, H. P., The Theory of Rotating Fluids, Cambridge University Press, 1980.
[2]Turner, J. S., Buoyancy Effects in Fluids, Cambridge at the University Press, 1973.

Advances in GFD theory have received great attention in the last few centuries, with emphasis given to a variety of environmental problems influenced by the earth's changing climate. As noted in the preface to GFD-I, the puzzle on the theory of ocean currents has first been challenged in the "Seas of the Old World" by actual measurements at sea by Marsili (1681).[3,4] These first observations of density and current variations in the Bosphorus Strait established the theory of buoyancy currents in the stratified environment, actually setting the foundations of ocean science more than three centuries before our present age, ignored until the last century or so, when these concepts first found their expression in fluid dynamics theory.

The two figures in the preceding page should serve to deliver the great impact of buoyancy effects on currents, as perceived by Marsili and elaborated by contemporary examples of stratified exchange at straits interconnecting ocean basins,[5] by numerous observations, analyses and model studies.

A review of earth's rotation effects already has been given in the preface to the first textbook (referred to as GFD-I hereafter).[6] An introduction to stratified fluid dynamics added on top of rotational effects is made in the present volume. Elements of basic processes in stratified fluids are described, with a list of recommended books and video sources supplied at the end of this book.

A large part of the presented material has been developed as part of graduate programs at the Institute of Marine Sciences and Eurasia Institute of Earth Sciences (respectively of the Middle East Technical University, Mersin and İstanbul Technical University, İstanbul), and also as part of a school at the International Center for Theoretical Physics, Trieste. The material presented in two volumes has been built on first principles and simplified theory. Enlightened peers and contemporaries contributed to our knowledge base. Foremost among peers, Ümit Ünlüata has provided continuous motivation, encouragement and contributions of review notes

---

[3]Marsili, L. F., Osservazioni Intorno al Bosforo Tracio overo Canale di Constantinopoli, Rappresentate in Lettera alla Sacra Real Maestá Cristina Regina di Svezia da Luigi Ferdinando Marsigli. Nicoló Angelo Tinassi, Roma, 124 p., 1681.

[4]Pinardi, N., Özsoy, E., Latif, M. A., Moroni, F., Grandi, A., Manzella, G., De Strobel, F., Lyubartsev, V., Measuring the Sea: the First Oceanographic Cruise (1679–1680) and the Roots of Oceanography, Physical Oceanography, 48(4), 845–860, 2018.

[5]Özsoy E., Di Iorio D., Gregg M., Backhaus J., Mixing in the Bosphorus Strait and the Black Sea Continental Shelf: Observations and a Model of the Dense Water Outflow, J. Mar. Sys., 31, 99–135, 2001.

[6]Özsoy, E. (2020). Geophysical Fluid Dynamics I: An Introduction to Atmosphere—Ocean Dynamics: Homogeneous Fluids, Springer, 296p.

(*e.g.* some sections with exercises in Chap. 3). The textbooks are hoped to motivate collaboration in sciences, primarily in Physical Oceanography, Climate and Marine Sciences programs in the Eastern Mediterranean, Black Sea and Caspian Seas region of the "Seas of the Old World", as in rest of the world.

Erdemli, Turkey Emin Özsoy
Mersin, Turkey
İstanbul, Turkey
Didyma, Turkey
Ankara, Turkey
December 2020

Contents

Hydro-Thermo-Dynamics of Stratified Fluids

1

1.1 Basic Hydrodynamics: Continuity and Momentum Equations

Continuity and momentum equations were developed [Eqs. (2.23.a,b) and (3.56) of the first volume GFD-I[1] of this two book series]:

$$\frac{D\rho}{Dt} + \rho \nabla \cdot \mathbf{u} = 0 \tag{1.1}$$

$$
\begin{aligned}
\frac{D\mathbf{u}}{Dt} + 2\boldsymbol{\Omega} \times \mathbf{u} &= \mathbf{g} + \frac{1}{\rho}\nabla \cdot \boldsymbol{\sigma} \\
&= \mathbf{g} - \frac{1}{\rho}\nabla p + \frac{1}{\rho}\nabla \cdot \mathbf{d}
\end{aligned}
\tag{1.2}
$$

where ρ, p, and $\mathbf{u}$ are the density, pressure and velocity of the fluid respectively, $\boldsymbol{\Omega}$ the earth's angular velocity, $\mathbf{d}$ the deviatoric stress tensor and $\mathbf{g}$ the gravitation vector including a centrifugal component added to gravity $\mathbf{g}'$,

$$\mathbf{g} = \mathbf{g}' - \boldsymbol{\Omega} \times \boldsymbol{\Omega} \times \mathbf{r} \tag{1.3}$$

with $\mathbf{r}$ measured from the earth's center, applying a small correction to Eq. (3.52) in GFD-I. The stress tensor $\boldsymbol{\sigma}$ in (1.2) is given by

$$\boldsymbol{\sigma} = -p\mathbf{I} + \mathbf{d} \tag{1.4}$$

---

[1]Özsoy, E. (2020). *Geophysical Fluid Dynamics I: An Introduction to Atmosphere—Ocean Dynamics: Homogeneous Fluids*, Springer Textbooks in Earth Sciences, Geography and Environment, 296p. (referred to as GFD-I in text)

© Springer Nature Switzerland AG 2021
E. Özsoy, *Geophysical Fluid Dynamics II*, Springer Textbooks in Earth Sciences, Geography and Environment,
https://doi.org/10.1007/978-3-030-74934-7_1

where $\mathbf{I} = \delta_{ij}$ is the identity matrix (tensor).

1.2   Thermodynamics

1.2.1   First Law of Thermodynamics

In the earlier GFD-I book on homogeneous fluids, density ρ has been taken as constant. In the present case of non-homogeneous fluids, variable density $\rho = \rho(x, y, z, t)$ is considered in general, so that the two Eqs. (1.1) and (1.2) are insufficient to solve for three unknowns $\mathbf{u}$, ρ, and p, implying additional equations needed to close the system.

The *First Law of Thermodynamics* is an energy conservation statement,

$$\frac{dE_T}{dt} = \frac{dH}{dt} + \frac{dW}{dt} \tag{1.5}$$

expressing the balance between rates of change in total energy (E_T), heat (H) supplied to the fluid and work (W) done on the fluid. Total energy for a material volume V bounded by material surface S is defined as

$$E_T = \int_V \rho \epsilon \, dV \tag{1.6.a}$$

where

$$\epsilon = e + \frac{1}{2}\mathbf{u} \cdot \mathbf{u} - \mathbf{g} \cdot \mathbf{x} \tag{1.6.b}$$

is the total energy per unit mass, including internal energy (or specific enthalpy) e, kinetic $\mathbf{u} \cdot \mathbf{u}/2$ and potential $-\mathbf{g} \cdot \mathbf{x}$ energy components. Positive contribution to potential energy is represented by inner product of $\mathbf{g}$ with position vector $\mathbf{x}$ relative to arbitrary datum.

1.2.2   Mechanical and Thermal Energy Equations

The rate of change of total energy, on the left hand side of Eq. (1.5) is expressed by making use of (1.6.a,b), Leibniz rule, divergence theorem (described in GFD-I) and continuity equation (1.1),

$$\frac{dE_T}{dt} = \frac{d}{dt} \int_V \rho \epsilon \, dV$$

$$= \int_V \frac{\partial \rho \epsilon}{\partial t} dV + \int_S \rho \epsilon \mathbf{u} \cdot \mathbf{n} \, dS$$

$$= \int_V \left(\frac{\partial \rho \epsilon}{\partial t} + \nabla \cdot \rho \epsilon \mathbf{u} \right) dV$$

$$= \int_V \left(\rho \frac{\partial \epsilon}{\partial t} + \epsilon \frac{\partial \rho}{\partial t} + \epsilon \rho \nabla \cdot \mathbf{u} + \epsilon \mathbf{u} \cdot \nabla \rho + \rho \mathbf{u} \cdot \nabla \epsilon \right) dV \qquad (1.7)$$

$$= \int_V \rho \frac{D\epsilon}{Dt} dV + \int_V \epsilon \left(\frac{D\rho}{Dt} + \rho \nabla \cdot \mathbf{u} \right) dV$$

$$= \int_V \rho \frac{D\epsilon}{Dt} dV$$

$$= \int_V \left\{ \rho \frac{De}{Dt} + \rho \frac{D}{Dt} \left(\frac{1}{2} \mathbf{u} \cdot \mathbf{u} - \mathbf{g} \cdot \mathbf{x} \right) \right\} dV.$$

On the other hand, the rate of change in heat input to a fluid element is

$$\frac{dH}{dt} = \int_V \rho Q \, dV - \int_S \mathbf{q} \cdot \mathbf{n} \, dS$$

$$= \int_V (\rho Q - \nabla \cdot \mathbf{q}) \, dV, \qquad (1.8)$$

$$= \int_V (\rho Q + \nabla \cdot K \nabla T) \, dV,$$

where Q is the rate of internal heating (per unit mass of fluid) delivered to the fluid volume V, the quantity $(-\mathbf{q})$ representing heat flux, such that $-\mathbf{q} \cdot \hat{\mathbf{n}}$ is flux per unit area through the enclosing surface S, in opposite direction to the outward normal $\hat{\mathbf{n}}$. The diffusive (conductive) heat flux is then expressed by *Fourier's Law*, which relates it to local gradient of temperature, following a linear relationship (e.g. Batchelor 1967):

$$\mathbf{q} = -K \nabla T \qquad (1.9)$$

with the negative sign implying heat flow from high to low temperatures. The constant of proportionality K is called *thermal conductivity*.

The rate of work done on the fluid by surroundings, due to surface stresses $\boldsymbol{\Sigma} = \boldsymbol{\sigma} \cdot \hat{\mathbf{n}}$ (from GFD-I) applied on the enclosing surface S, is used to deform it with velocity $\mathbf{u}$, which by the use of identities and divergence theorem (GFD-I) is obtained as

$$\frac{dW}{dt} = \int_S \boldsymbol{\Sigma} \cdot \mathbf{u} \, dS = \int_S (\boldsymbol{\sigma} \cdot \mathbf{n}) \cdot \mathbf{u} \, dS$$

$$= \int_S (\mathbf{u} \cdot \boldsymbol{\sigma}) \cdot \mathbf{n} \, dS = \int_V \nabla \cdot (\mathbf{u} \cdot \boldsymbol{\sigma}) \, dV = \int_V \rho R \, dV. \qquad (1.10.a)$$

where

$$R \equiv \nabla \cdot (\mathbf{u} \cdot \boldsymbol{\sigma})/\rho \qquad (1.10.b)$$

represents the redistribution (per unit mass) of external work supplied to volume V of the surface force $\boldsymbol{\Sigma}$. In fact, any external work on the fluid through surface forces $\boldsymbol{\Sigma}$ is integrated across surface S of fluid contact with any solid boundary. For example, a propeller (e.g. of a ship or aeroplane) would add kinetic energy to the fluid, by the action of pressure and viscous stresses applied at the fluid interface.

By making use of the momentum equation (1.2), the integrand (1.10) is written as

$$
\begin{aligned}
\nabla \cdot (\mathbf{u} \cdot \boldsymbol{\sigma}) &= \mathbf{u} \cdot (\nabla \cdot \boldsymbol{\sigma}) + (\boldsymbol{\sigma} \cdot \nabla) \cdot \mathbf{u} \\
&= \rho \mathbf{u} \cdot \left(\frac{D\mathbf{u}}{Dt} + 2\boldsymbol{\Omega} \times \mathbf{u} - \mathbf{g} \right) + (\boldsymbol{\sigma} \cdot \nabla) \cdot \mathbf{u} \\
&= \rho \mathbf{u} \cdot \frac{D\mathbf{u}}{Dt} - \rho \mathbf{u} \cdot \nabla (\mathbf{g} \cdot \mathbf{x}) + (\boldsymbol{\sigma} \cdot \nabla) \cdot \mathbf{u} \\
&= \rho \frac{D}{Dt} \left(\frac{1}{2} \mathbf{u} \cdot \mathbf{u} - \mathbf{g} \cdot \mathbf{x} \right) + (\boldsymbol{\sigma} \cdot \nabla) \cdot \mathbf{u}
\end{aligned}
\qquad (1.11)
$$

where we specifically note the product of momentum with total derivative to give

$$\rho \mathbf{u} \cdot \frac{D\mathbf{u}}{Dt} = \rho \mathbf{u} \cdot \frac{\partial \mathbf{u}}{\partial t} + \rho \mathbf{u} \cdot (\mathbf{u} \cdot \nabla) \mathbf{u} = \rho \frac{\partial}{\partial t} \frac{1}{2} \mathbf{u} \cdot \mathbf{u} + \rho \mathbf{u} \cdot \nabla \frac{1}{2} \mathbf{u} \cdot \mathbf{u} = \rho \frac{D}{Dt} \frac{1}{2} \mathbf{u} \cdot \mathbf{u}$$

and similarly for

$$-\rho \mathbf{u} \cdot \nabla (\mathbf{g} \cdot \mathbf{x}) = -\rho \frac{D}{Dt} (\mathbf{g} \cdot \mathbf{x})$$

so that (1.10) becomes

$$
\begin{aligned}
\frac{dW}{dt} &= \int_V \nabla \cdot (\mathbf{u} \cdot \boldsymbol{\sigma}) \, dV \\
&= \int_V (\boldsymbol{\sigma} \cdot \nabla) \cdot \mathbf{u} \, dV + \int_V \rho \frac{D}{Dt} \left(\frac{1}{2} \mathbf{u} \cdot \mathbf{u} - \mathbf{g} \cdot \mathbf{x} \right) dV.
\end{aligned}
\qquad (1.12)
$$

The second term of (1.12) is modified by making use of Eqs. (1.1) and (1.4), so that

$$
\begin{aligned}
\boldsymbol{\sigma} \cdot \nabla \cdot \mathbf{u} &= -p \nabla \cdot \mathbf{u} + (\mathbf{d} \cdot \nabla) \cdot \mathbf{u} \\
&= \frac{p}{\rho} \frac{D\rho}{Dt} + (\mathbf{d} \cdot \nabla) \cdot \mathbf{u} \\
&= \rho (\chi + \varPhi)
\end{aligned}
\qquad (1.13)
$$

where

$$\chi \equiv \frac{p}{\rho}\frac{D\rho}{Dt} \quad \text{and} \quad \varPhi \equiv \frac{1}{\rho}(\mathbf{d}\cdot\nabla)\cdot\mathbf{u} \qquad (1.14.a, b)$$

respectively represent work done by compression and heat generated by viscous mechanical energy dissipation per unit mass.

Substituting (1.11) and (1.12) into (1.10) and combining (1.6), (1.7), (1.10) yields the thermodynamic energy equation

$$\frac{De}{Dt} = \frac{1}{\rho}\nabla\cdot K\nabla T + Q + \chi + \varPhi \qquad (1.15)$$

From these source-sink terms on the right hand side of (1.15), the first three are quite straightforward: The first term represents Laplacian diffusion depending on the gradient of temperature which is a state variable, the second term represents sources of externally supplied heat (e.g. by radiation), the third term representing heat generation by compression as a function of state variables pressure and density.

The fourth term representing viscous dissipation in (1.15), described by (1.14.b) is a nonlinear function of velocity and deviatoric stress tensor, which further is a function of velocity, with unknown constant of viscosity incorporated. In fact, for a laminar flow it appears physically tractable (substituting from Eq. 3.29 of GFD-I). However, for most actual situations in the ocean and atmosphere, the fluid is in a turbulent state, whose eddy diffusive characteristics are either empirically specified or parameterized, so that the actual dependence of this term on flow variables depends on particular formulation.

The mechanical energy Eq. (1.11) is re-written by making use of (1.10.a,b) and (1.14.a,b) as

$$
\begin{aligned}
\frac{D}{Dt}\left(\frac{1}{2}\mathbf{u}\cdot\mathbf{u} - \mathbf{g}\cdot\mathbf{x}\right) &= \frac{1}{\rho}\mathbf{u}\cdot(\nabla\cdot\boldsymbol{\sigma}) \\
&= \frac{1}{\rho}\{\nabla\cdot(\boldsymbol{\sigma}\cdot\mathbf{u}) - \boldsymbol{\sigma}\cdot(\nabla\cdot\mathbf{u})\} \\
&= \frac{1}{\rho}\{\rho R - \frac{p}{\rho}\frac{D\rho}{Dt} - \mathbf{d}\cdot(\nabla\cdot\mathbf{u})\} \\
&= R - \chi - \varPhi
\end{aligned} \qquad (1.16)
$$

to describe conservation of mechanical energy.

Summing (1.15) and (1.16) gives the rate of change of total energy, including the internal, kinetic and potential energy components:

$$\frac{D\epsilon}{Dt} = \frac{D}{Dt}\left(e + \frac{1}{2}\mathbf{u}\cdot\mathbf{u} - \mathbf{g}\cdot\mathbf{x}\right) = \frac{1}{\rho}\nabla\cdot K\nabla T + Q + R. \qquad (1.17)$$

In conclusion, two separate equations for energy conservation have thus been developed, with corresponding rates of change respectively for heat (1.15) and mechanical energy (1.16). Adding these together, Eq. (1.17) expresses the conservation of total energy in the system. Note that the sum $\chi + \Phi$ represents heat generation by compression and viscous dissipation, lost from mechanical energy and gained by thermal part of internal energy.

The meaning of individual terms in these equations are clear: In the heat equation (1.15), the rate of change of internal energy (specific enthalpy) occurs due to the sum of heat diffusion, sources, compression and dissipation. In the mechanical energy equation (1.16), the rate of change of kinetic plus potential energy occurs due to work done on the fluid by the environment, with resulting conversions of mechanical energy into heat, by processes of compression and dissipation.

The interesting result is that when the rates of change for heat and mechanical energy are added together in (1.15) and (1.16), we observe that the compression and dissipation terms are canceled out resulting in rate of change in total energy Eq. (1.17). Since these opposing terms in thermal and mechanical equations are cancel out in the total energy equation, it can be concluded that what is lost/gained from mechanical energy is gained/lost from heat, by internal mechanisms of i.e. energy conversion.

With these considerations, the thermodynamic equation is written as

$$\frac{De}{Dt} - \frac{p}{\rho^2}\frac{D\rho}{Dt} = \frac{1}{\rho}\nabla \cdot K\nabla T + Q + \Phi \tag{1.18}$$

where Q and Φ are unknown functions at present, either externally specified or parameterized at best.

The mechanical energy Eq. (1.16), on the other hand, can simply be put into the following form:

$$\begin{aligned}\frac{D}{Dt}\left(\frac{1}{2}\mathbf{u}\cdot\mathbf{u} - \mathbf{g}\cdot\mathbf{x}\right) &= \frac{1}{\rho}\mathbf{u}\cdot(\nabla\cdot\boldsymbol{\sigma}) \\ &= \frac{1}{\rho}\{-(\mathbf{u}\cdot\nabla p) + \mathbf{u}\cdot(\nabla\cdot\mathbf{d})\}\end{aligned} \tag{1.19}$$

so that the change in mechanical energy is accounted for by the work done by the pressure gradient force (first term) with remaining dissipation term by viscous stresses (second term).

1.2.3   Equilibrium Thermodynamics

Classical (equilibrium) thermodynamics theory describes states of uniform matter, in which all local mechanical, physical and thermal quantities are assumed independent of position and time, e.g. as in the case of a homogeneous fluid at rest.

Despite the fact that non-equilibrium states such as moving fluids are inadequately addressed in classical theory, observations seem to indicate that the departure from equilibrium does not seem to have too great an influence on the thermodynamic quantities.

The *equilibrium thermodynamic state* of a fluid with *fixed composition* (i.e. constituents with fixed mixing ratios) is determined by at least three *parameters of state*, which are the temperature T, pressure p and density ρ.

An alternative variable v, is just the inverse of density ρ, which from now on, will be referred to as *specific volume*

$$v = \frac{1}{\rho} \tag{1.20}$$

for convenience. The state is therefore defined by a relation between these state variables i.e.

$$f(p, v, T) = 0. \tag{1.21}$$

Note that the *equation of state* for a fluid defines a surface with respect to p, v, T coordinates, alternately expressed by

$$p = f_1(v, T), \quad \text{or} \quad T = f_2(p, v), \quad \text{or} \quad v = f_3(p, T). \tag{1.22a − c}$$

in the thermodynamic equation (1.18), in which the right hand side terms are lumped together to represent heating rates by diffusion, internal heating and viscous dissipation processes

$$\frac{1}{\rho} \nabla \cdot K \nabla T + Q + \Phi \equiv Q^* \equiv \frac{Dq}{Dt}. \tag{1.23}$$

By making use of (1.20) and (1.23), Eq. (1.18) takes the following form:

$$\frac{De}{Dt} + p \frac{Dv}{Dt} = \frac{Dq}{Dt}. \tag{1.24}$$

In the case of an equilibrium state (T=constant and **u**=constant), $\nabla \cdot K \nabla T = 0$ and $\Phi = (\mathbf{d} \cdot \nabla) \cdot \mathbf{u} = 0$ implies uniform internal heating Q to be the only contribution to (1.23):

$$\frac{Dq}{Dt} = Q. \tag{1.25}$$

For vanishing internal heating $Q=0$, with other terms excluded in (1.23), the fluid is essentially said to be *adiabatic*, excluding any heat exchange. Equation (1.24) is alternatively written as

$$\delta e + p \, \delta v = \delta q \tag{1.26}$$

where δe is the change in specific internal energy, $-p\delta v$ the work done by compression leading to a change in volume, and δq the heat added (Fig. 1.1).

The above differentials are valid in the equilibrium case, since no temporal or spatial variations are present. The notation δ is then interpreted as the total change in the state of the fluid as a whole. For infinitesimal deviations, Eq. (1.26) describes *reversible* changes between neighboring states, i.e. the net change in internal energy δe is canceled out when returning to the original state after some change. In general,

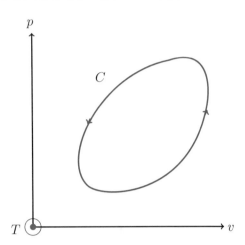

Fig. 1.1 Changes in the thermodynamic state of a fluid occurs along a closed path C, following coordinates of the basic state variables p, v and T, on an indicator diagram in the $p - v$ plane

whether the process is *reversible* or *irreversible* for a finite change is determined by the path taken. Since net changes occur along a contour in the p, v, T space, we evaluate the integral in $p - v$ plane (indicator diagram):

The state equation $f(p, v, T) = 0$ defines a surface in p, v, T coordinates for fluids of fixed composition. Then,

$$\oint_c de = -\oint_c p dv + \oint_c dq, \tag{1.27}$$

but since p is not a function of v alone (i.e. $p = p(v, T)$), the integral depends on path. If the process is *isothermal* ($T = $ constant) or $p = p(v)$ only, then the first term on the right hand side of (1.27) vanishes. If the process is *adiabatic*, the second term on the right hand side vanishes, yielding *reversible* changes in each of the above mutually exclusive cases. When the path is a special combination of these, the process may still be reversible. Another possibility for reversible change is when the first term cancels the second term on the right hand side, i.e. in the case of *compression heating*.

A practical quantity of importance is *specific heat c* of the fluid, i.e. the amount of heat given to unit mass of fluid during a small reversible change in temperature

$$c = \frac{\delta q}{\delta T}. \tag{1.28}$$

Specific heat is a function of the conditions under which the reversible change takes place, i.e. in any direction starting from point A in Fig. 1.2. Using the chain rule to express the changes in state variable e with respect to the other state variables in (1.26) gives

$$\delta q = \left(\frac{\partial e}{\partial p}\right)_v \delta p + \left(\frac{\partial e}{\partial v}\right)_p \delta v + p \delta v. \tag{1.29}$$

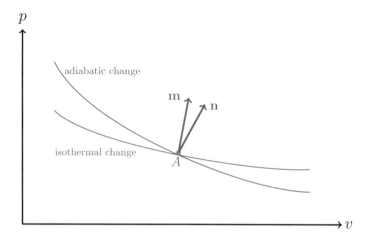

Fig. 1.2 Lines following adiabatic and isothermal changes projected on the $p-v$ plane with corresponding normal unit vectors

On the other hand, temperature differential based on the equation of state (1.21) is written as

$$\delta T = \left(\frac{\partial T}{\partial p}\right)_v \delta p + \left(\frac{\partial T}{\partial v}\right)_p \delta v \tag{1.30}$$

where the subscripts denote differentiation at constant value of the subscript variable, (e.g. $\left(\frac{\partial T}{\partial p}\right)_v$ denotes change in temperature with respect to pressure, while keeping the density ρ or specific volume v constant). Therefore (1.28) is expressed as

$$c = \frac{\left(\frac{\partial e}{\partial p}\right)_v \delta p + \left[\left(\frac{\partial e}{\partial v}\right)_p + p\right]\delta v}{\left(\frac{\partial T}{\partial p}\right)_v \delta p + \left(\frac{\partial T}{\partial v}\right)_p \delta v} \tag{1.31}$$

which depends on the ratio $\delta p/\delta v$ or on the choice of direction during change from point A. Two well-defined constants, specifying changes parallel to the axes of the $p-v$ diagram are defined as the *principal specific heats*. Definition of *specific heat at constant pressure* is

$$c_p = \left(\frac{\delta q}{\delta T}\right)_{\delta p=0} = \left(\frac{\delta e}{\delta T}\right)_p + p\left(\frac{\delta v}{\delta T}\right)_p \tag{1.32a}$$

and specific heat at constant volume is similarly defined as

$$c_v = \left(\frac{\delta q}{\delta T}\right)_{\delta v=0} = \left(\frac{\delta e}{\delta T}\right)_v. \tag{1.32b}$$

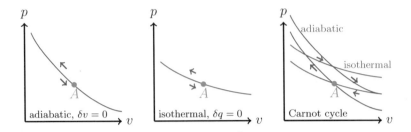

Fig. 1.3 Adiabatic and isothermal changes projected on the $p - v$ plane

If **m** and **n** are the unit normal vectors perpendicular to the isothermal ($\delta T = 0$) and adiabatic ($\delta q = 0$) lines respectively in Fig. 1.2, their components in p and v directions are then defined as $\mathbf{m} = (m_v, m_p)$, $\mathbf{n} = (n_v, n_p)$, yielding

$$c_p = \left(\frac{\delta q}{\delta T}\right)_{\delta p=0} = \frac{n_v(\delta q)_{max}}{m_v(\delta T)_{max}} \tag{1.33a}$$

$$c_v = \left(\frac{\delta q}{\delta T}\right)_{\delta v=0} = \frac{n_p\,(\delta q)_{max}}{m_p\,(\delta T)_{max}}. \tag{1.33b}$$

Since $-m_v/m_p$ and $-n_v/n_p$ are gradients of isothermal and adiabatic lines respectively, the ratio of principal specific heat component is found to be (Fig. 1.3)

$$\gamma = \frac{c_p}{c_v} = \frac{n_v/n_p}{m_v/m_p} = \frac{\left(\frac{\delta p}{\delta v}\right)_{\delta q=0}}{\left(\frac{\delta p}{\delta v}\right)_{\delta T=0}} = \frac{\left(\frac{\delta v}{\delta p}\right)_{\delta T=0}}{\left(\frac{\delta v}{\delta p}\right)_{\delta q=0}}. \tag{1.34}$$

1.2.4 Second Law of Thermodynamics

The *Second Law of Thermodynamics* relates *specific entropy s* to other state variables. In a reversible transition from one equilibrium state to another, the change in entropy is proportional to the heat supplied to the fluid. Expected change in state depends only on temperature, with proportionality constant chosen as $1/T$, specific entropy being defined as an equilibrium property of the fluid,

$$T\delta s = \delta q. \tag{1.35}$$

The above relation defines absolute scale for temperature T is in degrees Kelvin. An *adiabatic, reversible* change ($\delta q = 0$) is therefore also *isentropic* ($\delta s = 0$).

Considering *irreversible change q* (cf. Eq. 1.23), even in the case of an adiabatic process ($\nabla \cdot K\nabla T + \rho Q = 0$), heat dissipation Φ of mechanical energy is always positive, $\delta q > 0$, with absolute temperature T also defined as positive ($T > 0$), one arrives at the result that *entropy must unconditionally increase* ($\delta S > 0$) in irreversible changes.

A different way of expressing the Second Law of Thermodynamics follows from (1.26) and (1.35),

$$\delta e = T\delta s - p\delta v. \tag{1.36}$$

Other dependent properties (state variables) describe equilibrium thermodynamic changes in a fluid. These are *specific enthalpy, h*, *Helmholtz free energy (Helmholtz function), f*, and *Gibbs' function g*, defined respectively as

$$
\begin{aligned}
h &= e + pv \\
f &= e - Ts \\
g &= e + pv - Ts.
\end{aligned}
\tag{1.37.a - c}
$$

The above quantities described by (i.e. energy transforms) imply "thermodynamic equilibrium". Differentiating these energy transforms yields the following relations:

$$
\begin{aligned}
\delta h &= T\delta s + v\delta p \\
\delta f &= -p\delta v - s\delta T \\
\delta g &= v\delta p - s\delta T.
\end{aligned}
\tag{1.38.a - c}
$$

From (1.36) it follows that

$$\left(\frac{\delta e}{\delta v}\right)_{\delta s=0} = \left(\frac{\partial e}{\partial v}\right)_s = -p, \qquad \left(\frac{\delta e}{\delta s}\right)_{\delta v=0} = \left(\frac{\partial e}{\partial s}\right)_v = T. \tag{1.37}$$

Differentiating (1.39.a,b) in reverse order, we obtain two versions of the cross-derivatives of e:

$$
\begin{aligned}
\frac{\partial^2 e}{\partial v \partial s} &= \left[\frac{\partial}{\partial v}\left(\frac{\partial e}{\partial s}\right)_v\right]_s = \left(\frac{\partial T}{\partial v}\right)_s, \\
\frac{\partial^2 e}{\partial s \partial v} &= \left[\frac{\partial}{\partial s}\left(\frac{\partial e}{\partial v}\right)_s\right]_v = -\left(\frac{\partial p}{\partial s}\right)_v.
\end{aligned}
\tag{1.40a, b}
$$

since (1.40.a) and (1.40.b) should be equal, it follows that

$$\left(\frac{\partial p}{\partial s}\right)_v = -\left(\frac{\partial T}{\partial v}\right)_s. \tag{1.41.a}$$

This equality is the first one of the *Maxwell's thermodynamic relations*. Other Maxwell relations can be obtained by applying reciprocal differentiation of the energy transforms (1.38.a-c) to yield

$$\left(\frac{\partial v}{\partial s}\right)_p = \left(\frac{\partial T}{\partial p}\right)_s$$

$$\left(\frac{\partial v}{\partial T}\right)_p = -\left(\frac{\partial s}{\partial p}\right)_T \qquad (1.41.b - d)$$

$$\left(\frac{\partial p}{\partial T}\right)_v = \left(\frac{\partial s}{\partial v}\right)_T.$$

Alternative expressions can be written for the specific heat, making use of (1.35):

$$c = \frac{\delta q}{\delta T} = T\left(\frac{\delta s}{\delta T}\right), \qquad (1.42)$$

so that definitions of principal specific heats (1.32.a,b) are modified to become

$$c_p = T\left(\frac{\partial s}{\partial T}\right)_p, \qquad c_v = T\left(\frac{\partial s}{\partial T}\right)_v. \qquad (1.43)$$

Regarding s to be a function of T and v only, (related by the state equation $f_s(s, T, v) = 0$), one obtains

$$\delta s = \left(\frac{\partial s}{\partial T}\right)_v \delta T + \left(\frac{\partial s}{\partial v}\right)_T \delta v. \qquad (1.44)$$

Differentiating with respect to T at constant p and multiplying by T yields

$$T\left(\frac{\partial s}{\partial T}\right)_p = T\left(\frac{\partial s}{\partial T}\right)_v + T\left(\frac{\partial s}{\partial v}\right)_T \left(\frac{\partial v}{\partial T}\right)_p. \qquad (1.45)$$

Making repeated use of the Maxwell relations (1.41a-d), provides

$$
\begin{aligned}
c_p - c_v &= T\left(\frac{\partial p}{\partial T}\right)_v \left(\frac{\partial v}{\partial T}\right)_p \\
&= T\left(\frac{\partial s}{\partial v}\right)_T \left(\frac{\partial v}{\partial T}\right)_p \\
&= T\left(\frac{\partial p}{\partial v}\right)_T \left(\frac{\partial s}{\partial p}\right)_T \left(\frac{\partial v}{\partial T}\right)_p \\
&= -T\left(\frac{\partial p}{\partial v}\right)_T \left(\frac{\partial v}{\partial T}\right)_p^2.
\end{aligned}
\qquad (1.46)
$$

Entropy s is regarded a function of other state variables T and p, so that

$$\delta s = \left(\frac{\partial s}{\partial T}\right)_p \delta T + \left(\frac{\partial s}{\partial p}\right)_T \delta p$$
$$= \frac{c_p}{T} \delta T - \left(\frac{\partial v}{\partial T}\right)_p \delta p, \tag{1.47}$$

with the definition of *coefficient of thermal expansion* of the fluid

$$\alpha_T = \frac{1}{v}\left(\frac{\partial v}{\partial T}\right)_p = -\frac{1}{\rho}\left(\frac{\partial \rho}{\partial T}\right)_p, \tag{1.48}$$

Equation (1.47) can now be written as

$$T\delta s = c_p \delta T - \alpha_T T v \delta p. \tag{1.49}$$

1.2.5 Different Forms of the Thermodynamic Equation

Thermodynamic equation (1.18) has been derived in earlier sections:

$$\frac{De}{Dt} - \frac{p}{\rho^2}\frac{D\rho}{Dt} = \frac{1}{\rho}\nabla \cdot K\nabla T + (Q + \Phi) \equiv Q^*, \tag{1.50}$$

where the right hand side is re-defined as source function Q^*. Defining $v \equiv 1/\rho$, this equation is equivalent to (1.24)

$$\frac{De}{Dt} + p\frac{Dv}{Dt} = Q^*. \tag{1.51}$$

By making use of (1.36), a different version is obtained

$$T\frac{Ds}{Dt} = Q^*, \tag{1.52}$$

or by using (1.49), it takes either of the following forms

$$c_p\frac{DT}{Dt} - \alpha_T T v \frac{Dp}{Dt} = Q^* \tag{1.53}$$

$$c_p\frac{DT}{Dt} - \frac{\alpha_T T}{\rho}\frac{Dp}{Dt} = Q^*, \tag{1.53.b}$$

where Q^* represents non-isentropic sources of heat due to the combined effects of diffusion, internal and frictional heating, with rate $Q^* = Dq/Dt$. Note that equations (1.53.a,b) are convenient because they are in terms of measurable quantities p, ρ and T. The first term represents heat storage proportional to heat capacity (specific heat) and the second term represents compression effects.

1.3 Thermodynamic Equations Applied to the Atmosphere and the Ocean

1.3.1 Relative Roles of Compression in the Atmosphere and the Ocean

The relative importance of contributing terms on the left hand side of Eq. (1.53.a) are interpreted (using Eqs. 1.46 and 1.48) by the ratio

$$
-\frac{c_p \delta T}{\alpha_T v T \delta p} = -\frac{c_p \delta T}{\left(\frac{\partial v}{\partial T}\right)_p T \delta p} = \frac{c_p}{c_p - c_v}\left(\frac{\partial p}{\partial v}\right)_T \left(\frac{\partial v}{\partial T}\right)_p \frac{\delta T}{\delta p}
$$
$$
= \frac{c_p}{c_p - c_v}\frac{\left(\frac{\partial v}{\partial T}\right)_p \delta T}{\left(\frac{\partial v}{\partial p}\right)_T \delta p},
$$

(1.54)

measuring volume changes in proportion to temperature and pressure, multiplied by the factor $c_p/(c_p - c_v)$. Note that thermal expansion in Eq. (1.48)

$$
\left(\frac{\partial v}{\partial T}\right)_p = \alpha_T v
$$

(1.55)

and the *bulk modulus of elasticity M* of the fluid

$$
M = v\frac{\delta p}{\delta v} = -\rho\frac{\delta p}{\delta \rho}
$$

(1.56.a)

depends on state changes. In the case of isothermal changes, this becomes

$$
M_T = v\left(\frac{\partial p}{\partial v}\right)_T = -\rho\left(\frac{\partial p}{\partial \rho}\right)_T.
$$

(1.56.b)

The bulk modulus is related to sound velocity c_s through

$$
M_T = \rho c_s{}^2.
$$

Combining (1.55), (1.56.b,c) and noting $\gamma = c_p/c_v$ in Eqs. (1.34) and (1.55), one obtains the ratio of heat storage terms due to changes in specific heat and compression.

$$
-\frac{c_p \delta T}{v\alpha_T T \delta p} = \frac{\gamma}{\gamma - 1}\alpha_T M_T \frac{\delta T}{\delta p}
$$
$$
= \frac{\gamma}{\gamma - 1}\rho\alpha_T c_s{}^2\frac{\delta T}{\delta p}.
$$

(1.57)

If the atmosphere is assumed to be an ideal gas with $\gamma = 1.4$, the coefficient on the right hand side will be $\gamma/(\gamma - 1) = 3.5$. The maximum change in temperature from the earth's surface to top of the troposphere is $\delta T = 100\,°K = 100\,°C$, with pressure change $\delta p = 1000$ mbar $= 1$ bar $= 10^5$ Pa $= 10^5$ kg m^{-1} s^{-2}. Typical values are given as: sound speed 300 m s^{-1}, coefficient of thermal expansion $\alpha_T = 3.4 \times 10^{-3}\,°K^{-1}$ and air density $\rho = 1.2$ kg m^{-3}. Putting these typical values in (1.57) yields

$$-\frac{c_p \delta T}{v \alpha_T T \delta p} = 3.5 \times 1.2 \times 3.4 \times 10^3 \times 9 \times 10^4 \times \frac{100}{10^5} \cong 1,$$

so that the two terms on the left hand side of (1.53) are shown to be of the same order in the atmosphere.

For ocean water at typical conditions, c_p and c_v values are not too different, yielding a typical ratio of $\gamma = 0.997$, or $\gamma/(\gamma - 1) = 3.3 \times 10^3$. The density value is close to $\rho = 10^3$ kg/m^3, and the coefficient of expansion for ocean water is usually close to the pure water value of $\alpha_T = 1.5 \times 10^{-4}/°K$. The speed of sound is typically around $c_s = 1500$ m/s, and the temperature difference from the surface to the bottom in deep basins of the ocean is on the order of $\delta T = 10\,°C$ or more, within a pressure range of about $\delta p = 1000$ dbar $= 100$ bar $= 10^7$ Pa. With these values,

$$-\frac{c_p \delta T}{v \alpha_T T \delta p} = 3.3 \times 10^3 \times 10^3 \times 1.5 \times 10^{-4} \times 2.25 \times 10^6 \times \frac{10}{10^7} \cong 10^3,$$

shows the effects of compression heating (second term of 1.53) to be minimal, compared to specific-heat storage in the ocean.

Therefore two versions of the thermodynamic equation can be written; which, in the case of atmosphere

$$c_p \frac{DT}{Dt} - \frac{\alpha_T T}{\rho} \frac{Dp}{Dt} = \frac{1}{\rho} \nabla \cdot K \nabla T + (Q + \Phi) \tag{1.58}$$

and for the ocean is approximated as:

$$c_p \frac{DT}{Dt} = \frac{1}{\rho} \nabla \cdot K \nabla T + (Q + \Phi), \tag{1.59}$$

with corresponding equations of state applied to individual fluids.

1.3.2   Thermodynamic Equation for the Atmosphere

A review of governing equations is in order here. Continuity and momentum equations (1.1) and (1.2) involving unknowns ρ, p, and $\mathbf{u}$ are supplemented by thermodynamic equation (1.58), which introduces an additional variable T. Additionally, the equation of state is required to supplement the system:

$$f(\rho, p, T) = 0. \tag{1.60}$$

A further difficulty exists in the atmosphere, mainly due to wet-processes accounting for humidity, introducing an additional variable into the equations. Defining specific humidity

$$\sigma = \frac{\rho_v}{\rho_v + \rho_d} \qquad (1.61)$$

where ρ_d is density of dry air, ρ_v and σ respectively the density and concentration of water vapor in the atmosphere. Equation of state for the atmosphere is represented by ideal gas law

$$f(\rho, p, T, \sigma) = 0 \qquad (1.62.a)$$

modified as

$$p = \rho R(1 + 0.61\sigma)T = \rho R T_*, \qquad (1.62.b)$$

where the virtual temperature T_* incorporates effects of humidity. With an additional variable σ introduced, a conservation statement added for humidity

$$\frac{D\sigma}{Dt} = \nabla \cdot K_\sigma \nabla \sigma + \Sigma \qquad (1.63)$$

where K_σ is moisture diffusivity for small concentrations, $\sigma \ll 1$. In the atmosphere this is often violated since air becomes saturated ($\sigma \sim 1$) by phase change from vapor to water by precipitation. Releasing of latent heat is represented by a source function in the thermodynamic equation, and as a sink in the conservation of vapor, quantified by Σ in (1.63). This eventually becomes a highly nonlinear process, with switching of source/sink terms implied in the thermodynamic state.

Humidity effects are often ignored at first order to avoid above difficulties, thereby assuming the atmosphere to be an *ideal gas* of fixed composition, satisfying the ideal *equation of state*

$$p = \rho R T, \qquad (1.64)$$

where

$$R = c_p - c_v \qquad (1.65)$$

is the *universal gas constant*. Coefficient of thermal expansion becomes (cf. 1.48)

$$\alpha_T = -\frac{1}{\rho}\left(\frac{\partial \rho}{\partial T}\right)_p = -\frac{1}{\rho}\frac{\partial}{\partial T}\left(\frac{p}{RT}\right)_p = \frac{p}{\rho R}\frac{1}{T^2} = \frac{1}{T}, \qquad (1.66)$$

so that the thermodynamic equation (1.58) takes the following form

$$c_p \frac{DT}{Dt} - \frac{1}{\rho}\frac{Dp}{Dt} = Q^*. \qquad (1.67)$$

Furthermore, taking logarithms of both sides in (1.64)

$$\ln p = \ln \rho + \ln T + \ln R \tag{1.68.a}$$

followed by differentiation yields

$$\frac{1}{p}\frac{Dp}{Dt} = \frac{1}{\rho}\frac{D\rho}{Dt} + \frac{1}{T}\frac{DT}{Dt}. \tag{1.68.b}$$

Substituting (1.67) into (1.68) gives

$$c_p\frac{DT}{Dt} - \frac{p}{\rho^2}\frac{D\rho}{Dt} - \frac{p}{\rho T}\frac{DT}{Dt} = Q^*, \tag{1.68.c}$$

then, making use of (1.64) and (1.65) yields

$$c_v\frac{DT}{Dt} - \frac{p}{\rho^2}\frac{D\rho}{Dt} = Q^*. \tag{1.69}$$

Still another form of the thermodynamic equation is obtained by eliminating DT/Dt from (1.67) and (1.69)

$$\frac{1}{\gamma p}\frac{Dp}{Dt} - \frac{1}{\rho}\frac{D\rho}{Dt} = \frac{1}{c_p T}Q^* \tag{1.70}$$

where γ is defined by (1.34) for on ideal gas with a numerical value of

$$\gamma = \frac{c_p}{c_v} = 1.41. \tag{1.71}$$

Potential temperature is defined as

$$\theta = T\left(\frac{p_*}{p}\right)^{R/c_p} = T\left(\frac{p_*}{p}\right)^{(1-\frac{1}{\gamma})} \tag{1.72.a}$$

where p_* is an arbitrary (constant) reference pressure (e.g. using the sea-level pressure, $p_* = 1013$ mb). Based on the ideal gas law (1.64), potential temperature is defined as

$$\theta = \frac{p}{\rho R}\left(\frac{p_*}{p}\right)^{(1-\frac{1}{\gamma})} = \frac{p^{1/\gamma}}{\rho R}p_*^{(1-\frac{1}{\gamma})} = c_*\frac{p^{1/\gamma}}{\rho} \tag{1.72.b}$$

where c_* is a constant. Logarithmic differentiation of (1.72.a) yields

$$\ln \theta = \ln T - \frac{R}{c_p}\ln p + \frac{R}{c_p}\ln p_*$$

and

$$\frac{1}{\theta}\frac{D\theta}{Dt} = \frac{1}{T}\frac{DT}{Dt} - \frac{R}{c_p}\frac{1}{p}\frac{Dp}{Dt}. \tag{1.73}$$

Multiplying this expression by $\rho c_p T$ results in

$$\frac{\rho c_p T}{\theta}\frac{D\theta}{Dt} = \rho c_p \frac{DT}{Dt} - \frac{Dp}{Dt},$$

and comparing with (1.67) gives

$$\frac{D\theta}{Dt} = \frac{\theta}{c_p T}Q^*. \tag{1.74}$$

Further, by making use of (1.52), the last equation is interpreted as

$$T\frac{Ds}{Dt} = Q^*. \tag{1.75}$$

An *adiabatic* atmosphere ($Q^* = 0$) is also an *isentropic* one ($\delta s = 0$) by virtue of (1.75), in which case (1.74) becomes

$$\frac{D\theta}{Dt} = 0 \tag{1.76}$$

stating that θ should be conserved following a fluid particle, i.e. $\theta = \theta_0 = $ constant, whereupon (1.72) yields a simple relation between temperature and pressure corresponding to compression heating (i.e. temperature increase with pressure):

$$T = \theta_0 \left(\frac{p}{p_*}\right)^{R/c_p}. \tag{1.77}$$

1.3.3   Thermodynamic Equation for the Ocean

Compressibility effects are often considered negligible in the ocean, justified by the ratio $[\gamma/(\gamma - 1)]\rho\alpha_T c_s^2 \delta T/\delta p$ being on the order of 10^3. However, in the deep ocean, this ratio can be smaller and in fact become an $O(1)$ quantity, noticing that temporal and spatial variations of temperature (δT or $\frac{DT}{Dt}$) are often small. Thus the two terms on the left hand side of (1.49) can be of comparable magnitude,

$$c_p \frac{DT}{Dt} - \alpha_T \frac{T}{\rho}\frac{Dp}{Dt} = Q^*, \tag{1.78}$$

especially in the deep ocean, where the non-adiabatic term Q^* can be neglected, $Q^* = 0$.

Potential temperature in the ocean is defined as

$$\theta = T - \int_{p_0}^{p} \frac{\alpha_T T}{\rho c_p} dp \tag{1.79}$$

p_0 being a reference pressure (e.g. at the surface, $p_0 \simeq 0$), then the material derivative of (1.78) is

$$\begin{aligned}\frac{D\theta}{Dt} &= \frac{DT}{Dt} - \frac{Dp}{Dt}\frac{D}{Dp}\int_{p_0}^{p}\frac{\alpha_T T}{\rho c_p}dp \\ &= \frac{DT}{Dt} - \frac{\alpha_T T}{\rho c_p}\frac{Dp}{Dt}.\end{aligned} \tag{1.80}$$

Comparing (1.80) with (1.78) gives

$$\frac{D\theta}{Dt} = \frac{Q^*}{c_p}, \tag{1.81}$$

and since $Q^* = 0$ in the deep ocean, this becomes

$$\frac{D\theta}{Dt} = 0, \tag{1.82}$$

such that θ is conserved (i.e. $\theta = \theta_0 = $ constant) which then gives a relation between temperature and pressure through (1.79). Assuming hydrostatic pressure (with z pointing upwards)

$$\frac{\partial p}{\partial z} = -\rho g \tag{1.83}$$

then yields

$$\begin{aligned}T &= \theta_0 + \int_{p_0}^{p}\frac{\alpha_T T}{\rho c_p}dp \\ &= \theta_0 - \int_{z_0}^{z}\frac{g\alpha_T}{c_p}T dz\end{aligned} \tag{1.84}$$

or differentiating with respect to z yields adiabatic temperature increase with depth

$$\frac{\partial T}{\partial z} = -\frac{g\alpha_T}{c_p}T. \tag{1.85}$$

The solution to (1.85) is

$$T = T_0 e^{-(\frac{g\alpha_T}{c_p})(z-z_0)} \tag{1.86}$$

indicating an exponential increase in temperature with depth as a result of the compressibility of sea water. For typical values of $g = 9.81$ m s^{-2}, $\alpha_T = 1.5 \times 10^{-4}$ °C^{-1} and $c_p = 3.85 \times 10^3$ Jkg^{-1} °C^{-1}, the rate of increase is characterized by

$$\frac{g\alpha_T}{c_p} = \frac{9.81 \times 1.5 \times 10^{-4}}{3.85 \times 10^3} = 0.38 \times 10^{-6} m^{-1}$$

i.e. a one percent increase in temperature takes place at depth difference of $\Delta z \geq$ 28 km. Since the deepest part of the ocean has a depth of about 10 km, increase in temperature due to compressibility of sea-water is much less than 1% everywhere.

A thermodynamic relationship establishes the speed of sound as

$$c_s^2 = \left(\frac{\partial p}{\partial \rho}\right)_s, \tag{1.87}$$

leading to

$$\frac{Dp}{Dt} - c_s^2 \frac{D\rho}{Dt} = 0 \tag{1.88}$$

for isentropic conditions in the deep ocean (isentropic conditions). Definition of *potential density* is given by

$$\Delta = \rho - \int_{p_0}^{p} \frac{1}{c_s^2} dp = \rho + \int_{z_0}^{z} \frac{\rho g}{c_s^2} dz \tag{1.89}$$

so as to satisfy (1.88)

$$\begin{aligned}\frac{D\Delta}{Dt} &= \frac{D\rho}{Dt} - \frac{Dp}{Dt} \frac{D}{Dp} \int_{p_0}^{p} \frac{1}{c_s^2} dp \\ &= \frac{D\rho}{Dt} - \frac{1}{c_s^2} \frac{Dp}{Dt} = 0\end{aligned} \tag{1.90}$$

by virtue of (1.88), with potential density Δ conserved under isentropic conditions. Making further use of hydrostatic equation (1.83), isentropic density ($\Delta = \Delta_0 =$ constant) becomes

$$\rho = \Delta_0 - \int_{z_0}^{z} \frac{\rho g}{c_s^2} dz \tag{1.91}$$

or

$$\frac{\partial \rho}{\partial z} = -\frac{g}{c_s^2} \rho \tag{1.92.a}$$

or

$$\rho = \rho_0 e^{-\int_{z_0}^{z} \frac{g}{c_s^2} dz} \cong \rho_0 e^{-\frac{g}{c_s^2}(z-z_0)} \tag{1.92.b}$$

representing deep water density variations due to compressibility. Typical values for the ocean of rate of increase in density are estimated as

$$\frac{g}{c_s{}^2} = \frac{9.81}{(1600)^2} = 3.8 \times 10^{-6} m^{-1}$$

i.e. a one percent increase in in-situ density due to compressibility over a depth range of 2.5 km.

Having shown the negligible influence of compressibility in the ocean, we return to the simplified version of the thermodynamics equation (1.59). Dividing by c_p, and neglecting $Q^* + \Phi \simeq 0$,

$$\frac{DT}{Dt} = \nabla \cdot \kappa_T \nabla T \tag{1.93}$$

known as the heat diffusion equation, where $\kappa_T \equiv K/\rho c_p$ defined as *heat diffusivity*. Once again an additional equation for a new variable T has been introduced, in addition to the continuity and momentum equations which involve the unknowns p, ρ, **u**. To close the system, once again the equation of state $f(p, \rho, T)$ has to be invoked, for ocean water of fixed composition.

Further complication is to be expected, as the ocean density is actually determined by salinity as well as temperature. Since salts dissolved in water are of fixed composition but the solution is not, the total concentration of salt determines salinity, modifying *equation of state* for sea-water

$$f(\rho, p, T, S) = 0. \tag{1.94}$$

Since compressibility of seawater is negligible, the role of pressure in determining (1.94) is of minor importance. Non-linear empirical form of (1.94), linearized around some central values T_0, S_0, ρ_0 is given as

$$\rho = \rho_0(1 - \alpha_T(T - T_0) + \alpha_S(S - S_0)) \tag{1.95}$$

with extra variable S introduced in the equations, a conservation statement for salinity (i.e. a diffusion equation) is introduced to complement the system as

$$\frac{DS}{Dt} = \nabla \cdot \kappa_S \nabla S. \tag{1.96}$$

Solutions for **u**, ρ, p, T, S can therefore be obtained from the complete set of Eqs. (1.1), (1.2), (1.93), and (1.96).

Diffusivities κ_T and κ_S, generally speaking are not equal in the ocean. If they are assumed to be equal for the sake of simplicity, or if the effect of salinity in the equation of state is neglected altogether, then it is possible to combine (1.93) and (1.94) using (1.95)

$$\frac{D\rho}{Dt} = \nabla \cdot \kappa \nabla \rho \tag{1.97}$$

where $\kappa = \kappa_T = \kappa_S$.

Note that Eq. (1.97) is quite different in form and meaning from continuity equation (1.1), although both equations involve conservation statements for ρ. In continuity equation, $D\rho/Dt$ stands for density changes due to either compressibility or inhomogeneity, which create divergence of velocity; whereas compressibility neglected in the thermodynamic equation leads to Eq. (1.97) relating density changes to diffusion of heat and salt. Also note that by introducing (1.97), Eqs. (1.93), (1.95) and (1.96) become redundant. This is because Eq. (1.97) complements continuity and momentum equations (1.1, 1.2) to yield three equations for three unknowns ρ, p, **u**.

Appropriate versions of the thermodynamic equation (1.70) or (1.74) for the atmosphere and (1.97) for the ocean are used in further derivations of Chap. 2.

1.4   Vorticity Dynamics

Analyses of vorticity dynamics in rotating, stratified fluids will be based on earlier examination in GFD-I. *Relative vorticity* is defined as

$$\boldsymbol{\omega} = \nabla \times \mathbf{u} \tag{1.98}$$

with respect to inertial (rotating) coordinates. Velocity field in the absolute (fixed) coordinate system is given by (GFD-I), denoted by subscript A

$$\mathbf{u}_A = \mathbf{u} + \boldsymbol{\Omega} \times \mathbf{r}, \tag{1.99}$$

with $\boldsymbol{\Omega}$ representing earth's angular velocity and **r** the position vector in the rotating system, an *absolute vorticity* likewise can be defined as

$$\boldsymbol{\omega}_A = \nabla \times \mathbf{u} + \nabla \times \boldsymbol{\Omega} \times \mathbf{r} = (\boldsymbol{\omega} + 2\boldsymbol{\Omega}) \tag{1.100}$$

by virtue of vector differentiation rules given in GFD-I (Eq. 1.27.d).

By making use of vector identities, the momentum equation (1.2) is first written as

$$\frac{\partial \mathbf{u}}{\partial t} + \frac{1}{2}\nabla(\mathbf{u} \cdot \mathbf{u}) - \mathbf{u} \times \nabla \times \mathbf{u} + 2\boldsymbol{\Omega} \times \mathbf{u} = \mathbf{g} - \frac{1}{\rho}\nabla p + \mathbf{F} \tag{1.101}$$

where $\mathbf{F} = \frac{1}{\rho}\nabla \cdot \mathbf{d}$ is the net frictional force. Taking the curl of the modified momentum equation (1.101), and using the definition (1.98) and vector identities of GFD-I, it can be shown that $\boldsymbol{\omega}$ satisfies

$$\frac{\partial \boldsymbol{\omega}}{\partial t} + \nabla \times [(\boldsymbol{\omega} + 2\boldsymbol{\Omega}) \times \mathbf{u}] = -\nabla\frac{1}{\rho} \times \nabla p + \nabla \times \mathbf{F}. \tag{1.102}$$

For constant $\boldsymbol{\Omega}$, Eq. (1.100) is used to obtain

$$\frac{\partial \boldsymbol{\omega}_A}{\partial t} + \nabla \times (\boldsymbol{\omega}_A \times \mathbf{u}) = -\nabla\frac{1}{\rho} \times \nabla p + \nabla \times \mathbf{F}. \tag{1.103}$$

Upon making use of vector identities,

$$\nabla \cdot \boldsymbol{\omega}_A = \nabla \cdot (\boldsymbol{\omega} + 2\boldsymbol{\Omega}) = \nabla \cdot \nabla \times \mathbf{u}_A \equiv 0,$$

the second term is expanded to give

$$\frac{D\boldsymbol{\omega}_A}{Dt} = \frac{\partial \boldsymbol{\omega}_A}{\partial t} + \mathbf{u} \cdot \nabla \boldsymbol{\omega}_A = \boldsymbol{\omega}_A \cdot \nabla \mathbf{u} - \boldsymbol{\omega}_A \nabla \cdot \mathbf{u} - \nabla \frac{1}{\rho} \times \nabla p + \nabla \times \mathbf{F}. \quad (1.104)$$

Equations (1.102), (1.103) and (1.104) are different forms of the vorticity equation. In Eq. (1.104), the left hand side is the material derivative of absolute vorticity. The right hand side includes effects of vortex stretching (first term), vorticity change due to compressibility (second term), *vorticity induced by stratification* (third term) and torques generated by friction forces (fourth term). Here, the third term can be written as

$$-\nabla \frac{1}{\rho} \times \nabla p = \frac{1}{\rho^2} \nabla \rho \times \nabla p \qquad (1.105)$$

and in the absence of other effects (vortex stretching, compressibility and friction) this is the only term to induce changes in vorticity of a stratified fluid. If either the fluid is homogeneous ($\nabla \rho = 0$), or if the fluid is stratified such that density and pressure gradients depend on each other ($p = p(\rho)$), then the above term also vanishes, leaving absolute vorticity $\boldsymbol{\omega}_A$ to be conserved. If any of these conditions are not satisfied, (1.105) represents *overturning* tendency of a stratified fluid.

Next, consider some special quantity θ that is conserved following a fluid particle

$$\frac{D\theta}{Dt} = \frac{\partial \theta}{\partial t} + \mathbf{u} \cdot \nabla \theta = 0. \qquad (1.106)$$

For example, the conserved quantity θ could be salinity, temperature or any other conserved quantity, not being subject to effects of diffusion, friction or other internal sources of energy. Vector identities relate $\nabla \theta$, $\boldsymbol{\omega}_A$, and $\mathbf{u}$ by

$$\nabla \theta \times (\boldsymbol{\omega}_A \times \mathbf{u}) = (\boldsymbol{\omega}_A \cdot \nabla \theta) \mathbf{u} - (\mathbf{u} \cdot \nabla \theta) \boldsymbol{\omega}_A$$

and substituting from (1.106) gives

$$\nabla \theta \times (\boldsymbol{\omega}_A \times \mathbf{u}) = -\frac{\partial \theta}{\partial t} \boldsymbol{\omega}_A - (\boldsymbol{\omega}_A \cdot \nabla \theta) \mathbf{u}. \qquad (1.107)$$

If Eq. (1.103) is multiplied with $\nabla \theta$, considering (1.107) and applying vector identities results in the following:

$$\nabla \theta \cdot \frac{\partial \boldsymbol{\omega}_A}{\partial t} + \nabla \cdot \left[\boldsymbol{\omega}_A \frac{\partial \theta}{\partial t} + \mathbf{u} (\boldsymbol{\omega}_A \cdot \nabla \theta) \right] = -\nabla \theta \cdot (\nabla \frac{1}{\rho} \times \nabla p) + \nabla \theta \cdot (\nabla \times \mathbf{F}).$$
$$(1.108)$$

Further use of vector identities yields

$$\frac{D(\boldsymbol{\omega}_A \cdot \nabla\theta)}{Dt} + (\boldsymbol{\omega}_A \cdot \nabla\theta)\nabla \cdot \mathbf{u} = -\nabla\theta \cdot (\nabla\frac{1}{\rho} \times \nabla p) + \nabla\theta \cdot (\nabla \times \mathbf{F}). \quad (1.109)$$

Now, making use of the continuity equation (1.1) and dividing by ρ, the equation takes the following form, which is known as *Ertel's theorem*:

$$\frac{D}{Dt}\left[\frac{\boldsymbol{\omega}_A \cdot \nabla\theta}{\rho}\right] = \frac{1}{\rho}\nabla\theta \cdot \left[-\nabla\frac{1}{\rho} \times \nabla p + \nabla \times \mathbf{F}\right]. \quad (1.110)$$

In the absence of frictional effects (setting $\mathbf{F} = 0$), the left hand side bracketed quantity must be conserved if one of the following conditions is satisfied: either θ, p and ρ are related to each other, or ρ is constant (homogeneous fluids) or $\nabla\theta$ and ∇p are co-planar vectors.

If the flow is inviscid and incompressible but stratified (satisfying $\nabla \cdot \mathbf{u} = 0$ or $D\rho/Dt = 0$, by virtue of 1.1), ρ can replace θ in Eqs. (1.106) and (1.110), where the right hand side of (1.110) vanishes, yielding the following conservation statement:

$$\frac{D}{Dt}\left[\frac{\boldsymbol{\omega}_A \cdot \nabla\rho}{\rho}\right] = 0. \quad (1.111)$$

Next, consider a material surface S enclosed by a material curve C in a fluid. The quantity

$$\Gamma(t) = \oint_C \mathbf{u} \cdot d\mathbf{r} \quad (1.112)$$

is defined as the *circulation* in the non-inertial frame, and by virtue of Stokes' theorem (GFD-I)

$$\Gamma(t) = \int_S \nabla \times \mathbf{u} \cdot \mathbf{n} \, ds = \int_S \boldsymbol{\omega} \cdot \mathbf{n} \, ds \quad (1.113)$$

is the vorticity flux passing through the surface S. It is also possible to define circulation in the absolute frame of reference as

$$\Gamma_A(t) = \oint_C \mathbf{u}_A \cdot d\mathbf{r} = \Gamma(t) + \oint_C \boldsymbol{\Omega} \times \mathbf{r} \cdot d\mathbf{r} \quad (1.114)$$

by virtue of (1.99). Through the use of vector identities, and the Stokes' theorem, it can be shown that the second term equals

$$\oint_C \boldsymbol{\Omega} \times \mathbf{r} \cdot d\mathbf{r} = \int_S \nabla \times (\boldsymbol{\Omega} \times \mathbf{r}) \cdot \mathbf{n} \, ds = \int_S 2\boldsymbol{\Omega} \cdot \mathbf{n} \, ds, \quad (1.115)$$

yielding

$$\Gamma_A(t) = \Gamma(t) + \int_S 2\boldsymbol{\Omega} \cdot \mathbf{n} \, ds = \int_S (\boldsymbol{\omega} + 2\boldsymbol{\Omega}) \cdot \mathbf{n} \, ds = \int_S \boldsymbol{\omega}_A \cdot \mathbf{n} \, ds \quad (1.116)$$

where $\boldsymbol{\omega}_A$ is defined in (1.100). The integrand of the second term in (1.116) is the component of $2\boldsymbol{\Omega}$ normal to the surface S, or alternatively

$$\int_S 2\boldsymbol{\Omega} \cdot \mathbf{n}\, ds = 2\Omega S_p, \tag{1.117}$$

where vector $\boldsymbol{\Omega}$ is constant and $\Omega = |\boldsymbol{\Omega}|$ and S_p is the projection of the surface S on the plane perpendicular to $\boldsymbol{\Omega}$. The rate of change of Γ_A is then

$$\frac{d\Gamma_A}{dt} = \frac{d}{dt} \oint_C \mathbf{u}_A \cdot d\mathbf{r} = \oint_C \frac{D\mathbf{u}_A}{Dt} \cdot d\mathbf{r} + \oint_C \mathbf{u}_A \cdot \frac{D}{Dt}(d\mathbf{r}), \tag{1.118}$$

where the second term can be shown to vanish

$$\oint_C \mathbf{u}_A \cdot d\left(\frac{D\mathbf{r}}{Dt}\right) = \frac{1}{2} \oint_C d(\mathbf{u}_A \cdot \mathbf{u}_A) = 0 \tag{1.119}$$

for the integral is for a closed curve. Utilizing (1.116), (1.118) and (1.119), one obtains

$$\frac{d\Gamma}{dt} = \oint_C \frac{D\mathbf{u}_A}{Dt} \cdot d\mathbf{r} - 2\Omega \frac{dS_p}{dt}. \tag{1.120}$$

Making use of the momentum equation

$$\frac{D\mathbf{u}_A}{Dt} = \mathbf{g} - \frac{1}{\rho}\nabla p + \mathbf{F}, \tag{1.121}$$

and inserting in (1.120) yields

$$\frac{d\Gamma}{dt} = \oint_C \mathbf{g} \cdot d\mathbf{r} - \oint_C \frac{1}{\rho}\nabla p \cdot d\mathbf{r} + \oint_C \mathbf{F} \cdot d\mathbf{r} - 2\Omega \frac{dS_p}{dt}. \tag{1.122}$$

Next, by expressing $\mathbf{g} = -\nabla\phi$ and utilizing the Stokes theorem once more in (1.122)

$$\oint_C \mathbf{g} \cdot d\mathbf{r} = \oint_C \nabla\phi \cdot d\mathbf{r} = \int\int_S \nabla \times \nabla\phi \cdot \mathbf{n}\, dS = 0, \tag{1.123}$$

and simplification by making use of Stokes' theorem and vector identities leads to *Kelvin's circulation theorem*:

$$\frac{d\Gamma}{dt} = -\int_S \nabla\frac{1}{\rho} \times \nabla p \cdot \mathbf{n}\, dS + \oint_C \mathbf{F} \cdot d\mathbf{r} - 2\Omega \frac{dS_p}{dt}. \tag{1.124}$$

The first term on the right hand side is the rate of change in circulation due to *overturning* tendency in a stratified fluid, the second term by tangential shear stresses and the third term by changes in projected area (conservation of angular momentum). In short, these terms represent stratification, friction and rotation effects on the conservation of circulation. In the absence of all three effects, Kelvin's circulation theorem states that the circulation is conserved: i.e. an initially *irrotational fluid* will remain to be irrotational.

The first term represents creation or destruction of circulation (or vorticity) in a stratified fluid. For this term to vanish, the fluid is either (*i*) *homogeneous* (i.e. $\rho =$ constant, $\nabla \rho = 0$), or (*ii*) *barotropic*, (i.e. pressure is a function of density, $p = p(\rho)$), so that

$$-\nabla \frac{1}{\rho} \times \nabla p = \frac{1}{\rho^2} \nabla \rho \times \nabla p = \frac{1}{\rho^2} \frac{\partial p}{\partial \rho} \nabla \rho \times \nabla \rho = 0.$$

If either *inhomogeneous* (stratified) or *baroclinic* (not barotropic) conditions exist, the overturning term in (1.120) creates vorticity in the fluid.

Exercises

Exercise 1
In Chap. 1, equilibrium thermodynamics originally developed for fluids at rest consequently has been applied to fluids in motion. Compressible effects are expected to be significant in case of the atmosphere, while they are known to be rather small in the ocean.

The validity of assumptions regarding compressible effects in the ocean can be tested by considering thermodynamic behavior of sea water at great depths. Hydrophysical data obtained in the deepest pits of the earth's oceans, such as the Mariana Trench, provides an opportunity to test these basic elements of the theory. On the other hand, it is rather surprising that very few deep CTD casts have been performed to date in the Mariana Trench, providing evidence to test these assumptions.

Based on a literature review and evaluation of observations on seawater properties measured at the deepest oceanic regions, can you confirm compressibility of seawater as expressed in Chap. 1?

Basic State and Scales

2

2.1   Existence of a "Basic State"

Continuity and momentum equations (1.1), (1.2) applied to inhomogenous fluids were introduced in Chap. 1:

$$\frac{D\rho}{Dt} + \rho \nabla \cdot \mathbf{u} = 0 \tag{2.1.a}$$

$$\frac{D\mathbf{u}}{Dt} + 2\mathbf{\Omega} \times \mathbf{u} = \mathbf{g} - \frac{1}{\rho}\nabla p + \frac{1}{\rho}\nabla \cdot \mathbf{d} \tag{2.1.b}$$

followed by thermodynamic equations, respectively (1.58) for the atmosphere and in simplified form (1.59) for the ocean

$$c_p \frac{DT}{Dt} - \frac{\alpha_T T}{\rho}\frac{Dp}{Dt} = \frac{1}{\rho}\nabla \cdot K\nabla T + (Q + \Phi) \equiv Q^* \tag{2.1.c}$$

$$c_p \frac{DT}{Dt} = \frac{1}{\rho}\nabla \cdot K\nabla T + (Q + \Phi) \equiv Q^* \tag{2.1.d}$$

where $Q^* = Dq/Dt = \rho^{-1}\nabla \cdot K\nabla T + (Q + \Phi)$ represents non-isentropic sources of heat due to the combined effects of diffusion, internal and frictional heating. The above equations were then complemented by the equation of state (1.60)

$$f(\rho, p, T) = 0, \tag{2.1.e}$$

to solve for unknowns $(\rho, p, T, \mathbf{u})$ based on the above set of equations.

© Springer Nature Switzerland AG 2021
E. Özsoy, *Geophysical Fluid Dynamics II*, Springer Textbooks in Earth Sciences, Geography and Environment,
https://doi.org/10.1007/978-3-030-74934-7_2

In later development of quasi-geostrophic theory in Chap. 3, motions in the atmosphere and ocean will be expressed as perturbations with respect to a motionless *basic state*. The basic state of rest ($\mathbf{u} = 0$) assumes steady-state ($\partial/\partial t = 0$), implying ($D/Dt = 0$) for all basic state variables such as p, T and ρ under static forces.

Let us now investigate if such a state could actually exist on earth. The equations governing basic state of the fluid are then

$$\frac{1}{\rho}\nabla p + \mathbf{\Omega} \times \mathbf{\Omega} \times \mathbf{r} - \mathbf{g} = 0 \qquad (2.2.a)$$

$$\nabla^2 T = 0 \qquad (2.2.b)$$

$$\rho = \rho(p, T). \qquad (2.2.c)$$

The first of these is the 3-D momentum equation (1.2), where the gravity vector $\mathbf{g}$ is redefined according to (1.3). The second one is the thermodynamic equations (1.58) or (1.59) applied respectively to the atmosphere and ocean, assuming constant K. The last one is the equation of state for atmosphere or ocean, which for simplicity will be assumed linear in properties, excluding moisture effects in the atmosphere and salinity in the ocean.

Although one could investigate solutions for a geophysical fluid on earth, analogy to a container on a "rotating table" experiment (or even an infinite fluid under the influence of similar static forces for that matter) should be sufficient to demonstrate properties of the basic state. As the choice of coordinates is immaterial, $\mathbf{\Omega} = \Omega\mathbf{k}$ and $\mathbf{g} = -g\mathbf{k}$ are specified with unit vector $\mathbf{k}$ aligned with the axis of rotation

$$\nabla p = \rho\Omega^2\mathbf{r} + \rho g\mathbf{k}. \qquad (2.3)$$

Taking curl of (2.3) yields

$$\Omega^2 \nabla \times \rho\mathbf{r} + g\nabla \times \rho\mathbf{k} = 0 \qquad (2.4)$$

which in cylindrical coordinates (r, θ, z) reads as

$$g\frac{\partial \rho}{\partial r} + r\Omega^2\frac{\partial \rho}{\partial z} = 0 \qquad (2.5)$$

The heat equation (2.2.b) in the same coordinates is

$$\frac{\partial^2 T}{\partial z^2} + \frac{1}{r}\frac{\partial}{\partial r}\left(r\frac{\partial T}{\partial r}\right) + \frac{1}{r^2}\frac{\partial^2 T}{\partial \theta^2} = 0. \qquad (2.6)$$

It can be observed that the governing Eqs. (2.3) and (2.6) are obtained by simple extension of the limiting basic state. Equation (2.3) and its simplified form (2.5) are simply the hydro-static response to gravity, where centrifugal force due to earth's rotation have been included. The thermodynamic equation (2.6) is in adiabatic form, excluding any sources of heat. Although these two equations appear to be entirely independent of each other, the equation of state (2.2.c) links them together. We also note that the hydro-static Eq. (2.5) is of first order, while the heat equation (2.6) is of second order and Laplacian in form, while equation of state (2.2.c) in general, is a nonlinear equation, linking variables (ρ, p, T) in the basic state solution.

With the complete set of hydro and thermodynamic equations reviewed above, it can be immediately observed that a solution for basic state may not be possible. First of all, since density is a function of temperature and pressure, linked by (2.2.c), it would be impossible to obtain solutions for (2.5) and (2.6) simultaneously. For instance, a solution to (2.5) is of the form

$$\rho = \rho \left(z - \frac{\Omega^2 r^2}{2g} \right) \tag{2.7}$$

and consequently, by virtue of (2.2.c), temperature field should have the same dependence on coordinates

$$T = T \left(z - \frac{\Omega^2 r^2}{2g} \right). \tag{2.8}$$

It is clear that this functional form obtained from (2.5) contradicts with the typical solution that would satisfy (2.6). The contradiction in this simple hydro-thermo-static model shows that static equilibrium may not be achieved in a rotating-stratified fluid in the basic state.

To relieve incompatibility between thermal and hydro-static conditions of the basic state, it is needed to incorporate terms that have been neglected in the above simplified equations. To see which terms were neglected, comparison is needed with the original equations (2.1.a–e).

In the earlier part of this section a basic state solution has been searched, which is motionless $\mathbf{u} = 0$ and in steady state $D/Dt = 0$. The last term of (2.1.b) representing frictional effects also has been dropped in the motionless fluid. In fact, dropping all relevant terms in (2.1.a–e) has produced the set (2.2.a–c) bringing into focus the incompatibility in basic state, as reviewed. In the thermodynamic equation $\nabla \cdot K \nabla T = 0$, only the diffusion term is kept, which in fact is responsible for the incompatibility. Dropping diffusive term (and hence thermodynamics), the basic state would be the same as in a homogeneous fluid treated in GFD-I, in which case dependency similar to (2.7) could still be acceptable in terms of pressure.

There are two ways to eliminate and recover from incompatibility, so as to find an acceptable basic state. The situation is somewhat similar to geostrophic degeneracy reviewed in GFD-I. Making small corrections by either including friction and heating sources other than diffusion, introducing time dependence would automatically bring about motion, that however would contradict with the existence of a basic state.

The search is made to develop perturbations around a basic state of rest. A way out seems to exist, only by considering centrifugal forces to be typically much smaller than gravity, producing small changes in horizontal density compared to hydro-static forces in the vertical i.e.

$$\frac{\Omega^2 r^2}{2gz} \ll 1.$$
(2.9)

With this simplification, (2.7) and (2.8) reduce to

$$\rho = \rho(z) \text{ and } T = T(z)$$
(2.10.a, b)

and therefore an approximate solution of (2.7) is obtained

$$T = T_0 + \alpha z.$$
(2.11)

As a result, the basic state is prescribed by state variables which are functions of z alone, neglecting centrifugal acceleration in comparison to gravity. Furthermore, with isentropic conditions often imposed later, static equilibrium can exist in the fluid, under these assumptions. Basic state variables of the form $\rho_r(z)$, $p_r(z)$, $T_r(z)$ *etc.* therefore are allowed, implying $\mathbf{u}_r = 0$ in this state. Approximated static conditions without motion allow only vertical variations resulting from gravitation being the only force acting on the fluid. Therefore (2.2.a) yields

$$\frac{1}{\rho_r} \nabla p_r = \mathbf{g}$$
(2.12)

and making tangent-plane approximations with $\mathbf{g} = -g\mathbf{k}$, $\mathbf{k}$ being the unit vector in z-direction yields "hydro-static pressure" which is the accepted basic state

$$\frac{dp_r}{dz} = -\rho_r g.$$
(2.13)

2.2   Scale Heights

Considering simple models, estimates of vertical scales can be given for state variables.

For the atmosphere, ideal gas law $p_r = \rho_r R T_r$ is used. Upon using logarithmic differentiation

$$\frac{1}{p_r}\frac{dp_r}{dz} = \frac{1}{\rho_r}\frac{d\rho_r}{dz} + \frac{1}{T_r}\frac{dT_r}{dz}$$
(2.14)

and substituting (2.13) then gives

$$-\frac{1}{\rho_r}\frac{d\rho_r}{dz} = \frac{g}{RT_r} + \frac{1}{T_r}\frac{dT_r}{dz}.$$
(2.15)

A simple relation assuming *isothermal* atmosphere ($T_r = \bar{T}$ = constant) as the basic state in (2.15) yields

$$\rho_r = \rho_0 e^{-(z-z_0)/H_s} \tag{2.16}$$

where $H_s = R\bar{T}/g$ is the *density scale height*. For typical values of $R = 2.9 \times 10^2$ m^{2}s^{-2} $^\circ$K^{-1}, $\bar{T} = 293\,^\circ$K, $g = 9.81$ ms^{-2}, scale height is estimated as $H_s = 9$ km which is almost the same order as the thickness of troposphere.

For the ocean, we obtain a simple relation in the isentropic case (potential density $\Delta = 0$). By virtue of (1.88.a)

$$\rho_r(z) = \rho_0 e^{-g \int_{z_0}^{z} \frac{1}{c_s^2} dz} \cong \rho_0 e^{-(z-z_0)/H_s} \tag{2.17}$$

where $H_s = \bar{c}_s^2/g$ is the density scale height, $\bar{c}_s$ is the average speed of sound ($\bar{c}_s$ = constant). For typical values of $\bar{c}_s = 1500$ ms^{-1} and $g = 9.81$ ms^{-2}, a scale height of $H_s \cong 200$ km is obtained. Since the deepest part of the ocean has a depth of only about 10 km, the basic-state density change in the vertical is in fact shown to be negligible.

Approximating (2.17) with these estimates, density scale height H_s is obtained as

$$H_s = \left(\frac{-1}{\rho_r} \frac{d\rho_r}{dz} \right)^{-1} \tag{2.18}$$

where H_s is typically 10 km for the atmosphere and 200 km for the ocean, as seen from the above simplified models.

2.3   Perturbation Equations

Considering motion of geophysical fluids composed of a basic state with superimposed perturbations, state variables are broken into corresponding components

$$\begin{Bmatrix} p \\ \rho \\ T \\ \theta \\ \mathbf{u} \end{Bmatrix} = \begin{Bmatrix} p_r(z) \\ \rho_r(z) \\ T_r(z) \\ \theta_r(z) \\ 0 \end{Bmatrix} + \begin{Bmatrix} \tilde{p}(x,y,z,t) \\ \tilde{\rho}(x,y,z,t) \\ \tilde{T}(x,y,z.t) \\ \tilde{\theta}(x,y,z.t) \\ \mathbf{u}(x,y,z.t) \end{Bmatrix}. \tag{2.19}$$

Neglecting friction in the momentum equation (1.2) applied on a *tangent plane* at earth's surface results in

$$\frac{\partial \mathbf{u}_h}{\partial t} + \mathbf{u}_h \cdot \nabla_h \mathbf{u}_h + w \frac{\partial \mathbf{u}_h}{\partial z} + f\mathbf{k} \times \mathbf{u}_h = -\frac{1}{\rho} \nabla_h p$$

$$\frac{\partial w}{\partial t} + \mathbf{u}_h \cdot \nabla_h w + w \frac{\partial w}{\partial z} = -\frac{1}{\rho} \frac{\partial \rho}{\partial z} - g \tag{2.20.a, b}$$

These equations are the horizontal and vertical components of (1.2), with three dimensional velocity $\mathbf{u}$ decomposed into horizontal ($\mathbf{u}_h = (u, v)$) and vertical (w) components

$$\mathbf{u} = \mathbf{u}_h + w\mathbf{k} \tag{2.21}$$

such that horizontal $\nabla_h = \left(\frac{\partial}{\partial x}, \frac{\partial}{\partial y} \right)$ and vertical $\frac{\partial}{\partial z}$ operators add up to form the three dimensional gradient

$$\nabla = \nabla_h + \mathbf{k}\frac{\partial}{\partial z}. \tag{2.22}$$

The gravity vector is in the vertical direction $\mathbf{g} = -g\mathbf{k}$ with rotational contribution in (2.20.a) approximated as $f\mathbf{k} \times \mathbf{u}_h$ with Coriolis parameter $f = 2\Omega \sin \phi$, with vertical components being much smaller than $\mathbf{g}$ (in GFD-I). With the same notation, continuity equation is written as

$$\frac{\partial \rho}{\partial t} + \nabla_h \cdot \rho \mathbf{u}_h + \frac{\partial}{\partial z}\rho w = 0. \tag{2.23}$$

The above equations are complemented by thermodynamic counterparts. Thermodynamics equation (1.74) for the atmosphere has a source term Q^* representing non-adiabatic terms

$$\frac{\partial \theta}{\partial t} + \mathbf{u}_h \cdot \nabla_h \theta + w\frac{\partial \theta}{\partial z} = \frac{\theta}{c_p T}Q^*. \tag{2.24}$$

The equation of state is given as (1.68.b)

$$\theta = C_* \frac{p^{1/\gamma}}{\rho}, \tag{2.25}$$

where $C_* = p_*^{\left(1-\frac{1}{\gamma}\right)/R}$ is an appropriate constant.

For the ocean, full version of thermodynamic equation (1.78) is used, neglecting compressible and the non-adiabatic terms in quasi-geostrophic theory. It suffices to adopt the form of thermodynamic equation (1.97) written in terms of density

$$\frac{\partial \rho}{\partial t} + \mathbf{u}_h \cdot \nabla_h \rho + w\frac{\partial \rho}{\partial z} = R^*, \tag{2.26}$$

with non-adiabatic terms R^* involving thermal diffusion and external heating effects. This form of the density (mass) conservation is therefore sufficient, without the need for a separate equation of state for the ocean.

In the following development of the simple theory, all non-isentropic terms are neglected by setting $Q^* = R^* = 0$ in (2.24) and (2.26). Perturbed variables described in (2.19) are substituted in the above equations.

The horizontal momentum equation (2.11.a) becomes, (hereafter dropping the $()_h$ notation),

$$(\rho_r + \tilde{\rho}) \left\{ \frac{\partial \mathbf{u}}{\partial t} + \mathbf{u} \cdot \nabla \mathbf{u} + w \frac{\partial \mathbf{u}}{\partial z} + f \mathbf{k} \times \mathbf{u} \right\} = -\nabla (p_r + \tilde{p}) = -\nabla \tilde{p} \quad (2.27)$$

noting that the right hand side term applies to basic state pressure $p_r = p_r(z)$. Making use of the hydro-static approximation (2.13), the vertical momentum equation (2.21.b) then reads as

$$(\rho_r + \tilde{\rho}) \left\{ \frac{\partial w}{\partial t} + \mathbf{u} \cdot \nabla w + w \frac{\partial w}{\partial z} \right\} = -\frac{\partial (p_r + \tilde{p})}{\partial z} - (\rho_r + \tilde{\rho}) g$$
$$= -\frac{\partial \tilde{p}}{\partial z} - \tilde{\rho} g. \quad (2.28)$$

Since $\rho_r = \rho_r(z)$ only, the continuity equation (2.23) simplifies to

$$\frac{\partial \tilde{\rho}}{\partial t} + \mathbf{u} \cdot \nabla \tilde{\rho} + (\rho_r + \tilde{\rho}) \nabla \cdot \mathbf{u} + \frac{\partial}{\partial z} \rho_r w + \frac{\partial}{\partial z} \tilde{\rho} w = 0. \quad (2.29)$$

The thermodynamic equation for the atmosphere (2.24)becomes

$$\frac{\partial \tilde{\theta}}{\partial t} + \mathbf{u} \cdot \nabla \tilde{\theta} + w \frac{\partial \theta_r}{\partial z} + w \frac{\partial \tilde{\theta}}{\partial z} = 0 \quad (2.30)$$

for which the equation of state (2.25) can be written as

$$(\theta_r + \tilde{\theta}) = C_* (p_r + p)^{1/\gamma} (\rho_r + \tilde{\rho})^{-1}. \quad (2.31.a)$$

For the ocean, (2.26) takes the form

$$\frac{\partial \tilde{\rho}}{\partial t} + \mathbf{u} \cdot \nabla \tilde{\rho} + w \frac{\partial \rho_r}{\partial z} + w \frac{\partial \tilde{\rho}}{\partial z} = 0. \quad (2.31.b)$$

2.4   Order of Magnitude Analysis

The following scales are selected for an order of magnitude analysis of the equations

$$\begin{aligned}
\mathbf{x} &\sim L, \quad z \sim H, \quad t \sim \frac{L}{U} \\
\mathbf{u} &\sim U, \quad w \sim \alpha \frac{H}{L} U, \quad f \sim f_0 \\
p_r &\sim p_0, \quad \rho_r \sim \rho_0, \quad \theta_r \sim \theta_0 \\
\tilde{p} &\sim p_*, \quad \tilde{\rho} \sim \rho_*, \quad \tilde{\theta} \sim \theta_*,
\end{aligned} \quad (2.32.a-k)$$

where the time scale represents travel time with speed U over an excursion length
of L. Alternatively,

$$t \sim \frac{L}{U} = \frac{1}{f_0}\left(\frac{f_0 L}{U}\right) = \frac{1}{f_0 \delta} \tag{2.33}$$

where f_0 is the Coriolis parameter and

$$\delta = \frac{U}{f_0 L} \tag{2.34}$$

is the *Rossby number*. Equation (2.33) implies that the time scale of the motion is
much larger than the inertial period $2\pi/f_0$, corresponding to small values of Rossby
number, $\delta \ll 1$. The imposed scaling inherently assumes that the vertical velocity is
much smaller than horizontal, in proportion to a small parameter

$$\lambda = H/L \tag{2.35}$$

justifying the shallow water approximation $\lambda \ll 1$. In fact, this is implied by the
continuity equation (2.23), especially when density effects are small. As have been
shown in GFD-I, vertical motion in a rotating homogeneous fluid appears to be
decoupled from the horizontal, when $\lambda \ll 1$. Vertical motion is typically imposed by
external or boundary effects such as Ekman pumping or bottom topography. Vertical
velocity imposed by topographic height h_b is expected to be of the order

$$w = O\left(U\frac{h_b}{L}\right) = O\left(\alpha\lambda U\right) \tag{2.36}$$

where $\alpha = h_b/H$ is a small parameter ($\alpha \ll 1$) representing small deviations of
bottom topography compared to total depth. In the development of quasi-geostrophic
theory for homogeneous fluids in GFD-I, it has been assumed that $\alpha \ll 1$. Imposing
large deviations of topography $\alpha = O(1)$ can be selected within limits of the theory.
Vertical motions are represented by replacing the small parameter α with δ, to yield

$$w \sim \delta\lambda U. \tag{2.37}$$

A final dimensionless parameter which appears after scaling is the (external)
divergence parameter

$$\mu = \frac{f_0^2 L^2}{gH} = \left(\frac{L}{R}\right)^2 \tag{2.38}$$

where $R = \sqrt{gH}/f_0$ is defined as the *external radius of deformation*.

Typical scales of pressure, density and temperature are obtained as follows. Small
perturbations are applied to pressure $p_* \ll p_0$, similarly for density and temperature.
To correctly scale the terms $\tilde{p}$ and $\tilde{\rho}$ one must have

$$p_* \sim \rho_0 f_0 UL. \tag{2.39}$$

and similarly for

$$\rho_* \sim \frac{f_0 U L}{g H} \rho_0 = \mu \delta \rho_0. \tag{2.40}$$

where it is noted that

$$\frac{f_0 U L}{g H} = \frac{f_0^2 L^2}{g H} \frac{U}{f_0 L} = \mu \delta.$$

It can also be noted that hydro-static balance (2.13) for the basic state gives

$$p_0 \sim \rho_0 g H \tag{2.41.a}$$

yielding

$$p_* \sim \frac{f_0 U L}{g H} \rho_0 g H = \mu \delta p_0. \tag{2.41.b}$$

Scaling (2.27) by (2.32.a–k) and (2.37) yields

$$\delta \left(\rho_r + \delta \mu \tilde{\rho}\right) \left(\mathbf{u}_t + \mathbf{u} \cdot \nabla \mathbf{u} + \delta \, w \mathbf{u}_z\right) + \mathbf{k} \times \mathbf{u} = -\nabla \tilde{p}, \tag{2.42}$$

With small factors multiplying inertial terms, the Coriolis term balances pressure gradient to $O(1)$. The small perturbation terms involving $\tilde{\rho}$ are two orders of magnitude smaller than the basic state density $\rho_r(z)$. Leaving out these small terms in total density and including only the basic state component in pressure gradient term appropriately gives the horizontal momentum equation

$$\delta(\mathbf{u}_t + \mathbf{u} \cdot \nabla \mathbf{u} + \delta \, w \mathbf{u}_z) + \mathbf{k} \times \mathbf{u} = -\nabla \frac{\tilde{p}}{\rho_r}. \tag{2.43}$$

Similarly, scaling (2.28) and ignoring the perturbation density yields the vertical component of the momentum equation

$$\lambda^2 \delta \left(\frac{\partial w}{\partial t} + \mathbf{u} \cdot \nabla w + \delta w \frac{\partial w}{\partial z}\right) = -\frac{1}{\rho_r} \frac{\partial \tilde{p}}{\partial z} - \frac{\tilde{\rho}}{\rho_r}. \tag{2.44}$$

Since the left hand side is much smaller than the right hand side terms, hydro-static pressure is applicable also in perturbation field, in the same way as the basic state. By similar arguments about density, the potential temperature anomaly is scaled as

$$\theta_* \sim \mu \delta \theta_0. \tag{2.45}$$

Since ρ and θ are related, (2.31) is scaled as in the above, yielding

$$\left(\theta_r + \delta \mu \tilde{\theta}\right) = \frac{c_* (\rho_0 g H)^{1/\gamma}}{\rho_0 \theta_0} (p_r + \delta \mu \tilde{p})^{1/\gamma} (\rho_r + \delta \mu)^{-1} \tag{2.46}$$

The constant $c_* = p_*^{1-\frac{1}{\gamma}}/R$, with p_* is noted to be an arbitrary reference pressure. Selecting $p_* = \rho_0 g H = p_0$, the constant in (2.46) is evaluated as

$$\frac{c_*(\rho_0 g H)^{1/\gamma}}{\rho_0 \theta_0} = \frac{p_*^{1-\frac{1}{\gamma}}}{R} \frac{(\rho_0 g H)^{1/\gamma}}{\rho_0 \theta_0} = \frac{(\rho_0 g H)^{1-\frac{1}{\gamma}+\frac{1}{\gamma}}}{\rho_0 R \theta_0} = \frac{g H}{R \theta_0}. \qquad (2.47)$$

On the other hand, a relationship can be established between arbitrary scales θ_0, p_0 and ρ_0, by using relation (2.25)

$$\theta_0 = \frac{p_*^{1-\frac{1}{\gamma}}}{R} \frac{p_0^{1/\gamma}}{\rho_0} = \frac{p_0}{\rho_0 R} = \frac{g H}{R} \qquad (2.48)$$

so that the constant in (2.47) i.e. the right hand side of (2.48) becomes unity, yielding (in 2.46)

$$\left(1 + \delta\mu \frac{\tilde{\theta}}{\theta_r}\right) = \left(1 + \delta\mu \frac{\tilde{p}}{p_r}\right)^{1/\gamma} \left(1 + \delta\mu \frac{\tilde{\rho}}{\rho_r}\right)^{-1}. \qquad (2.49)$$

Since $\delta\mu \ll 1$, expansion in Taylor Series yields

$$\left(1 + \delta\mu \frac{\tilde{\theta}}{\theta_r}\right) = \left(1 + \frac{1}{\gamma}\delta\mu \frac{\tilde{p}}{p_r} + \cdots\right)\left(1 - \delta\mu \frac{\tilde{\rho}}{\rho_r} + \cdots\right)$$
$$= 1 + \delta\mu \left(\frac{\tilde{p}}{\gamma p_r} - \frac{\tilde{\rho}}{\rho_r}\right) + (\delta\mu)^2 () + \cdots \qquad (2.50)$$

which gives, to $O(\delta\mu)$

$$\frac{\tilde{\theta}}{\theta_r} = \frac{1}{\gamma}\frac{\tilde{p}}{p_r} - \frac{\tilde{\rho}}{\rho_r}. \qquad (2.51)$$

Similarly expanding the ideal gas law $p = \rho R T$ and scaling as $T_r \sim T_0$ and $\tilde{T} \sim T_* = \delta\mu T_0$,

$$p_0 (p_r + \delta\mu \tilde{p}) = \rho_0 R T_0 (\rho_r + \delta\mu \tilde{\rho})(T_r + \delta\mu T) \qquad (2.52.a)$$

and noting $p_0 = \rho_0 R T_0$, to $O(\delta\mu)$ gives

$$\frac{\tilde{p}}{p_r} = \frac{\tilde{\rho}}{\rho_r} + \frac{\tilde{T}}{T_r}. \qquad (2.52.b)$$

These relations establish the conversion between perturbation variables $\tilde{p}$, $\tilde{\rho}$, $\tilde{\theta}$ and $\tilde{T}$.

Finally to develop non-dimensional form of the Coriolis parameter using the tangent plane approximation, the scales $f \sim f_0$, $y \sim L$, $\beta \sim \beta_0$ are utilized, so that

$$f = f_0 f' = f_0 + \beta_0 y = f_0 \left(1 + \frac{\beta_0 L}{f_0} \beta' y' \right) = f_0 (1 + \delta \beta' y') \qquad (2.53.a)$$

$$\beta_0 = \frac{2\Omega \cos \phi_0}{r_0} = \frac{f_0 \cot \phi_0}{r_0} \qquad (2.53.b)$$

where Ω is the earth's rotation rate, r_0 its radius, ϕ_0 is the latitude and primes indicate the non-dimensional variables. The dimensionless parameter $\delta = \beta_0 L / f_0$ can thus be estimated from

$$\delta = \frac{\beta_0 L}{f_0} = \cot \phi_0 \left(\frac{L}{r_0} \right) \qquad (2.54)$$

Taking $\cot \phi_0 \simeq O(1)$ for mid-latitudes, $r_0 = 6000$ km, and $L = 100$ km (ocean) or 1000 km (atmosphere) as appropriate scales, we find $\delta = \beta_0 L / f_0 = O(10^{-2} - 10^{-1})$, this small parameter to be the same order as the Rossby number. Therefore, appropriately scaled, non-dimensional tangent plane function (with primes dropped) is

$$f = 1 + \delta \beta y. \qquad (2.55)$$

Making use of (2.32.a–k), (2.34), (2.35), (2.37), (2.38), (2.40), (2.41), (2.42), (2.44), (2.45), (2.51) and (2.55), Eqs. (2.27)–(2.31) are scaled as follows:

$$(\rho_r + \delta \mu \tilde{\rho}) \delta \left\{ \frac{\partial \mathbf{u}}{\partial t} + \mathbf{u} \cdot \nabla \mathbf{u} + \delta w \frac{\partial \mathbf{u}}{\partial z} \right\} + (\rho_r + \delta \mu \tilde{\rho})(1 + \delta \beta y) \mathbf{k} \times \mathbf{u} = -\nabla \tilde{p}$$

$$(\rho_r + \delta \mu \tilde{\rho}) \lambda^2 \delta^2 \left\{ \frac{\partial w}{\partial t} + (\mathbf{u} \cdot \nabla) w + \delta w \frac{\partial w}{\partial z} \right\} = -\frac{\partial \tilde{p}}{\partial z} - \tilde{\rho}$$

$$\delta \mu \left\{ \frac{\partial \tilde{\rho}}{\partial t} + \mathbf{u} \cdot \nabla \tilde{\rho} + \delta \frac{\partial}{\partial z} \tilde{\rho} w \right\} + (\rho_r + \delta \mu \tilde{\rho}) \nabla \cdot \mathbf{u} + \delta \frac{\partial}{\partial z} \rho_r w = 0$$

$$\frac{\partial \tilde{\theta}}{\partial t} + \mathbf{u} \cdot \nabla \tilde{\theta} + \delta w \frac{\partial \tilde{\theta}}{\partial z} + \frac{1}{\mu} w \frac{\partial \theta_r}{\partial z} = 0 \quad \text{(atmosphere)}$$

$$\frac{\tilde{\theta}}{\theta_r} = \frac{1}{\gamma} \frac{\tilde{p}}{p_r} - \frac{\tilde{\rho}}{\rho_r} \qquad \text{(atmosphere)}$$

$$\frac{\partial \tilde{\rho}}{\partial t} + \mathbf{u} \cdot \nabla \tilde{\rho} + \delta w \frac{\partial \tilde{\rho}}{\partial z} + \frac{1}{\mu} w \frac{\partial \rho_r}{\partial z} = 0. \quad \text{(ocean)}$$

$$(2.56.a - f)$$

The first three are the momentum and continuity equations valid for both the ocean and the atmosphere. The thermodynamic and state equations for the atmosphere are given by (2.56.d, e) and for the ocean is given by (2.56.f).

2.5   Scale Analyses

Typical estimates of the small parameters δ, λ and μ can be obtained by taking $f_0 = 10^{-4}\,\mathrm{s}^{-1}$, $g = 10\,\mathrm{m s}^{-1}$, $U = 10\,\mathrm{m/s}$, $L = 1000\,\mathrm{km}$, $H = 10^4\,\mathrm{m}$ for the atmosphere, and $U = 1\,\mathrm{m/s}$, $L = 100\,\mathrm{km}$, $H = 10^3\,\mathrm{m}$ for the ocean,

$$\delta \sim 10^{-1}$$
$$\lambda \sim 10^{-2}$$
$$\mu \sim 10^{-1} - 10^{-2}$$

which can be reduced for both fluids, i.e.

$$\mu = O(\delta) - O(\delta^2)$$
$$\lambda = O(\delta^2) \tag{2.57}$$

so that δ is the largest of the small parameters. In fact, if we take slightly lower velocities or larger horizontal scales this parameter may become $\delta = O(1)$. Note also that the parameters appear in groups of δ, μ, $\delta\mu$, or $\delta\lambda$, the latter being always smaller. Therefore only those terms of $O(\delta)$ kept in the equations,

$$\delta\rho_r \left\{ \mathbf{u}_t + \mathbf{u} \cdot \nabla \mathbf{u} + \beta \mathbf{k} \times \mathbf{u} \right\} + \rho_r \mathbf{k} \times \mathbf{u} = -\nabla \tilde{p}$$

$$0 = -\frac{\partial \tilde{p}}{\partial z} - \tilde{\rho}$$

$$\nabla \cdot \mathbf{u} + \frac{\delta}{\rho_r} \frac{\partial}{\partial z}(\rho_r w) = 0$$

$$\frac{\partial \tilde{\theta}}{\partial t} + \mathbf{u} \cdot \nabla \tilde{\theta} + \frac{1}{\mu} \frac{\partial \theta_r}{\partial z} w = 0 \quad \text{(atmosphere)} \tag{2.58.a - f}$$

$$\frac{\tilde{\theta}}{\theta_r} = \frac{1}{\gamma} \frac{\tilde{p}}{p_r} - \frac{\tilde{\rho}}{\rho_r} \quad \text{(atmosphere)}$$

$$\frac{\partial \tilde{\rho}}{\partial t} + \mathbf{u} \cdot \nabla \tilde{\rho} + \frac{1}{\mu} \frac{\partial \rho_r}{\partial z} w = 0 \quad \text{(ocean)}$$

Note that those terms with $O\left(\frac{1}{\mu}\right) \gg 1$ are dominant, while smaller terms of $O(1)$, and $O(\delta)$ are kept in (2.56.a–f). Dividing equation (2.56.c) by μ allows to keep terms of $O(\mu) \simeq O(\delta^2)$ and $O(\mu\delta)$, otherwise leading to dynamically inconsistent result $w = 0$.

The set of equations to be used for the atmosphere are (2.58.a–e), while for the ocean they are (2.58.a–c) and (2.58.f), with new variables defined as

$$\phi = \frac{\tilde{p}}{\rho_r}$$

$$r = \frac{\tilde{\rho}}{\rho_r} \qquad (2.59.a - c)$$

$$s = \frac{\tilde{\theta}}{\theta_r}$$

Dividing (2.58.b) by ρ_r

$$\frac{1}{\rho_r}\frac{\partial \tilde{p}}{\partial z} + \frac{\tilde{\rho}}{\rho_r} = 0 \qquad (2.60)$$

and rearranging,

$$\frac{\partial \left(\frac{\tilde{p}}{\rho_r}\right)}{\partial z} + \frac{1}{\rho_r}\frac{\partial \rho_r}{\partial z}\frac{\tilde{p}}{\rho_r} + \frac{\tilde{\rho}}{\rho_r} = 0 \qquad (2.61)$$

and defining

$$K(z) = \frac{1}{\rho_r}\frac{d\rho_r}{dz} \qquad (2.62)$$

enables this to be rewritten as

$$\frac{\partial \phi}{\partial z} + K(z)\phi + r = 0. \qquad (2.63)$$

Secondly, (2.58.e) is written as

$$s = L(z)\phi - r \qquad (2.64)$$

where

$$L(z) = \frac{1}{\gamma}\frac{\rho_r}{p_r}. \qquad (2.65)$$

Finally, noting that θ_r and ρ_r are functions of z alone, and defining

$$M(z) = \frac{1}{\theta_r}\frac{d\theta_r}{dz}, \qquad (2.66)$$

Equations (2.58.d) and (2.58.f) take the forms

$$\left(\frac{\partial}{\partial t} + \mathbf{u}\cdot\nabla\right)s + \frac{1}{\mu}M(z)w = 0$$

$$\left(\frac{\partial}{\partial t} + \mathbf{u}\cdot\nabla\right)r - \frac{1}{\mu}K(z)w = 0. \qquad (2.67a, b)$$

With these notations, equations (2.58.a–f) become

$$\delta \left\{ \frac{\partial \mathbf{u}}{\partial t} + \mathbf{u} \cdot \nabla \mathbf{u} + \beta y \mathbf{k} \times \mathbf{u} \right\} + \mathbf{k} \times \mathbf{u} = -\nabla \phi$$

$$\frac{\partial \phi}{\partial z} + K(z)\phi + r = 0$$

$$\nabla \cdot \mathbf{u} + \delta \frac{1}{\rho_r} \frac{\partial}{\partial z}(\rho_r w) = 0 \qquad\qquad\qquad (2.68.a - f)$$

$$\frac{\partial s}{\partial t} + \mathbf{u} \cdot \nabla s + \frac{1}{\mu} M(z)w = 0 \qquad \text{(atmosphere)}$$

$$s = L(z)\phi - r \qquad \text{(atmosphere)}$$

$$\frac{\partial r}{\partial t} + \mathbf{u} \cdot \nabla r + \frac{1}{\mu} K(z)w = 0 \qquad \text{(ocean).}$$

These steps allow atmospheric and oceanic components to be unified in the next sections.

2.6   Brunt–Väisälä (Stratification) Parameter

2.6.1   Atmosphere

The dimensionless stratification parameter for the atmosphere is defined as

$$S(z) = \frac{1}{\mu} \frac{1}{\theta_r} \frac{d\theta_r}{dz} = \frac{1}{\mu} M(z) \qquad\qquad (2.69)$$

which in dimensional (primed) variables, gives

$$S(z) = \frac{H}{\left(\frac{f_0^2 L^2}{gH}\right)} \frac{1}{\theta_r'} \frac{d\theta_r'}{dz'} = \frac{H^2}{f_0^2 L^2} \frac{g}{\theta_r'} \frac{d\theta_r'}{dz'}$$

$$= \left(\frac{H}{L}\right)^2 \left(\frac{N'(z)}{f_0}\right)^2 = \left(\frac{H}{L}\right)^2 \left(\frac{N_0}{f_0}\right)^2 N^2(z) \qquad (2.70)$$

$$= S_0 N^2(z)$$

with Brunt Väisälä frequency defined as

$$N'^2(z) \equiv N_0^2 \, N^2(z) = \frac{g}{\theta_r'} \frac{d\theta_r'}{dz'}. \qquad\qquad (2.71)$$

For typical values of $H = 10^4 \, \text{m}$, $L = 10^6 \, \text{m}$, $f_0 = 10^{-4} \, \text{s}^{-1}$ and $N_0 = 10^{-2} \, \text{s}^{-1}$ for the atmosphere,

$$S_0 = \left(\frac{H}{L}\right)^2 \left(\frac{N_0}{f_0}\right)^2 = O(1) \qquad\qquad (2.72)$$

so that $\frac{1}{\mu}M(z) = S(z) = O(1)$ in equation (2.68.d). Making use of potential temperature (2.25) for the basic state, first taking logarithms of both sides,

$$\ln \theta'_r = \frac{1}{\gamma} \ln p'_r - \ln \rho'_r + \ln c_*, \tag{2.73}$$

Equation (2.71) is evaluated as

$$
\begin{aligned}
N'^2(z) = g \frac{d \ln \theta'_r}{dz} &= \frac{g}{\gamma} \frac{1}{p'_r} \frac{dp'_r}{dz} - \frac{g}{\rho'_r} \frac{d\rho'_r}{dz} \\
&= g \left(-\frac{1}{\rho'_r} \frac{d\rho'_r}{dz} - \frac{g\rho'_r}{\gamma p'_r} \right)
\end{aligned} \tag{2.74}
$$

where use has been made of the hydro-static relation (2.13). Non-dimensional form with scales $N' \sim N_0$, $\rho_r \sim \rho_0$, $p_r \sim \rho_0 g H$, $z \sim H$ yields

$$
\begin{aligned}
\mu S_0 N^2(z) &= -\frac{1}{\rho_r} \frac{d\rho_r}{dz} - \frac{g}{\gamma} \frac{\rho_r}{p_r} \\
&= -K(z) - L(z).
\end{aligned} \tag{2.75}
$$

and since $\mu \ll 1$, if follows that

$$K(z) \cong -L(z) \tag{2.76}$$

with this approximation it follows from (2.68.e) that

$$s = L(z)\phi - r \cong -K(z)\phi - r \tag{2.77}$$

and comparing these with (2.68.a–e), the governing equations for the atmosphere become

$$
\begin{aligned}
&\delta \left\{ \frac{\partial \mathbf{u}}{\partial t} + \mathbf{u} \cdot \nabla \mathbf{u} + \beta y \mathbf{k} \times \mathbf{u} \right\} + \mathbf{k} \times \mathbf{u} = -\nabla \phi \\
&\frac{\partial \phi}{\partial z} = s \\
&\nabla \cdot \mathbf{u} + \delta \frac{1}{\rho_r} \frac{\partial}{\partial z}(\rho_r w) = 0 \\
&\frac{\partial s}{\partial t} + \mathbf{u} \cdot \nabla s + S_0 N^2(z) w = 0
\end{aligned} \tag{2.78.a - d}
$$

2.6.2 Ocean

The stratification parameter in dimensionless form is given as

$$S(z) = -\frac{1}{\mu}\frac{1}{\rho_r}\frac{d\rho_r}{dz} = -\frac{1}{\mu}K(z) \tag{2.79}$$

which in dimensional variables reads as

$$S(z) = \frac{H^2}{f_0 L^2}\left(-\frac{g}{\rho_r}\frac{d\rho_r}{dz}\right) = S_0 N^2(z) \tag{2.80}$$

with N^2 defined as

$$N'^2(z) = N_0^2 N^2(z) = -\frac{g}{\rho_r}\frac{d\rho_r}{dz} - \frac{g^2}{c_s^2} \tag{2.81}$$

and

$$S_0 = \left(\frac{H}{L}\right)^2\left(\frac{N_0}{f_0}\right)^2. \tag{2.82}$$

Taking typical values of $H = 10^3$ m, $L = 10^5$ m, $N_0 = 10^{-2}\,\mathrm{s}^{-1}$, $f_0 = 10^{-4}\,\mathrm{s}^{-1}$, it is found to be of the order $S_0 = O(1)$. The second term $+g^2/c_s^2$ is by an order of magnitude smaller than the first term.

On the other hand, by virtue of (2.79)

$$K(z) = O(\mu) \tag{2.83}$$

can be neglected in equation (2.68.b), yielding the following governing equations

$$\delta\left\{\frac{\partial \mathbf{u}}{\partial t} + \mathbf{u}\cdot\nabla\mathbf{u} + \beta y\mathbf{k}\times\mathbf{u}\right\} + \mathbf{k}\times\mathbf{u} = -\nabla\phi$$

$$\frac{\partial\phi}{\partial z} = -r$$

$$\nabla\cdot\mathbf{u} + \delta\frac{1}{\rho_r}\frac{\partial}{\partial z}(\rho_r w) = 0 \tag{2.84.a - d}$$

$$\frac{\partial r}{\partial t} + \mathbf{u}\cdot\nabla r - S_0 N^2(z)w = 0$$

2.6.3   Combined Equations for the Atmosphere and Ocean

Note that the same final set of equations for the atmosphere and the ocean have been arrived at, in above derivations of the governing equations, the only difference being in how variables in each set are interpreted. The set of equations are basically of the following form, either taking (2.78.a–d) or (2.84.a–d) into consideration:

$$\delta \left\{ \frac{\partial \mathbf{u}}{\partial t} + \mathbf{u} \cdot \nabla \mathbf{u} + \beta y \mathbf{k} \times \mathbf{u} \right\} + \mathbf{k} \times \mathbf{u} = -\nabla \phi$$

$$\frac{\partial \phi}{\partial z} = s$$

$$\nabla \cdot \mathbf{u} + \delta \frac{1}{\rho_r} \frac{\partial}{\partial z}(\rho_r w) = 0$$

$$\frac{\partial s}{\partial t} + \mathbf{u} \cdot \nabla s + S_0 N^2(z) w = 0$$

$$(2.85.a - d)$$

The above set directly corresponds to (2.78.a–d) developed for the atmosphere. It is only needed to replace the variable s with $-r$ to obtain the set of equations (2.84.a–d) for the ocean. The only difference between the equations for the atmosphere and the ocean in interpreting the variable $s = \tilde{\theta}/\theta_r$ as the normalized perturbation potential temperature for the atmosphere, and the variable $s = -r = -\tilde{\rho}/\rho_r$ as the normalized perturbation density in case of the ocean. Further, the Brunt–Väisälä stratification parameter is defined in terms of basic state potential temperature (2.71) in the case of the atmosphere, and in terms of basic state density (2.81) in case of the ocean.

With these simplifications it is time to move on to the development of quasi-geostrophic theory in the next chapter.

Quasi-geostrophic Theory

3

3.1 Development of Quasi-geostrophic Theory

Based on developments in Chap. 2, it is possible to arrive at sets of governing equations unifying GFD theory for the atmosphere and ocean, by appropriately defining the state variable s for each of these fluid domains. Briefly, a normalized potential temperature $s = \tilde{\theta}/\theta_r$ has been defined in equation (2.78.a–d) for the atmosphere, and normalized perturbation density $s = -r = \tilde{\rho}/\rho_r$ in Eq. (2.84.a–d) for the ocean. Further, the Brunt–Väisälä stratification parameter is defined in terms of the basic state potential temperature (2.71) in the atmospheric case, and in terms of the basic state density (2.81) in the oceanic case.

With these replacements, the following unified set of equations valid for either atmospheric or oceanic fluid domains are obtained

$$
\delta \left\{ \frac{\partial \mathbf{u}}{\partial t} + \mathbf{u} \cdot \nabla \mathbf{u} + \beta y \mathbf{k} \times \mathbf{u} \right\} + \mathbf{k} \times \mathbf{u} = -\nabla \phi
$$

$$
\frac{\partial \phi}{\partial z} = s
$$

$$
\nabla \cdot \mathbf{u} + \delta \frac{1}{\rho_r} \frac{\partial}{\partial z} (\rho_r w) = 0 \qquad (3.1.a - d)
$$

$$
\frac{\partial s}{\partial t} + \mathbf{u} \cdot \nabla s + S_0 N^2(z) w = 0.
$$

It is noted however, that these equations become *degenerate* for geostrophic flows (i.e. Rossby number in the limit $\delta = U/f_0 L \to 0$), as demonstrated earlier for homogeneous fluids in GFD-I. This is simply shown by considering what happens when $\delta = 0$ in (3.1.a), with geostrophic velocity $\mathbf{u} = \mathbf{k} \times \nabla \phi$, automatically satisfying (3.1.c), $\nabla \cdot \mathbf{u} = 0$. This results in redundancy of the solutions. To overcome this

© Springer Nature Switzerland AG 2021   45
E. Özsoy, *Geophysical Fluid Dynamics II*, Springer Textbooks in Earth Sciences,
Geography and Environment,
https://doi.org/10.1007/978-3-030-74934-7_3

problem of *geostrophic degeneracy*, the first remedy is to seek power series solutions expanded in terms of the small parameter δ, as follows:

$$
\begin{aligned}
\mathbf{u} &= \mathbf{u}_0 + \delta \mathbf{u}_1 + \delta^2 \mathbf{u}_2 + \cdots \\
w &= w_0 + \delta w_1 + \delta^2 w_2 + \cdots \\
\phi &= \phi_0 + \delta \phi_1 + \delta^2 \phi_2 + \cdots \\
s &= s_0 + \delta s_1 + \delta^2 s_2 + \cdots .
\end{aligned}
\qquad (3.2.a-d)
$$

Inserting these in Eqs. (3.1.a–d) and by grouping terms in equal powers of δ, perturbed sets of equations are obtained. First order terms of $O(\delta^0)$ are

$$
\begin{aligned}
&\mathbf{k} \times \mathbf{u}_0 = \nabla \phi_0 \\
&\frac{\partial \phi_0}{\partial z} = s_0 \\
&\nabla \cdot \mathbf{u}_0 = 0 \\
&\frac{\partial s_0}{\partial z} + \mathbf{u}_0 \cdot \nabla s_0 + S_0 N^2(z) w_0 = 0,
\end{aligned}
\qquad (3.3.a-d)
$$

which, on the other hand, continue to give degeneracy at this order, though the can be used for diagnostic purposes. The first order velocity defines a stream-function $\mathbf{u}_0 = \mathbf{k} \times \nabla \phi_0$ and a first order vorticity based on Eq. (3.3.a),

$$
\boldsymbol{\omega}_0 = \nabla \times \mathbf{u}_0 = \nabla \times \mathbf{k} \times \nabla \phi_0 \equiv \mathbf{k} \nabla^2 \phi_0. \qquad (3.4)
$$

alternatively, scalar vorticity is obtained as $\zeta_0 = \nabla^2 \phi_0$ in two-dimensional flow $\boldsymbol{\omega}_0 = \mathbf{k} \zeta_0$, by making use of vector identities provided in GFD-I,

$$
\nabla \times \mathbf{k} \times \nabla \phi_0 = \underbrace{(\nabla \phi_0 \cdot \nabla)\mathbf{k}}_{0} - \underbrace{(\mathbf{k} \cdot \nabla)\nabla \phi}_{0} - \underbrace{\nabla \phi (\nabla \cdot \mathbf{k})}_{0} + \mathbf{k}(\nabla \cdot \nabla \phi_0) = \mathbf{k} \nabla^2 \phi_0.
$$

Examining equations at next order $O(\delta^1)$ involves terms carried over from $O(\delta^0)$,

$$
\begin{aligned}
&\frac{\partial \mathbf{u}_0}{\partial t} + \mathbf{u}_0 \cdot \nabla \mathbf{u}_0 + \beta y \mathbf{k} \times \mathbf{u}_0 + \mathbf{k} \times \mathbf{u}_1 = -\nabla \phi_1 \\
&\frac{\partial \phi_1}{\partial z} = s_1 \\
&\nabla \cdot \mathbf{u}_1 + \frac{1}{\rho_r} \frac{\partial}{\partial z}(\rho_r w_0) = 0 \\
&\frac{\partial s_1}{\partial t} + \mathbf{u}_1 \cdot \nabla s_0 + \mathbf{u}_0 \cdot \nabla s_1 + S_0 N^2(z) w_1 = 0.
\end{aligned}
\qquad (3.5.a-d)
$$

By eliminating terms of $O(\delta^1)$, closure at $O(\delta^0)$ is achieved. By taking curl of Eq. (3.3.a) one obtains

$$\frac{\partial \nabla \times \mathbf{u}_0}{\partial t} + \nabla \times (\mathbf{u}_0 \cdot \nabla \mathbf{u}_0) + \nabla \times (\beta y \mathbf{k} \times \mathbf{u}_0) + \nabla \times \mathbf{k} \times \mathbf{u}_1 = -\nabla \times \nabla \phi_1 \equiv 0.$$

(3.6)

Then, vector identities from GFD-I are used to simplify various terms

$$\mathbf{u}_0 \cdot \nabla \mathbf{u}_0 \equiv \frac{1}{2}\nabla(\mathbf{u}_0 \cdot \mathbf{u}_0) - \mathbf{u}_0 \times \nabla \times \mathbf{u}_0 = \frac{1}{2}\nabla|\mathbf{u}_0|^2 + \zeta_0 \mathbf{k} \times \mathbf{u}_0$$

$$\nabla \times (\mathbf{u}_0 \cdot \nabla \mathbf{u}_0) = \frac{1}{2}\nabla \times \nabla|\mathbf{u}_0|^2 + \nabla \times \zeta_0 \mathbf{k} \times \mathbf{u}_0 = \mathbf{k}\nabla \cdot (\zeta_0 \mathbf{u}_0)$$

$$= \mathbf{k}(\mathbf{u}_0 \cdot \nabla \zeta_0 + \zeta_0 \nabla \cdot \mathbf{u}_0)$$

$$= \mathbf{k}(\mathbf{u}_0 \cdot \nabla \zeta_0)$$

$$\nabla \times \beta y \mathbf{k} \times \mathbf{u}_0 = \mathbf{k} \cdot \nabla \beta y \mathbf{u}_0 = \mathbf{k}\beta y \nabla \cdot \mathbf{u}_0 + \mathbf{k}\mathbf{u}_0 \cdot \nabla \beta y$$

$$= \mathbf{k}\beta v_0$$

$$\nabla \times \mathbf{k} \times \mathbf{u}_1 = \mathbf{k}(\nabla \cdot \mathbf{u}_1).$$

Substituting these in Eq. (3.6) yields

$$\frac{\partial \zeta_0}{\partial t} + \mathbf{u}_0 \cdot \nabla \zeta_0 + \beta v_0 + \nabla \cdot \mathbf{u}_1 = 0,$$

and by making further use of (3.5.c), one obtains

$$\left(\frac{\partial}{\partial t} + \mathbf{u}_0 \cdot \nabla\right)\zeta_0 + \beta v_0 = \frac{1}{\rho_r}\frac{\partial}{\partial z}(\rho_r w_0).$$

(3.7)

On the other hand, combining (3.3.b) and (3.3.d) leads to

$$\left(\frac{\partial}{\partial t} + \mathbf{u}_0 \cdot \nabla\right)\left(\frac{\partial \phi_0}{\partial z}\right) + S_0 N^2(z)w_0 = 0,$$

(3.8)

which is re-written as follows

$$\rho_r w_0 = -\left(\frac{\partial}{\partial t} + \mathbf{u}_0 \cdot \nabla\right)\left(\frac{\rho_r}{S_0 N^2}\frac{\partial \phi_0}{\partial z}\right).$$

(3.9)

Differentiating with respect to z, the term on the right hand side of (3.7) is expressed as

$$\frac{\partial \rho_r w_0}{\partial z} = -\left(\frac{\partial}{\partial t} + \mathbf{u}_0 \cdot \nabla\right)\frac{\partial}{\partial z}\left(\frac{\rho_r}{S_0 N^2}\frac{\partial \phi_0}{\partial z}\right) - \frac{\partial \mathbf{u}_0}{\partial z} \cdot \nabla\left(\frac{\rho_r}{S_0 N^2}\frac{\partial \phi_0}{\partial z}\right).$$

(3.10)

Equation (3.10) can further be simplified if one is to make use of the geostrophic relation

$$\mathbf{u}_0 = \mathbf{k} \times \nabla \phi_0$$

leading to

$$\mathbf{u}_0 \cdot \nabla \phi_0 = (\mathbf{k} \times \nabla \phi_0) \cdot \nabla \phi_0 \equiv 0.$$

The last term in (3.10) is a product of the following form

$$L_z(\mathbf{u}_0) \cdot K_z(\nabla \phi_0) = L_z(\mathbf{k} \times \nabla \phi_0) \cdot K_z \nabla(\phi_0) \equiv 0$$

where L_z and K_z are derivatives depending on vertical coordinate z, so as to make the last term in (3.10) vanish. Noting that the remaining term is of total derivative form, combination of Eqs. (3.7) and (3.10) produces

$$\left(\frac{\partial}{\partial t} + \mathbf{u}_0 \cdot \nabla \right) \left[\zeta_0 + \frac{1}{\rho_r} \frac{\partial}{\partial z} \left(\frac{\rho_r}{S_0 N^2} \frac{\partial \phi_0}{\partial z} \right) \right] + \beta v_0 = 0. \qquad (3.11)$$

Finally, substituting equalities $\mathbf{u}_0 = \mathbf{k} \times \nabla \phi_0$, $\zeta_0 = \nabla^2 \phi_0$ and $v_0 = \partial \phi_0 / \partial x$ into (3.11), further manipulations in perturbation analysis leads to a new form of vorticity equation

$$\left(\frac{\partial}{\partial t} + \mathbf{k} \times \nabla \phi_0 \cdot \nabla \right) \left[\nabla^2 \phi_0 + \frac{1}{\rho_r} \frac{\partial}{\partial z} \left(\frac{\rho_r}{S_0 N^2} \frac{\partial \phi_0}{\partial z} \right) \right] + \beta \frac{\partial \phi_0}{\partial x} = 0. \qquad (3.12)$$

First order dynamics represented by the *quasi-geostrophic vorticity equation* (3.12) allows closed form solutions for ϕ_0, with supporting boundary conditions. The solution for other dynamic variables p_0, $\mathbf{u}_0$, w_0, $\tilde{\rho}_0$, $\tilde{\theta}_0$ and $\tilde{T}_0$ can be obtained from

$$\mathbf{u}_0 = \mathbf{k} \times \nabla \phi_0$$

$$w_0 = -\frac{1}{S_0 N^2 (z)} \left(\frac{\partial}{\partial t} + \mathbf{u}_0 \cdot \nabla \right) \frac{\partial \phi_0}{\partial z}$$

$$\frac{\tilde{p}_0}{\rho_r} \equiv \phi_0$$

$$\frac{\tilde{\theta}_0}{\theta_r} \equiv s_0 = \frac{\partial \phi_0}{\partial z} \qquad\qquad (3.13.e - f)$$

$$\frac{\tilde{\rho}_0}{\rho_r} \equiv r_0 = \frac{1}{\gamma} \frac{\rho_r}{p_r} \phi_0 - s_0$$

$$\frac{\tilde{T}_0}{T_r} \equiv \tau_0 = \frac{\rho_r}{p_r} \phi_0 - r_0.$$

Remembering that the above equations were derived from dimensionless perturbations, a first step is to convert to dimensional equivalents

$$
\left(\frac{\partial}{\partial t} + \frac{1}{f_0} \mathbf{k} \times \nabla \phi_0 \cdot \nabla \right) \left[\nabla^2 \phi_0 + \frac{1}{\rho_r} \frac{\partial}{\partial z} \frac{f_0^2 \rho_r}{N^2} \frac{\partial \phi_0}{\partial z} \right] + f_0 \beta \frac{\partial \phi_0}{\partial x} = 0 \quad (3.14)
$$

by setting

$$
\mathbf{u}_0 = f_0^{-1} \mathbf{k} \times \nabla \phi_0
$$

$$
s_0 = g^{-1} \frac{\partial \phi_0}{\partial z}
$$

$$
w_0 = -\frac{g}{N^2(z)} \left(\frac{\partial}{\partial t} + \mathbf{u}_0 \cdot \nabla \right) s_0 \qquad (3.15.a-e)
$$

$$
r_0 = \frac{1}{\gamma} \frac{\rho_r}{p_r} \phi_0 - s_0
$$

$$
\tau_0 = \frac{\rho_r}{p_r} \phi_0 - r_0.
$$

Defining variables $\psi_0 = \phi_0 / f_0 = \tilde{p}_0 / f_0 \rho_r$ and also noting that

$$
\beta \frac{\partial \psi_0}{\partial x} = \left(\frac{\partial}{\partial t} + k \times \nabla \psi_0 \cdot \nabla \right) (\beta y)
$$

the dimensional equations are written as follows,

$$
\left(\frac{\partial}{\partial t} + \mathbf{k} \times \nabla \psi_0 \cdot \nabla \right) \left[\nabla^2 \psi_0 + f_0^2 \frac{1}{\rho_r} \frac{\partial}{\partial z} \frac{\rho_r}{N^2} \frac{\partial \psi_0}{\partial z} \right] + f_0 \beta \frac{\partial \psi_0}{\partial x} = 0 \quad (3.16)
$$

where interrelated variables are specified as

$$
\frac{\tilde{p}_0}{p_r} = f_0 \frac{\rho_r}{p_r} \psi_0
$$

$$
\mathbf{u}_0 = \mathbf{k} \times \nabla \psi_0
$$

$$
\frac{\tilde{\theta}_0}{\theta_r} = \frac{f_0}{g} \frac{\partial \psi_0}{\partial z}
$$

$$
w_0 = -\frac{f_0}{N^2} \left(\frac{\partial}{\partial t} + \mathbf{u}_0 \cdot \nabla \right) \frac{\partial \psi_0}{\partial z} \qquad (3.17.a-f)
$$

$$
\frac{\tilde{\rho}_0}{\rho_r} = \frac{1}{\gamma} \frac{\tilde{p}_0}{p_r} - \frac{\tilde{\theta}_0}{\theta_r}
$$

$$
\frac{\tilde{T}_0}{T_r} = \frac{\tilde{p}_0}{p_r} - \frac{\tilde{\rho}_0}{\rho_r}.
$$

Also to be noted is the fact that (3.14) can be written in the following form:

$$\left(\frac{\partial}{\partial t} + \mathbf{k} \times \nabla \psi_0 \cdot \nabla \right) \left[f_0(1 + \beta y) + \nabla^2 \psi_0 + \frac{f_0^2}{\rho_r} \frac{\partial}{\partial z} \frac{\rho_r}{N^2} \frac{\partial \psi_0}{\partial z} \right] = 0. \quad (3.18)$$

In the above equations the stratification parameter $S(z)$ can be interpreted as follows. The non-dimensional function

$$S(z) = S_0 N^2(z) = \left(\frac{N(z)}{f_0}\right)^2 \left(\frac{H}{L}\right)^2, \quad (3.19)$$

defines *internal radius of deformation L_I*

$$L_I = \frac{N(z)H}{f_0} = \sqrt{\frac{g}{\theta_r} \frac{\partial \theta_r}{\partial z}} \frac{H}{f_0} = \sqrt{\frac{1}{\theta_r} \frac{\partial \theta_r}{\partial z}} \frac{\sqrt{gH}\sqrt{H}}{f_0} = \frac{1}{f_0} \sqrt{\frac{gH}{\theta_r} \frac{\partial \theta_r}{\partial z}} H = \frac{\sqrt{g'H}}{f_0}, \quad (3.20)$$

and *reduced gravity*

$$g' \equiv \frac{gH}{\theta_r} \frac{\partial \theta_r}{\partial z} \quad (3.21.a)$$

to set

$$S(z) = \left(\frac{L_I}{L}\right)^2, \quad (3.21.b)$$

i.e. the stratification parameter being proportional to the ratio between internal radius of deformation and horizontal scale. In comparison, the ratio between internal radius of deformation L_I and the external one $L_E = \sqrt{gH}/f_0$ (GFD-I) is obtained as

$$\frac{L_I}{L_E} = \sqrt{\frac{g'}{g}} = \left(\frac{H}{\theta_r} \frac{\partial \theta_r}{\partial z}\right)^{1/2}. \quad (3.22)$$

In dimensionless terms it is noted that

$$\frac{H}{\theta_r} \frac{\partial \theta_r}{\partial z} = \frac{N^2 H}{g} = \frac{N_0^2 H}{g} N^2(z) = \frac{N_0^2 H^2}{gH}$$

$$= \left(\frac{N_0}{f_0}\right)^2 \left(\frac{H}{L}\right)^2 \left(\frac{f_0^2 L^2}{gH}\right) = S_0 \mu, \quad (3.23)$$

with $S_0 = N_0 H / f_0 L = O(1)$ and $\mu = f_0^2 L^2 / gH \ll 1$, it can be shown that

$$\frac{L_I}{L_E} = (\mu S_0)^{1/2} \ll 1. \quad (3.24)$$

Equation (3.18) is a conservation statement of the form

$$\frac{Dq}{Dt} = 0, \qquad (3.25.a)$$

$$q = f_0(1 + \beta y) + \nabla^2 \psi_0 + \frac{f_0^2}{\rho_r} \frac{\partial}{\partial z} \frac{\rho_r}{N^2} \frac{\partial \psi_0}{\partial z} \qquad (3.25.b)$$

where q is called the *quasi-geostrophic quasi-potential vorticity*. Individual terms of vorticity are

$$f = f_0(1 + \beta y) = f_0 + \hat{\beta} y = \text{planetary vorticity}$$
$$\nabla^2 \psi_0 + \frac{f_0^2}{\rho_r} \frac{\partial}{\partial z} \frac{\rho_r}{N^2} \frac{\partial \psi_0}{\partial z} = \text{fluid vorticity.} \qquad (3.26.a, b)$$

where $\hat{\beta} = \beta_0$ as defined in (2.53.b). The second term of (3.26.b) is termed as *thermal vorticity*, representing stratification effects in vorticity. Note that Eq. (3.25) can be written in either of the following ways:

$$\frac{\partial q}{\partial t} + \mathbf{u}_0 \cdot \nabla q = 0,$$

$$\text{or} \quad \frac{\partial q}{\partial t} + \mathbf{k} \times \nabla \psi_0 \cdot \nabla q = 0, \qquad (3.27.a - c)$$

$$\text{or} \quad \frac{\partial q}{\partial t} + \frac{\partial \psi_0}{\partial x} \frac{\partial q}{\partial y} - \frac{\partial \psi_0}{\partial y} \frac{\partial q}{\partial x} = 0.$$

The non-linear part of the quasi-geostrophic vorticity equation could also be written using the *Jacobian* of ψ_0 and q defined as

$$J(\psi_0, q) = \begin{vmatrix} \frac{\partial \psi_0}{\partial x} & \frac{\partial \psi_0}{\partial y} \\ \frac{\partial q}{\partial x} & \frac{\partial q}{\partial y} \end{vmatrix}. \qquad (3.28)$$

The Jacobian also establishes whether the two functions ψ_0 and q are independent of each other or not. If $J(\psi_0, q) = 0$, then vorticity and stream-function are independent of each other (similar to *Wronskian* of differentiable functions in calculus), in which case the nonlinear terms in the equations have to vanish.

3.2 Quasi-geostrophic, Barotropic Motions

3.2.1 Barotropic Rossby Waves

To obtain barotropic version of the quasi-geostrophic vorticity equation, vertical derivatives are simply canceled out to yield

$$\left(\frac{\partial}{\partial t} + \frac{\partial \psi_0}{\partial x} \frac{\partial}{\partial y} - \frac{\partial \psi_0}{\partial y} \frac{\partial}{\partial x} \right) \nabla^2 \psi_0 + \hat{\beta} \frac{\partial \psi_0}{\partial x} = 0 \qquad (3.29)$$

where $\hat{\beta} = f_0\beta$. Plane-wave solutions of this equation are sought in the form

$$\psi_0 = Re\left\{A\, e^{i(kx+ly-\omega t)}\right\}. \tag{3.30}$$

Substituting this solution, and evaluating terms, it is shown that

$$\nabla^2\psi_0 = -(k^2 + l^2)\psi_0, \quad \text{then} \quad J = (\psi_0, \nabla^2\psi_0) = 0, \tag{3.31}$$

leaving

$$\frac{\partial}{\partial t}\nabla^2\psi_0 + \hat{\beta}\frac{\partial\psi_0}{\partial x} = 0 \tag{3.32}$$

which then gives

$$\left\{-(k^2 + l^2)(-i\omega) + \hat{\beta}(ik)\right\}\psi_0 = 0. \tag{3.33}$$

The above relationship yields the *dispersion relation* for *barotropic Rossby waves*,

$$\omega = -\frac{\hat{\beta}k}{k^2 + l^2} \tag{3.34}$$

and defining

$$\hat{\beta} = f_0\beta \quad \text{and} \quad \boldsymbol{\kappa} = k\mathbf{i} + l\mathbf{j} \tag{3.35}$$

$$|\boldsymbol{\kappa}|^2 = k^2 + l^2 \quad k = \boldsymbol{\kappa}\cdot\mathbf{i}, \quad l = \boldsymbol{\kappa}\cdot\mathbf{j}, \tag{3.36}$$

dispersion relation takes the form

$$\omega = -\frac{\hat{\beta}}{|\boldsymbol{\kappa}|}\cos\alpha = \hat{\beta}\frac{\boldsymbol{\kappa}\cdot\mathbf{i}}{|\boldsymbol{\kappa}|}. \tag{3.37}$$

Excluding the *beta-effect* by setting $\hat{\beta} = 0$ leads to $\omega \to 0$, so that wave solutions can not exist. It is therefore clear that *Rossby waves* owe their existence to beta-effect.

Phase velocity vector is defined as

$$\mathbf{C} = \left(\frac{\omega}{k}, \frac{\omega}{l}\right) = \left(C_x, C_y\right) \tag{3.38}$$

with components

$$C_x = \frac{\omega}{k} = -\left(\frac{\hat{\beta}}{k^2 + l^2}\right) < 0 \tag{3.39.a}$$

$$C_y = \left(\frac{k}{l}\right)C_x. \tag{3.39.b}$$

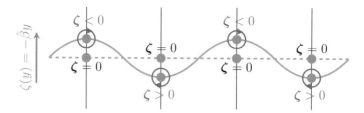

Fig. 3.1 A line of fluid particles oriented east-west with initial zero vorticity, deformed into a wave form will travel to the west by the β effect

These results show basic characteristics of Rossby waves. The zonal phase speed C_x is always negative, indicating phase propagation to the west (i.e. $\omega > 0$, $k < 0$ or *vice versa*). The longitudinal phase speed C_y can either be positive or negative.

Conservation of potential vorticity requires that

$$\left(\frac{\partial}{\partial t} + \frac{\partial \psi_0}{\partial x}\frac{\partial}{\partial y} - \frac{\partial \psi_0}{\partial y}\frac{\partial}{\partial x}\right)(f_0 + \hat{\beta}y + \nabla^2 \psi_0) = 0 \qquad (3.40)$$

equivalently written as

$$\left(\frac{\partial}{\partial t} + \mathbf{u} \cdot \nabla\right)(f + \zeta) = \frac{D}{Dt}(f + \zeta) = 0. \qquad (3.41)$$

The mechanism of Rossby waves is explained bu considering zonally oriented chains of fluid with initial vorticity $\zeta = 0$ at $y = 0$, later deformed into a sinusoidal form, as shown in Fig. 3.1.

Conservation of vorticity in the quasi-geostrophic approximation requires that $f + \zeta = f_0 + \hat{\beta}y + \zeta =$ constant with vorticity change $\zeta(y) = -\hat{\beta}y +$ constant, as a function of y, following the motion of material elements in longitude direction. In the northern hemisphere, particles moving north ($y > 0$) will gain negative vorticity $\zeta < 0$, while those moving south ($y < 0$) will gain positive vorticity $\zeta > 0$. The resulting street of vortices will advect themselves with their own field velocity resulting in the the entire wave moving towards the left (west) as discovered in Fig. 3.1.

The group velocity $\mathbf{C}_g$ is defined as

$$\mathbf{C}_g = \left(\frac{\partial \omega}{\partial k}, \frac{\partial \omega}{\partial l}\right) = \mathbf{i}\frac{\partial \omega}{\partial k} + \mathbf{j}\frac{\partial \omega}{\partial l} = \mathbf{i}C_{gx} + \mathbf{j}C_{gy} \qquad (3.42)$$

and its components for Rossby waves can be calculated from the dispersion relation as follows:

$$C_{gx} = \hat{\beta}\frac{k^2 + l^2}{\left(k^2 - l^2\right)}$$
$$C_{gy} = 2\hat{\beta}\frac{kl}{\left(k^2 + l^2\right)^2} \qquad (3.43.a, b)$$

It is noted that components of group velocity for shorter waves (in terms of east-west wave orientation) are positive (energy propagation to east), while longer waves are negative (energy propagation the west). For Rossby waves in the north hemisphere, energy is always propagated north.

$$C_{gx} = > 0 \text{ if } |k| > |l|$$
$$< 0 \text{ if } |k| < |l|. \tag{3.44}$$

3.2.2 Barotropic Rossby Waves in Basic Zonal Flows

The case of Rossby waves superposed on basic state of constant zonal flow is considering next. Starting with the vorticity equation

$$\frac{\partial}{\partial t}(\nabla^2 \psi_0) + J(\psi_0, \nabla^2 \psi_0) + \hat{\beta}\frac{\partial \psi_0}{\partial x} = 0, \tag{3.45}$$

the flow is assumed to be a superposition of basic and perturbed states

$$\psi_0 = \Psi(y) + \epsilon \psi \tag{3.46}$$

in the range $\epsilon \ll 1$, with Ψ allowed to be a function of y only

$$\Psi = \Psi(y), \quad U(y) = -\frac{d\Psi}{dy}. \tag{3.47}$$

Substitution in the vorticity equation gives

$$\frac{\partial}{\partial t}(\nabla^2 \Psi) + \epsilon\frac{\partial}{\partial t}(\nabla^2 \psi) + J\left(\Psi + \epsilon\psi, \nabla^2(\Psi + \epsilon\psi)\right) + \hat{\beta}\overset{0}{\cancel{\frac{\partial \Psi}{\partial x}}} + \epsilon\hat{\beta}\frac{\partial \psi}{\partial x} = 0. \tag{3.48}$$

The term $\hat{\beta}\partial\Psi/\partial x$ drops out because $\Psi = \Psi(y)$ only. Expanding the Jacobian,

$$J\left(\Psi + \epsilon\psi, \nabla^2(\Psi + \epsilon\psi)\right) = \frac{\partial(\Psi + \epsilon\psi)}{\partial x}\frac{\partial}{\partial y}\nabla^2(\Psi + \epsilon\psi)$$
$$- \frac{\partial(\Psi + \epsilon\psi)}{\partial y}\frac{\partial}{\partial x}\nabla^2(\Psi + \epsilon\psi)$$
$$= \frac{\partial\Psi}{\partial x}\frac{\partial\nabla^2\Psi}{\partial y} - \frac{\partial\nabla^2\Psi}{\partial x}\frac{\partial\Psi}{\partial y}$$
$$+ \epsilon\left(\frac{\partial\psi}{\partial x}\frac{\partial}{\partial y}\nabla^2\Psi - \frac{\partial}{\partial x}\nabla^2\Psi\overset{0}{\cancel{\frac{\partial\psi}{\partial y}}}\right) \tag{3.49}$$
$$+ \epsilon\left(\overset{0}{\cancel{\frac{\partial\Psi}{\partial x}\frac{\partial}{\partial y}\nabla^2\psi}} - \frac{\partial}{\partial x}\nabla^2\psi\frac{\partial\Psi}{\partial y}\right)$$
$$+ \epsilon^2(\dots)$$

and dropping x derivatives applied to Ψ terms, the Jacobian is expressed in Taylor series

$$J = \epsilon \left(\frac{\partial \psi}{\partial x} \frac{\partial}{\partial y} \nabla^2 \Psi - \frac{\partial}{\partial x} \nabla^2 \psi \frac{\partial \Psi}{\partial y} \right) + \epsilon^2(\ldots). \tag{3.50}$$

Substituting in vorticity equation and collecting equal powers of ϵ gives

$$\frac{\partial}{\partial t}(\nabla^2 \Psi) + \hat{\beta}\cancel{\frac{\partial \Psi}{\partial x}}^{\,0} + \epsilon \left\{ \frac{\partial \psi}{\partial x} \frac{\partial}{\partial y} \nabla^2 \Psi - \frac{\partial}{\partial x} \nabla^2 \psi \frac{\partial \Psi}{\partial y} + \frac{\partial}{\partial t}(\nabla^2 \psi) + \hat{\beta}\frac{\partial \psi}{\partial x} \right\} + \epsilon^2(\ldots) + \cdots = 0. \tag{3.51}$$

To first order $(O(\epsilon^0))$, one obtains the simple equation

$$\frac{\partial}{\partial t} \nabla^2 \Psi + \hat{\beta}\cancel{\frac{\partial \Psi}{\partial x}}^{\,0} = 0, \tag{3.52}$$

and since $\Psi = \Psi(y)$ is constant in time, the above equation is identically satisfied. It is also to be noted that

$$\frac{\partial \Psi}{\partial y} = -U(y)$$

$$\nabla^2 \Psi = -\frac{dU}{dy} \tag{3.53}$$

$$\frac{\partial}{\partial y} \nabla^2 \Psi = -\frac{d^2 U}{dy^2}$$

and therefore, equation obtained to next order $(O(\epsilon^1))$ is

$$\frac{\partial}{\partial t}(\nabla^2 \psi) + U \frac{\partial}{\partial x}(\nabla^2 \psi) + \left(\hat{\beta} - \frac{d^2 U}{dy^2} \right) \frac{\partial \psi}{\partial x} = 0. \tag{3.54}$$

This equation will serve as the vorticity equation for the perturbation stream-function ψ on a basic flow U. The notation is simplified by writing

$$\left(\frac{\partial}{\partial t} + U \frac{\partial}{\partial x} \right) \nabla^2 \psi + (\hat{\beta} - U'') \frac{\partial \psi}{\partial x} = 0 \tag{3.55}$$

with second derivative of the basic flow velocity denoted as

$$U'' \equiv \frac{d^2 U}{dy^2}. \tag{3.56}$$

To obtain plane wave solution for the above equation, let

$$\psi = Re \left(A e^{i(kx + ly - \omega t)} \right) \tag{3.57.a}$$

with substitution

$$(-i\omega + ikU)(-k^2 - l^2) + (\hat{\beta} - U'')ik = 0 \tag{3.57.b}$$

yields the dispersion relation

$$\omega = -\frac{(\hat{\beta} - U'')k}{(k^2 + l^2)} + kU. \tag{3.58}$$

Comparing with the earlier case without basic flow, the zonal phase speed can be calculated as

$$C_x = \frac{\omega}{k} = -\frac{(\hat{\beta} - U'')}{k^2 + l^2} + U. \tag{3.59}$$

The above is valid for a zonal basic flow $U(y)$. For simplicity, suppose a uniform basic current $U = U_0$=constant, $U'' = 0$. Then we see that the phase speed $C_x = 0$ for a velocity of $U_0 = \hat{\beta}/|\kappa|^2$. With $|\kappa| = \kappa = 2\pi/\lambda_x$, λ_x being the zonal wavelength, it is found that this condition occurs when

$$\lambda_x = \lambda_* \equiv 2\pi\sqrt{U_0/\hat{\beta}}. \tag{3.60}$$

The wavelength λ_* characterizes a *standing Rossby wave*, without propagation along east-west, at wavelength $\lambda_x = \lambda_*$. For typical values of $U_0 \simeq 25$ ms^{-1}, $\hat{\beta} = 1.2 \times 10^{-11}$ m^{-1} s^{-1}, one finds $\lambda_* \sim 9000$ km. The Rossby wave propagates from west to east for $\lambda_x < \lambda_*$, and from east to west for $\lambda_x > \lambda_*$. In mid-latitudes, (e.g. the European area) the observed wavelengths of storm systems are usually less than 7000 km, so that the atmospheric waves are often observed to travel from west to east.

Note that for $U \leq 0$ (for instance, easterly jets such as the *TEJ*—Tropical Easterly Jet), setting $U = -U_0$, it is shown that

$$C_x = -\frac{\hat{\beta}}{k^2 + l^2} - U_0 < 0 \tag{3.61}$$

propagation will be towards the west. In the tropics, such waves are often found to be superposed on *Easterlies*.

3.2.3   Barotropic Instability

To study barotropic stability of basic zonal flows, non-dimensional version of the vorticity equation developed in the last section is considered

$$\left(\frac{\partial}{\partial t} + U\frac{\partial}{\partial x}\right)\nabla^2\psi + (\hat{\beta} - U'')\frac{\partial\psi}{\partial x} = 0 \tag{3.62}$$

where $U = U(y)$ is the zonal velocity profile.

For simplicity, an idealized case of zonal flow confined between two rigid vertical walls at latitudes $y = 0$ and $y = 1$ (channel flow) is considered, where the following boundary conditions are applicable

$$v_0 = \frac{\partial \psi}{\partial x} = 0 \text{ on } y = 0, 1. \tag{3.63}$$

A trial solution $\psi(x, y, t) = P(y)e^{ik(x-ct)}$, allowing wave propagation along the x-direction, $c = C_x$ being the zonal phase speed and $P(y)$ the amplitude variation across y. Substituting these into Eq. (3.62) gives

$$(-ikc + Uik)\left(\frac{d^2 P}{dy^2} - k^2 P\right) + (\hat{\beta} - U'')ikP = 0, \tag{3.64}$$

or

$$\left\{\frac{d^2}{dy^2} - k^2 + \frac{\hat{\beta} - U''}{U - c}\right\} P(y) = 0. \tag{3.65}$$

The corresponding boundary conditions at the walls are

$$\begin{cases} P(0) = 0, \\ P(1) = 0. \end{cases} \tag{3.66}$$

Re-defining phase speed as a complex number $c = c_r + ic_i$ gives the possibility of time dependent wave solutions. In particular, positive imaginary part will represent wave growth or decay in time

$$\psi = P(y)e^{+kc_i t}e^{ik(x-c_r t)}. \tag{3.67}$$

since the wave number is constrained to negative values $k < 0$ for Rossby waves, the factor $e^{+kc_i t}$ in the proposed solution corresponds to a growing wave, enabling to investigate stability by substituting (3.67) into (3.65),

$$\left\{\frac{d^2}{dy^2} - k^2 + \frac{\hat{\beta} - U''}{(U - c_r) - ic_i}\right\} P(y) = 0 \tag{3.68}$$

alternatively written as

$$\left\{\frac{d^2}{dy^2} - k^2 + \frac{(\hat{\beta} - U'')\left[(U - c_r) + ic_i\right]}{(U - c_r)^2 + c_i^2}\right\} P(y) = 0. \tag{3.69}$$

Defining

$$\lambda \equiv (\beta - U'')(U - c_r)$$

$$\mu \equiv (\hat{\beta} - U'')c_i$$

gives

$$\left[\frac{d^2}{dy^2} - k^2 + \frac{\lambda + i\mu}{|U - c|^2} \right] P = 0. \tag{3.70}$$

Splitting the equation into real and imaginary parts, respective terms of the complex amplitude $P = P_r + P_i$ are evaluated as

$$\left[\frac{d^2}{dy^2} - k^2 + \frac{\lambda}{|U - c|^2} \right] P_r - \frac{\mu}{|U - c|^2} P_i = 0$$

$$\left[\frac{d^2}{dy^2} - k^2 + \frac{\lambda}{|U - c|^2} \right] P_i + \frac{\mu}{|U - c|} P_r = 0. \tag{3.71.a, b}$$

By multiplying Eq. (3.71.a) with P_i and multiplying Eq. (3.71.b) with $-P_r$, the two equations are then combined as

$$P_i \frac{d^2 P_r}{dy^2} - P_r \frac{d^2 P_i}{dy^2} - \frac{\mu}{|U - c|^2} \left(P_r^2 + P_i^2 \right) = 0. \tag{3.72}$$

Integrating across the channel,

$$\int_0^1 P_i \frac{d^2 P_r}{dy^2} dy - \int_0^1 P_r \frac{d^2 P_i}{dy^2} dy - \int_0^1 \frac{\mu}{|U - c|^2} \left(P_r^2 + P_i^2 \right) dy = 0, \tag{3.73}$$

and carrying out integration by parts, gives

$$\left(P_i \frac{dP_r}{dy} \Big|_0^1 - \int_0^1 \frac{dP_r}{dy} \frac{dP_i}{dy} dy \right) - \left(P_r \frac{dP_i}{dy} \Big|_0^1 - \int_0^1 \frac{dP_i}{dy} \frac{dP_r}{dy} dy \right)$$

$$- \int_0^1 \frac{\mu}{|U - c|^2} (P_r^2 + P_i^2) dy = 0. \tag{3.74}$$

With boundary conditions $P_r(0) = P_i(0) = 0$, $P_r(1) = P_i(1) = 0$, some terms in brackets vanish, with cancellations, leaving

$$\int_0^1 \frac{\mu}{|U - c|^2} (P_r^2 + P_i^2) dy = 0 \tag{3.75.a}$$

then replacing $\mu \equiv (\hat{\beta} - U'')c_i$ gives

$$c_i \int_0^1 \left(\hat{\beta} - U'' \right) \left\{ \frac{P_r^2 + P_i^2}{|U - c|^2} \right\} dy = 0. \qquad (3.75.b)$$

Note that, since the term in braces is always positive, and since $c_i \neq 0$ the only way that an instability can occur is to have $(\hat{\beta} - U'')$ *change sign at least once in the domain* $0 < y < 1$. This however is a *necessary* (but not *sufficient*) condition for *barotropic instability*. In case the specific integral would be identical to zero, only stable solutions with $c_i = 0$ could still possible. In the absence of β-effect, the necessary condition for instability is to have an inflection point in $U(y)$, which is the typical stability criterion for shear flows in the absence of geophysical scale β effects.

3.3 Quasi-geostrophic, Baroclinic Motions

3.3.1 Rossby Waves in Quasi-geostrophic Stratified Fluids

To study Rossby waves in a quasi-geostrophic stratified fluid, the equations are simplified a little to have the z-dependent coefficients in the original vorticity equations. In the present simple case, a specific type of basic state stratification is considered, with constant Brunt–Väisälä frequency

$$N^2(z) = N_0^2 = \text{constant}. \qquad (3.76)$$

The scale height would then be defined as constant in Eq. (2.74) for the atmosphere and (2.81) for the ocean as

$$H_s^{-1} \equiv -\frac{1}{\rho_r} \frac{\partial \rho_r}{\partial z} = \frac{N_0^2}{g} = \text{constant}, \qquad (3.77)$$

so that it is possible to write

$$\frac{1}{\rho_r} \frac{\partial}{\partial z} \left\{ \rho_r \frac{f_0^2}{N_0^2} \frac{\partial \psi_0}{\partial z} \right\} = \frac{f_0^2}{N_0^2} \left\{ \frac{\partial^2 \psi_0}{\partial z^2} + \frac{1}{\rho_r} \frac{\partial \rho_r}{\partial z} \frac{\partial \psi_0}{\partial z} \right\}$$
$$= r_0 \left\{ \frac{\partial^2 \psi_0}{\partial z^2} - \frac{1}{H_s} \frac{\partial \psi_0}{\partial z} \right\}, \qquad (3.78)$$

where $r_0 = f_0^2 / N_0^2$, and quasi-geostrophic vorticity equation then reads as

$$\left\{ \frac{\partial}{\partial t} + \frac{\partial \psi_0}{\partial x} \frac{\partial}{\partial y} - \frac{\partial \psi_0}{\partial y} \frac{\partial}{\partial x} \right\} \left\{ \frac{\partial^2 \psi_0}{\partial x^2} + \frac{\partial^2 \psi_0}{\partial y^2} + r_0 \frac{\partial^2 \psi_0}{\partial z^2} - \frac{r_0}{H_s} \frac{\partial \psi_0}{\partial z} \right\} + \hat{\beta} \frac{\partial \psi_0}{\partial x} = 0. \qquad (3.79)$$

The above equation is now a simplified form of the more general one, with constant coefficients. A plane-wave periodic solution with zonal and meridional components is proposed, with exponential amplitude varying with height

$$\psi_0 = A e^{\alpha z} \cos(kx + ly + mz - \omega t)$$
$$= Re \left\{ A e^{\alpha z} e^{i(kx + ly + mz - \omega t)} \right\}, \tag{3.80}$$

which then allows to obtain a basic solution, inserting it in the definition of fluid vorticity in (3.79)

$$Q \equiv \frac{\partial^2 \psi_0}{\partial x^2} + \frac{\partial^2 \psi_0}{\partial y^2} + r_0 \frac{\partial^2 \psi_0}{\partial z^2} - \frac{r_0}{H_s} \frac{\partial \psi_0}{\partial z}$$
$$= \left\{ -k^2 - l^2 - r_0 (\alpha + im)^2 - \frac{r_0}{H_s} (\alpha + im) \right\} \psi_0 \tag{3.81}$$
$$= \left\{ -k^2 - l^2 - r_0 \left[\alpha^2 + 2\alpha im - m^2 - \frac{1}{H_s}\alpha - im\frac{1}{H_s} \right] \right\} \psi_0$$

which is simplified by selecting the coefficient $\alpha = 1/(2H_s)$ to make imaginary terms (second and fifth terms in bracket) cancel each other

$$Q = \left\{ -k^2 - l^2 - r_0 m^2 - \frac{r_0}{4H_s^2} \right\} \psi_0. \tag{3.82}$$

The nonlinear Jacobian with interdependent arguments vanish, according to a basic rule of calculus, $J(\psi_0, Q(\psi_0)) = 0$,

$$\frac{\partial \psi_0}{\partial x} \frac{\partial Q}{\partial y} - \frac{\partial \psi_0}{\partial y} \frac{\partial Q}{\partial x} = - \left(k^2 + l^2 + r_0 m^2 + \frac{r_0}{4H_s^2} \right) \left\{ \frac{\partial \psi_0}{\partial x} \frac{\partial \psi_0}{\partial y} - \frac{\partial \psi_0}{\partial y} \frac{\partial \psi_0}{\partial x} \right\} = 0. \tag{3.83}$$

which in this case gives a linear equation of vorticity

$$\frac{\partial Q}{\partial t} + \hat{\beta} \frac{\partial \psi_0}{\partial x} = 0. \tag{3.84}$$

Substituting from (3.80) yields

$$-i\omega \left\{ -k^2 - l^2 - r_0 m^2 - \frac{r_0}{4H_s^2} \right\} + i\hat{\beta} k = 0$$

which is the dispersion relation for quasi-geostrophic Rossby waves valid in uniformly stratified environment with constant N_0^2

$$\omega = - \frac{\beta k}{k^2 + l^2 + r_0 m^2 + \frac{r_0}{4H_s^2}}. \tag{3.85}$$

The ratio $r_0 = f_0^2/N_0^2$ compares inertial (Coriolis) and buoyancy (Brunt–Väisälä) frequencies. In the limit $r_0 \to 0$, the dispersion relation (3.85) corresponds to Eq. (3.34) for barotropic Rossby waves propagating only in the horizontal direction. Otherwise, the third term in the denominator allows wave motion in the vertical with wave-number m, the ratio r_0 measuring the relative roles of horizontal and vertical components. The fourth term in the denominator decreases the frequency of motion by a small amount, obtained from (3.77) as

$$\frac{r_0}{4H_s^2} = \frac{1}{4}\frac{f_0^2}{N_0^2}\left(\frac{N_0^2}{g}\right)^2 = \frac{f_0^2 N_0^2}{4g^2}$$

$$= \frac{1}{4}\frac{f_0^2 L^2}{gH}\frac{N_0^2}{gH}\left(\frac{H}{L}\right)^2 = \frac{1}{4}\frac{f_0^2 L^2}{gH}\left(\frac{H}{L}\right)^2\frac{1}{HH_s} = \frac{1}{4}\mu\lambda^2\frac{1}{HH_s}$$

where μ, λ are small parameters consistent with quasi-geostrophic scaling, multiplying wave-number contribution $1/(HH_s)$ in (3.85).

3.3.2 Boussinesq Approximations

Returning to quasi-geostrophic dimensionless equations (3.1),

$$\delta\left\{\frac{\partial \mathbf{u}}{\partial z} + \mathbf{u}\cdot\nabla\mathbf{u} + \beta y\mathbf{k}\times\mathbf{u}\right\} + \mathbf{k}\times\mathbf{u} = \nabla\phi$$

$$\frac{\partial\phi}{\partial z} = s$$

$$\nabla\cdot\mathbf{u} + \delta\frac{1}{\rho_r(z)}\frac{\partial}{\partial z}\rho_r(z)w = 0 \qquad (3.86.a-d)$$

$$\frac{\partial s}{\partial t} + \mathbf{u}\cdot\nabla s + S_0 N^2(z)w = 0,$$

remembering *Rossby number* $\delta = U/f_0 L$, perturbed state variables for pressure, potential temperature (used for the atmosphere) and density (for the ocean) are defined with respect to mean (rest or basic) state variables denoted by the subscript r,

$$\phi = \frac{\tilde{p}}{\rho_r}, \quad s = \frac{\tilde{\theta}}{\theta_r} = -\frac{\tilde{\rho}}{\rho_r}. \qquad (3.87a-c)$$

Further approximations are then made here, by ignoring relatively small compressible effects in the perturbed fields, both in the atmosphere and ocean.

Since perturbation density is much smaller than the basic state density, the latter is further split into a constant part and a superposed component varying with depth as shown in Fig. 3.2,

$$\rho_r(z) = \rho_0 + \rho_1(z), \qquad (3.88)$$

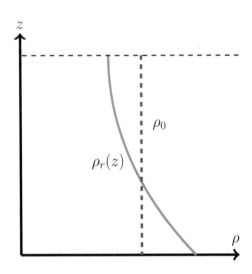

Fig. 3.2 Basic state density and its vertical mean

establishing the vertical average (constant) and the deviation from this average

$$\rho_0 = \int \rho_r(z)dz = \overline{\rho_r(z)}, \qquad \rho_1(z) = \rho_r(z) - \rho_0. \qquad (3.89.a, b)$$

More specifically, approximations $\rho_1/\rho_0 < O(\delta)$ and $\rho_r(z) \simeq \rho_0$ are applied in the equations. Equation of state is simplified, respectively for the atmosphere and ocean, canceling out the compressible terms

$$\frac{\tilde{\theta}}{\theta_r} = \frac{1}{\gamma}\frac{\cancel{\tilde{p}}^{\,0}}{p_r} - \frac{\tilde{\rho}}{\rho_r}, \qquad \frac{\cancel{\tilde{p}}^{\,0}}{p_r} = \frac{\tilde{\rho}}{\rho_r} + \frac{\tilde{T}}{T_r}. \qquad (3.90)$$

Furthermore, compressibility effects are expected to be even smaller for the perturbation field, neglecting $\tilde{p}/p_r$ in the state equations,

$$\frac{\tilde{\theta}}{\theta_r} = \frac{\tilde{T}}{T_r} = -\frac{\tilde{\rho}}{\rho_r}. \qquad (3.91.a)$$

In addition, variables $\theta_r(z)$, $T_r(z)$, $\rho_r(z)$ in denominator are respectively replaced by constants θ_0, T_0, ρ_0 for small deviations, in similar way as (3.88)

$$\frac{\tilde{\theta}}{\theta_0} = \frac{\tilde{T}}{T_0} = -\frac{\tilde{\rho}}{\rho_0}. \qquad (3.91.b)$$

In a series of simplifications often referred to as *Boussinesq approximations*, basic state stratification and compression are ignored in the continuity, momentum, and state equations, but kept in the thermodynamic equation, lumped into the variable parameter $N^2(z)$. The essence of Boussinesq approximations is that, stratification constitutes the dynamic restoring force, rather than compression. With these approximations, the governing Boussinesq equations are written in dimensional form as

$$\frac{\partial \mathbf{u}}{\partial t} + \mathbf{u} \cdot \nabla \mathbf{u} + f \mathbf{k} \times \mathbf{u} = \frac{1}{\rho_0} \nabla \tilde{p}$$

$$\frac{\partial \tilde{p}}{\partial z} = -g \tilde{\rho}$$

$$\nabla \cdot \mathbf{u} + \frac{\partial w}{\partial z} = 0 \qquad (3.92.a-d)$$

$$\frac{\partial \tilde{\rho}}{\partial t} + \mathbf{u} \cdot \nabla \tilde{\rho} - \frac{\rho_0}{g} N^2 w = 0$$

where it is remembered that $f = f_0(1 + \beta y) = f_0 + \hat{\beta} y$. The Boussinesq version of the vorticity equation turns out to be

$$\left[\frac{\partial}{\partial t} + (\mathbf{k} \times \nabla \psi) \cdot \nabla \right] \left[\nabla^2 \psi + \frac{\partial}{\partial z} \left(\frac{f_0^2}{N^2} \frac{\partial \psi}{\partial x} \right) \right] + \hat{\beta} \frac{\partial \psi}{\partial x} = 0, \qquad (3.93)$$

where

$$\tilde{\rho} = -\left(\frac{\rho_0}{T_0} \right) \tilde{T}, \quad \text{and} \quad \psi = \frac{\tilde{p}}{\rho_0 f_0} \qquad (3.94.a, b)$$

have been used. An alternative form of the Boussinesq vorticity equation is

$$\left(\frac{\partial}{\partial t} + \mathbf{u} \cdot \nabla \right) \left(\nabla^2 \psi + f_0^2 \frac{\partial}{\partial z} \frac{1}{N^2} \frac{\partial \psi}{\partial z} \right) + \hat{\beta} v = 0. \qquad (3.95)$$

In obtaining vorticity equation, use has been made of the non-dimensional Eq. (3.7). In dimensional form, the second equation reads

$$\left(\frac{\partial}{\partial t} + \mathbf{u} \cdot \nabla \right) \zeta + \hat{\beta} v = \frac{f_0}{\rho_r} \frac{\partial}{\partial z} (\rho_r w), \qquad (3.96)$$

where $\zeta = \nabla^2 \psi$. The Boussinesq version of vorticity equation therefore reads as

$$\left(\frac{\partial}{\partial t} + \mathbf{u} \cdot \nabla \right) \zeta + \hat{\beta} v = f_0 \frac{\partial w}{\partial z}. \qquad (3.97)$$

Fig. 3.3 Horizontal stream-function and velocity fields

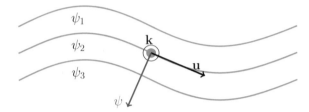

3.3.3 Physics Represented by Quasi-geostrophic Equations

To leading order in Eq. (3.86.a), horizontal velocity $\mathbf{u} = \hat{k} \times \nabla\psi$ is related to *stream-function* $\psi = \tilde{p}/\rho_0 f_0$ defined by (3.94.b) at any vertical level z of atmosphere and ocean, as shown in Fig. 3.3.

Near ground level in the atmosphere, and at the surface and bottom of the ocean, inviscid approximation is not necessarily valid, because of frictional effects which are mostly confined in Ekman boundary layers of typical thickness about $\sim$1 km in the atmosphere and $\sim$50 m in the ocean. Vorticity conservation states that

$$\frac{D\zeta}{Dt} \equiv \left(\frac{\partial}{\partial t} + \mathbf{u} \cdot \nabla\right)\zeta = -\hat{\beta}v + f_0 \frac{\partial w}{\partial z}$$

$$= -\hat{\beta}v - f_0 \nabla \cdot \mathbf{u} \tag{3.98}$$

vorticity $\zeta = \nabla^2\psi = \mathbf{k} \cdot (\nabla \times \mathbf{u})$ is changed either by (*i*) *beta-effect* (first term) or by (*ii*) *divergence effect* (second term):

(*i*) The *beta-effect* contributes to vorticity when fluid elements move to a position with different planetary vorticity from their earlier ambient value. For example, for northerly transport $v > 0$ (fluid moving to a region with higher planetary vorticity), we find that the fluid vorticity decreases ($D\zeta/Dt < 0$). Similarly, for southerly transport $v < 0$ (fluid moving to a region with lower planetary vorticity), we find that the fluid vorticity increases ($D\zeta/Dt > 0$). The role of the beta-effect in vorticity conservation has been envisioned in Fig. 3.1.

(*ii*) The *divergence effect* changes the vorticity or equivalently the angular momentum of a finite fluid volume. In the case of a fluid cylinder with fixed volume, this has following consequences: when there is convergence ($\partial w/\partial z > 0$ or $\nabla \cdot \mathbf{u} < 0$), the fluid column gets thinner and rotates faster, so that $D\zeta/Dt > 0$. Similarly, when there is divergence ($\partial w/\partial z < 0$ or $\nabla \cdot \mathbf{u} > 0$), the fluid column gets fatter and rotates slower, leading to $D\zeta/Dt < 0$.

In case of a fluid column of given height undergoing convergence of fluid from its periphery, the cylinder of fluid will contract and therefore become elongated in the vertical, with relative increase in vertical velocity with height, thus rotating faster as shown in Fig. 3.4 (top). This leads to the *ballerina effect*, simulating a ballerina controlling her speed of rotation by extending or contracting arms. The opposite tendency occurs in the case of divergence displayed in Fig. 3.4 (bottom).

The effects of top and bottom boundaries can be visualized as shown in Fig. 3.5. For simplicity, consider flat boundaries, where vertical velocity has to vanish. In the

Fig. 3.4 Changes in vorticity of an isolated fluid column undergoing convergence (top) and divergence (bottom)

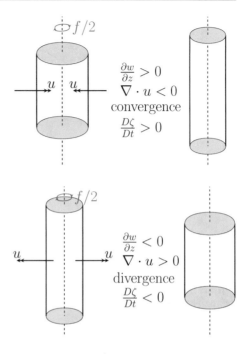

Fig. 3.5 Fluid column divergence/convergence with rates of change of vertical velocity and vorticity in domains limited by top and bottom boundaries, corresponding to cases of positive (top) and negative (bottom) vertical velocity

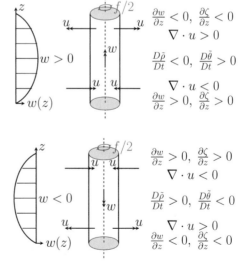

case of ascending motion one must have $\partial w/\partial z < 0$ or $\nabla \cdot \mathbf{u} > 0$ in the upper part, and the opposite in the lower part. As a result, vorticity decreases $D\zeta/Dt < 0$ in the upper part, and increases in the lower part. In the case of descending motion one must have the opposite case for upper and lower parts, with vorticity increasing in the upper part, and decreasing in the lower part.

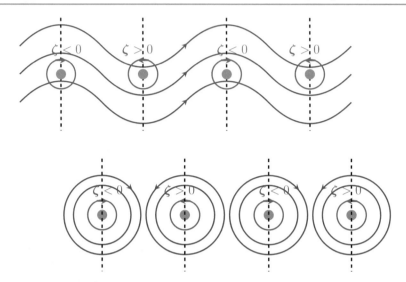

Fig. 3.6 Streamfunction and cyclonic/anti-cyclonic fields of vorticity in upper (top) and lower (bottom) level atmospheric waves

Idealized streamline patterns for the upper and lower parts of the domain, for instance in the mid-latitude troposphere are sketched in Fig. 3.6. For a basic zonal flow bounded by two horizontal surfaces, at the surface we often encounter cyclones/anticyclones that make up a stream of vortices with alternating sign (Fig. 3.4, lower part), rather looked upon as part of unstable motions of the fluid. In the atmosphere and in most places in the ocean, the upper air and the upper ocean currents are often better organized leading to structures like the jet-stream or major ocean currents which undulate in meanders, hiding vortices within them (Fig. 3.4, upper part), associated with an increase of the mean current. We expect and observe *phase shifts* between motions near upper and lower parts of the fluid. This *slanting* motion pattern results from different rates of the change of vorticity at different vertical levels. While the vorticity of wave motion decreases with time near the bottom, it would tend to increase near the upper part, and *vica versa*, as often observed in the troposphere.

Finally, density changes during the motion can be studied by making use of the thermodynamic equation

$$\frac{D\tilde{\rho}}{Dt} = \left(\frac{\rho_0}{g}N^2\right)w. \tag{3.99}$$

This equation shows that the perturbation density following a fluid particle changes only by vertical motions in relation to basic state density stratification. In particular, for or a fluid parcel moving upward ($w > 0$), the density increases $D\tilde{\rho}/Dt > 0$ (temperature decreases) as indicated by Eq. (3.99). Similarly, for a fluid parcel moving downward ($w < 0$), the density decreases $D\tilde{\rho}/Dt < 0$ (temperature increases) as shown in Fig. 3.5. Note however that this result runs contrary to the

Fig. 3.7 Warm and cold advection respectively before (top) and after (bottom) a passing storm. The meridional gradients $\nabla \tilde{T}$ and $\nabla \tilde{\rho}$ are represented in the y-z plane, while the horizontal components of vectors **u** and d $\nabla \tilde{\rho}$ are in a plane parallel to the x-y plane

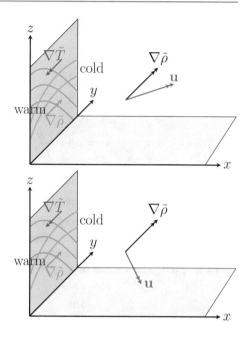

popular belief that a warming parcel should ascend, and a cooling parcel should descend!

What happens near the top and bottom boundaries? For instance, consider what actually would happen near the bottom of the atmosphere. It is expected that $w \sim 0$ near the boundary, so that the above thermodynamic equation yields

$$\frac{\partial \tilde{\rho}}{\partial t} \simeq -\mathbf{u} \cdot \nabla \tilde{\rho}. \tag{3.100}$$

Therefore, in a warming situation ($\partial \tilde{T}/\partial t > 0$ or $\partial \tilde{\rho}/\partial t < 0$) at fixed position, it is expected that $\mathbf{u} \cdot \nabla \tilde{\rho} > 0$, i.e., the vectors **u** and $\nabla \tilde{\rho}$ to have components pointing in the same direction as shown in Fig. 3.7 (top). Because $\nabla \tilde{\rho}$ must have a component towards north (it is usually warmer in the south than north), this means that the flow also must have a component from south to north, favoring *warm advection*, such as during the period preceding approaching cyclone.

On the other hand if there is a cooling situation ($\partial \tilde{\rho}/\partial t > 0$), it is expected that $\mathbf{u} \cdot \nabla \tilde{\rho} < 0$, i.e. the vectors **u** and $\nabla \tilde{\rho}$ having components pointing in the opposite direction as shown in Fig. 3.7 (bottom). This means that the flow must have a component from north to south, i.e. *cold advection*, such as during the period following a cyclone.

One of the most basic relationships reminded here is the *thermal wind*, simply combining the geostrophic and hydrostatic relations:

$$\mathbf{u} = \mathbf{k} \times \nabla \psi = \mathbf{k} \times \nabla \left(\frac{\tilde{\rho}}{f_0 \rho_0} \right), \quad \frac{\partial \tilde{p}}{\partial z} = -\tilde{\rho} g \tag{3.101}$$

yielding, with the earlier definitions of (3.91),

$$\frac{\partial \mathbf{u}}{\partial z} = \frac{1}{f_0 \rho_0} \mathbf{k} \times \nabla \left(\frac{\partial \tilde{p}}{\partial z} \right) = -\frac{g}{f_0 \rho_0} \mathbf{k} \times \nabla \tilde{\rho} = \frac{g}{f_0 T_0} \mathbf{k} \times \nabla \tilde{T}. \qquad (3.102)$$

Integrating in the vertical gives

$$\mathbf{u} = \mathbf{u}_0 + \frac{g}{f_0 T_0} \int_0^z \mathbf{k} \times \nabla \tilde{T} \, dz$$
$$= \mathbf{u}_0 - \frac{g}{f_0 \rho_0} \int_0^z \mathbf{k} \times \nabla \tilde{\rho} \, dz \qquad (3.103.a)$$

where $\mathbf{u}_0$ is the velocity at $z = 0$.

The so called thermal wind establishes a relationship between horizontal velocity and the vertical integral of the horizontal gradient of temperature or density. In summary, the vertical gradient of horizontal velocity is related to the horizontal gradient of temperature and density,

$$\frac{\partial \mathbf{u}}{\partial z} = +\frac{g}{f_0 T_0} \mathbf{k} \times \nabla \tilde{T}$$
$$= -\frac{g}{f_0 \rho_0} \mathbf{k} \times \nabla \tilde{\rho} \qquad (3.103.b)$$

3.3.4   Quasi-geostrophic Motion Linearly Superposed on Uniform Current

Closed form quasi-geostrophic equations based on the Boussinesq approximations promote the development of quasi-geostrophic theory for studying motions of geophysical fluids. In the earlier section, perturbations have been linearly superposed on top of a mean basic state assumed to be at rest. In this section, a further change is made to the by adding a uniform zonal current, relative to the perturbed variables. Former state variables are denoted by $\tilde{()}$ on top, allowing them to be split into the prescribed components. Equations of motion are composed of perturbations superimposed on a basic (zonal) flow component along the x-direction:

$$\tilde{p} = \bar{p} + p'$$
$$\tilde{\rho} = \bar{\rho} + \rho'$$
$$\tilde{T} = \bar{T} + T'$$
$$\tilde{\theta} = \bar{\theta} + \theta' \qquad (3.104.a - g)$$
$$\tilde{\mathbf{u}} = U(y, z)\mathbf{i} + \mathbf{u}'$$
$$\tilde{w} = w'$$
$$\tilde{\psi} = \Psi(y, z) + \psi$$

Fig. 3.8 Uniform zonal flow
jet velocity $U(y, z)$ in a
basic state

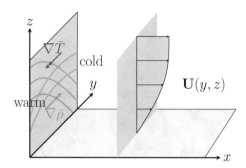

The mean zonal flow $U(y, z)$ remains in balance with meridional gradients of
mean temperature $\bar{T}$, potential temperature $\bar{\theta}$ and density $\bar{\rho}$, as shown in Fig. 3.8,
according to the thermal wind relationship (3.103.b) established between these mean
properties

$$
\begin{aligned}
\frac{\partial U(y, z)}{\partial z} &= -\frac{g}{f T_0} \frac{\partial \bar{T}(y, z)}{\partial y} \\
&= +\frac{g}{f \rho_0} \frac{\partial \bar{\rho}(y, z)}{\partial y}.
\end{aligned}
\tag{3.104.b}
$$

Substituting these, and using the momentum Eq. (3.92.a) for total flow perturbation velocity is given as $\tilde{u}$

$$
\tilde{\mathbf{u}}_t + \tilde{\mathbf{u}} \cdot \nabla \tilde{\mathbf{u}} + f \mathbf{k} \times \tilde{\mathbf{u}} = -\frac{1}{\rho_r} \nabla \tilde{p}
\tag{3.105}
$$

where the nonlinear terms are expanded as

$$
\begin{aligned}
\tilde{\mathbf{u}} \cdot \nabla \tilde{\mathbf{u}} &= (U\mathbf{i} + \mathbf{u}') \cdot \nabla (U\mathbf{i} + \mathbf{u}') \\
&= U \left(\mathbf{i} \overset{0}{\frac{\partial U}{\partial x}} + \frac{\partial \mathbf{u}'}{\partial x} \right) + \mathbf{u}' \cdot \nabla (U\mathbf{i} + \mathbf{u}') \\
&= U \frac{\partial \mathbf{u}'}{\partial x} + \mathbf{i} v' \frac{\partial U}{\partial y} + \mathbf{u}' \cdot \nabla \mathbf{u}'.
\end{aligned}
\tag{3.106}
$$

Note that the x-dependent first term in the above equation disappears because
$U = U(y, z)$ only. The Coriolis and pressure gradient terms are

$$
\begin{aligned}
f \mathbf{k} \times \tilde{\mathbf{u}} &= f \mathbf{k} \times (\mathbf{i} U) + f \mathbf{k} \times \mathbf{u}' \\
-\frac{1}{\rho_r} \nabla \tilde{p} &= -\frac{1}{\rho_r} \nabla \bar{p} - \frac{1}{\rho_r} \nabla p'
\end{aligned}
\tag{3.107.a, b}
$$

and the first terms on the right hand side of (3.107.a, b) cancel out in (3.105)

$$f\mathbf{k} \times (\mathbf{i}U) = -\frac{1}{\rho_r}\nabla\bar{p} \tag{3.107.c}$$

yielding a balance between Coriolis and pressure gradient forces in the basic state, also noting $\bar{p} = \bar{p}(y, z)$,

$$U = -\frac{1}{f\rho_r}\frac{\partial\bar{p}}{\partial y} = -\frac{\partial\Psi}{\partial y} \tag{3.108}$$

defining a relationship between pressure gradient and stream-function of the mean flow. With above simplifications, the remaining terms in the equation of motion (3.105) are

$$\mathbf{u}'_t + U\frac{\partial\mathbf{u}'}{\partial x} + \mathbf{i}v'\frac{\partial U}{\partial y} + \mathbf{u}'\cdot\nabla\mathbf{u}' + f\mathbf{k}\times\mathbf{u}' = -\frac{1}{\rho_r}\nabla p'. \tag{3.109}$$

By making use of Eqs. (3.91.a) and (3.94.b), hydrostatic balance is expressed as

$$g\frac{\tilde{\theta}}{\theta_r} = f\frac{\partial\tilde{\psi}}{\partial z} \tag{3.110}$$

then by splitting basic state and perturbed variables as in (3.104.a–g) the basic state $\bar{\theta}$ must satisfy

$$\tilde{\psi} = \Psi + \psi \quad\text{and}\quad \tilde{\theta} = \bar{\theta} + \theta',$$

$$g\frac{\bar{\theta}}{\theta_r} = f\frac{\partial\Psi}{\partial z}, \tag{3.111}$$

and subtracting the basic state equation from the total gives

$$g\frac{\theta'}{\theta_r} = f\frac{\partial\psi}{\partial z}. \tag{3.112}$$

Differentiating equation (3.111)

$$\frac{g}{\theta_r}\frac{\partial\bar{\theta}}{\partial y} = f\frac{\partial^2\Psi}{\partial y\partial z} \tag{3.113}$$

and replacing from (3.108) results in the familiar *thermal wind* relationship

$$\frac{g}{f\theta_r}\frac{\partial\bar{\theta}}{\partial y} = -\frac{\partial U}{\partial z}. \tag{3.114}$$

On the other hand, through the use of (3.91. a, b), (3.99) and (3.110), vertical velocity is given by

$$w = -\frac{f}{N^2}\left(\frac{\partial}{\partial t} + \tilde{\mathbf{u}}\cdot\nabla\right)\frac{\partial\psi}{\partial z}. \tag{3.115}$$

Expanding terms in (3.115) and neglecting the small nonlinear term once again yields vertical velocity w

$$\begin{aligned} w &= -\frac{f}{N^2}\left(\frac{\partial}{\partial t} + \tilde{\mathbf{u}}\cdot\nabla\right)\frac{\partial(\Psi+\psi)}{\partial z} \\ &= -\frac{f}{N^2}\left\{\frac{\partial}{\partial t}\left(\frac{\partial\psi}{\partial z}\right) + v'\frac{\partial}{\partial z}\left(\frac{\partial\Psi}{\partial y}\right) + U\frac{\partial}{\partial x}\frac{\partial\psi}{\partial z} + \cancel{\mathbf{u}'\cdot\nabla\frac{\partial\psi}{\partial z}}\right\} \end{aligned} \tag{3.116.a}$$

$$w = w' = -\frac{f}{N^2}\left\{\frac{\partial}{\partial t}\frac{\partial\psi}{\partial z} - v'\frac{\partial U}{\partial z} + U\frac{\partial}{\partial x}\frac{\partial\psi}{\partial z}\right\} \tag{3.116.b}$$

and re-organizing (3.116.b) gives back the simplified momentum equation

$$\frac{\partial}{\partial t}\frac{\partial\psi}{\partial z} + U\frac{\partial}{\partial x}\frac{\partial\psi}{\partial z} - v'\frac{\partial U}{\partial z} + \frac{N^2}{f}w' = 0. \tag{3.117}$$

Alternative derivation of vertical velocity w, based on Eq. (3.115) is obtained by making use of (3.99) and (3.110)

$$\begin{aligned} w &= -\frac{g}{N^2\theta_r}\left(\frac{\partial\tilde{\theta}}{\partial t} + \tilde{\mathbf{u}}\cdot\nabla\tilde{\theta}\right) \\ &= -\frac{g}{N^2\theta_r}\left(\frac{\partial\theta'}{\partial t} + U\frac{\partial\theta'}{\partial x} + v'\frac{\partial\bar{\theta}}{\partial y} + \cancel{\mathbf{u}'\cdot\nabla\theta'}\right) \end{aligned} \tag{3.118}$$

neglecting the expected small nonlinear contribution, the simplified form of the thermal equation is obtained.

$$\frac{\partial\theta'}{\partial t} + v'\frac{\partial\bar{\theta}}{\partial y} + U\frac{\partial\theta'}{\partial x} + \frac{N^2\theta_r}{g}w' = 0. \tag{3.119}$$

3.3.5 Energetics of Linear Motions Superposed on Uniform Zonal Flow

The evolution of perturbed states were given by Eqs. (3.109) and (3.119) in the last section. Excluding nonlinear terms involving perturbed variables, the equivalent momentum and thermal equations are obtained for motions superposed on a uniform flow

$$\frac{\partial \mathbf{u}'}{\partial t} + U\frac{\partial \mathbf{u}'}{\partial x} + i v'\frac{\partial U}{\partial y} + \mathbf{u}' \cdot \nabla \mathbf{u}' + f\mathbf{k} \times \mathbf{u}' = -\frac{1}{\rho_r}\nabla p'$$

$$\frac{\partial \theta'}{\partial t} + U\frac{\partial \theta'}{\partial x} + v'\frac{\partial \bar{\theta}}{\partial y} + \mathbf{u}' \cdot \nabla \theta' + \frac{N^2\theta_r}{g}w' = 0$$

$$(3.120.a, b)$$

To investigate mechanical energy conversions, dot product of $\rho_r\mathbf{u}$ is formed with Eq. (3.120.a),

$$\left(\frac{\partial}{\partial t} + U\frac{\partial}{\partial x}\right)\left(\frac{1}{2}\rho_r\mathbf{u}'\cdot\mathbf{u}'\right) + \rho_r u'v'\frac{\partial U}{\partial y} + \overbrace{\rho_r f\mathbf{u}'\cdot(\mathbf{k}\times\mathbf{u}')}^{=0} = -\mathbf{u}'\cdot\nabla p'$$

$$(3.121)$$

Note however, that the right hand side of this equation is interpreted, using continuity (3.92.c) and hydrostatic (3.92.b) equations, to yield

$$-\mathbf{u}'\cdot\nabla p' = -\nabla\cdot p'\mathbf{u}' + p'\nabla\cdot u'$$

$$= -\nabla\cdot p'\mathbf{u}' - p'\frac{\partial w'}{\partial z}$$

$$= -\nabla\cdot p'\mathbf{u}' - \frac{\partial p'w'}{\partial z} + w'\frac{\partial p'}{\partial z} \qquad (3.122)$$

$$= -\nabla\cdot p'\mathbf{u}' - \frac{\partial p'w'}{\partial z} - g\rho'w'$$

$$= -\nabla_3\cdot p'\mathbf{u}'_3 - g\rho'w'$$

where various terms are expanded and combined, with earlier notations $\nabla = \nabla_h$ representing horizontal gradient operator, $\mathbf{u}' = \mathbf{u}'_h$ and w' respectively are the horizontal and vertical components of velocity. We let notations $\nabla_3 = \nabla_h + \mathbf{k}\frac{\partial}{\partial z}$ and $\mathbf{u}'_3 = \mathbf{u}'_h + w'\mathbf{k}$ define three dimensional gradient and velocity. Finally substituting these, the Kinetic Energy equation is obtained

$$\underbrace{\left(\frac{\partial}{\partial t} + U\frac{\partial}{\partial x}\right)}_{\text{rate of change}}\underbrace{\left(\frac{1}{2}\rho_r\mathbf{u}'\cdot\mathbf{u}'\right)}_{KE} = \underbrace{-\rho_r u'v'\frac{\partial U}{\partial y}}_{\substack{SP \\ \text{shear} \\ \text{production}}} - \underbrace{g\rho'w'}_{\substack{BP \\ \text{buoyancy} \\ \text{production}}} - \underbrace{\nabla_3\cdot p'\mathbf{u}'_3}_{\substack{PP \\ \text{pressure} \\ \text{production}}} \qquad (3.123)$$

The left hand side of the above equation constitutes the rate of change of Kinetic Energy $KE = \frac{1}{2}\rho_r\mathbf{u}'\cdot\mathbf{u}'$ per unit of mass, while the right hand side terms respectively are the production of kinetic energy by shear, buoyancy and pressure effects.

To obtain the Thermal Energy equation, (3.120.b) is multiplied with $\rho_r \frac{g^2}{N^2\theta_r^2}\theta'$ to yield

$$\underbrace{\left(\frac{\partial}{\partial t} + U\frac{\partial}{\partial x}\right)}_{\text{rate of change}} \underbrace{\left[\frac{1}{2}\rho_r\frac{g^2}{N^2}\left(\frac{\theta'}{\theta_r}\right)^2\right]}_{PE} = \underbrace{-\frac{\rho_r g^2}{N^2\theta_r^2}\left(\overline{v'\theta'}\frac{\partial\bar\theta}{\partial y}\right)}_{\substack{TP \\ \text{thermal} \\ \text{production}}} - \underbrace{\frac{\rho_r g}{\theta_r}(\overline{\theta'w'})}_{\substack{BP \\ \text{buoyancy} \\ \text{production}}} . \quad (3.124)$$

The left hand side of the above equation is the rate of change of Potential Energy $PE = \frac{1}{2}\rho_r\frac{g^2}{N^2}\left(\frac{\theta'}{\theta_r}\right)^2$ per unit mass, while the terms on the right hand side respectively are the production of potential energy by thermal (TP) and buoyancy (BP) effects. By using Eq. (3.91.a) temperature can be exchanged with density to give

$$\underbrace{\left(\frac{\partial}{\partial t} + U\frac{\partial}{\partial x}\right)}_{\text{rate of change}} \underbrace{\left[\frac{1}{2}\rho_r\frac{g^2}{N^2}\left(\frac{\rho'}{\rho_r}\right)^2\right]}_{PE} = \underbrace{-\frac{g^2}{N^2\rho_r}\left(\overline{v'\rho'}\frac{\partial\bar\rho}{\partial y}\right)}_{\substack{TP \\ \text{thermal} \\ \text{production}}} + \underbrace{g\overline{\rho'w'}}_{\substack{BP \\ \text{buoyancy} \\ \text{production}}} . \quad (3.125)$$

An important quantity that appears in both of the above Eqs. (3.124) and (3.125) is the *buoyancy production* shown to be equivalent to the following, appearing with change of sign in either equation.

$$BP = g\overline{\rho'w'} = \frac{\rho_r g}{\theta_r}(\overline{\theta'w'}). \quad (3.126)$$

The BP term acts as a sink in mechanical energy and source in potential energy, therefore canceled out in total energy. Adding kinetic and thermal parts together, the equation for Total Energy TE is obtained

$$\left(\frac{\partial}{\partial t} + U\frac{\partial}{\partial x}\right)\underbrace{\frac{1}{2}\rho_r\left[\mathbf{u'}\cdot\mathbf{u'} + \frac{g^2}{N^2}\left(\frac{\rho'}{\rho_r}\right)^2\right]}_{TE}$$

$$= \underbrace{-\rho_r\overline{u'v'}\frac{\partial U}{\partial y}}_{\substack{SP \\ \text{shear} \\ \text{production}}} - \underbrace{\frac{g^2}{N^2\rho_r}\left(\overline{v'\rho'}\frac{\partial\bar\rho}{\partial y}\right)}_{\substack{TP \\ \text{thermal} \\ \text{production}}} - \underbrace{\nabla_3\cdot\overline{p'\mathbf{u'}_3}}_{\substack{PP \\ \text{pressure} \\ \text{production}}} \quad (3.127)$$

where it is noted that the buoyancy production term $-g\overline{\rho'w'}$ in kinetic and thermal energy Eqs. (3.123) and (3.125) cancel each other, yielding conservation of Total Energy $TE = KE + PE$.

Mechanical and thermal energy conversions can be analyzed based on (3.123)–(3.127). The 3D motion superposed on uniform flow $U(y, z)$ is specified in the x-direction, envisioned in a channel of width Y in latitude and depth H in the vertical. The solution is assumed to be cyclic with period X

$$F(0, y, z) = F(X, y, z). \tag{3.128.a}$$

and total integration over a full period X yields

$$\int_0^X \frac{\partial}{\partial x} F(x, y, z) dx = F(X, y, z) - F(0, y, z) = 0. \tag{3.128.b}$$

Equation (3.123) is integrated over fixed volume $V = HYX$ in 3D domain denoted as

$$\langle () \rangle = \int_0^H \int_0^Y \int_0^X () \, dx dy dz$$

and with time derivative $\partial/\partial t$ carried out, fluxes canceling as a result of periodic conditions and integrals of total derivative terms yielding zero,

$$\left(\frac{\partial}{\partial t} + U \frac{\partial}{\partial x} \right) \langle KE \rangle$$

$$= \frac{\partial}{\partial t} \iiint \frac{1}{2} \rho_r \mathbf{u}' \cdot \mathbf{u}' \, dx dy dz + \iint U \int_0^X \overset{0}{\cancel{\frac{\partial}{\partial x} \frac{1}{2} \rho_r \mathbf{u}' \cdot \mathbf{u}'}} \, dx dy dz$$

$$= - \iiint \rho_r u' v' \frac{\partial U}{\partial y} \, dx dy dz - \iiint g \rho' w' \, dx dy dz - \iiint \overset{0}{\cancel{\nabla_3 \cdot p' \mathbf{u}'_3}} \, dx dy dz. \tag{3.129.a}$$

simplified to give rate of change (denoted by a dot placed over the variable) of volume averaged Kinetic Energy

$$\dot{K} \equiv \frac{d}{dt} \langle KE \rangle = \frac{d}{dt} \iiint \frac{1}{2} \rho_r \mathbf{u}' \cdot \mathbf{u}' \, dx dy dz$$

$$= - \iiint u' v' \frac{\partial U}{\partial y} \, dx dy dz - \iiint g \rho' w' \, dx dy dz \tag{3.129.b}$$

$$= - \left\langle \rho_r u' v' \frac{\partial U}{\partial y} \right\rangle - \left\langle g \rho' w' \right\rangle$$

Similarly, the integration of thermal equation (3.125) yields for the fixed volume integral

$$
\left(\frac{\partial}{\partial t} + U \frac{\partial}{\partial x}\right) \langle PE \rangle
$$

$$
= \frac{\partial}{\partial t} \iiint \frac{1}{2} \rho_r \frac{g^2}{N^2} \left(\frac{\theta'}{\theta_r}\right)^2 \, dxdydz \qquad (3.130.a)
$$

$$
= -\iiint \frac{\rho_r g^2}{\theta_r^2 N^2} v' \theta' \frac{\partial \bar\theta}{\partial y} \, dxdydz \; - \; \iiint \frac{\rho_r g}{\theta_r} \theta' w' \, dxdydz = 0.
$$

to give rate of change of volume averaged Potential Energy

$$
\dot P \equiv \frac{d}{dt} \langle PE \rangle = \frac{d}{dt} \iiint \frac{1}{2} \rho_r \frac{g^2}{N^2} \left(\frac{\rho'}{\rho_r}\right)^2 \, dxdydz
$$

$$
= -\iiint \frac{g^2}{\rho_r N^2} v' \rho' \frac{\partial \bar\rho}{\partial y} \, dxdydz \; + \; \iiint g\rho' w' \, dxdydz
$$

$$
= -\left\langle v' \rho' \frac{g^2}{\rho_r N^2} \frac{\partial \bar\rho}{\partial y} \right\rangle + \left\langle g\rho' w' \right\rangle
$$

$$
(3.130.b)
$$

The rate of change of Total Energy is then

$$
\dot T = \dot K + \dot P \equiv \frac{d}{dt} \equiv \frac{d}{dt} \left(\langle KE \rangle + \langle PE \rangle \right)
$$

$$
= \frac{d}{dt} \iiint \left[\frac{1}{2} \rho_r \mathbf{u}' \cdot \mathbf{u}' + \frac{1}{2} \rho_r \frac{g^2}{N^2} \left(\frac{\rho'}{\rho_r}\right)^2 \right] \, dxdydz
$$

$$
= -\iiint u' v' \frac{\partial U}{\partial y} \, dxdydz - \iiint \frac{g^2}{\rho_r N^2} v' \rho' \frac{\partial \bar\rho}{\partial y} \, dxdydz
$$

$$
= -\iiint \rho_r u' v' \frac{\partial U}{\partial y} \, dxdydz - \iiint \frac{\rho_r g^2}{\theta_r^2 N^2} v' \theta' \frac{\partial \bar\theta}{\partial y} \, dxdydz
$$

$$
= -\left\langle \rho_r u' v' \frac{\partial U}{\partial y} \right\rangle - \left\langle v' \rho' \frac{g^2}{\rho_r N^2} \frac{\partial \bar\rho}{\partial y} \right\rangle
$$

$$
= -\left\langle \rho_r u' v' \frac{\partial U}{\partial y} \right\rangle + \left\langle g v' \rho' \left(\frac{\partial \bar\rho}{\partial z}\right)^{-1} \frac{\partial \bar\rho}{\partial y} \right\rangle.
$$

$$
(3.131)
$$

Here, once again the buoyancy production term $\langle -g\rho' w' \rangle$ in kinetic and thermal energy equations (3.129.b) and (3.130.b) cancel each other, yielding conservation of the rate of change of Total Energy $\dot T = \dot K + \dot P$.

3.3.6 Interpretation of Quasi-geostrophic Energetics

The volume integrals of Total Energy is a *positive definite* quantity by definition, under conditions of static stability ($N^2 > 0$) of the stratified fluid. Note also that both the kinetic and potential energy components (each of the terms below) individually are positive definite quantities,

$$\langle TE \rangle = \langle KE \rangle + \langle PE \rangle = \left\langle \frac{1}{2} \rho_r \mathbf{u}' \cdot \mathbf{u}' + \frac{1}{2} \rho_r \frac{g^2}{N^2} \left(\frac{\rho'}{\rho_r} \right)^2 \right\rangle > 0, \qquad (3.132)$$

Dynamic stability of the fluid, however, is determined by the rate of change of energy, or power, depending on exchange between different forms of energy,

$$\begin{aligned}
\dot{T} = \dot{K} + \dot{P} &\equiv \frac{d}{dt} \left(\langle KE \rangle + \langle PE \rangle \right) \\
&= \left\langle -\rho_r u'v' \frac{\partial U}{\partial y} \right\rangle + \left\langle g v' \rho' \left(\frac{\partial \bar{\rho}}{\partial z} \right)^{-1} \frac{\partial \bar{\rho}}{\partial y} \right\rangle = SP + TP.
\end{aligned} \qquad (3.133)$$

The stability of quasi-geostrophic flow is determined by the various production terms. Unstable conditions result if on the average, a positive rate of change $\dot{T} > 0$ is obtained for the studied volume of fluid. Contributions to $\dot{T}$ are broken down into kinetic and potential energy conversion terms in the following

$$\begin{aligned}
\dot{K} &\equiv \frac{d}{dt} \left(\langle KE \rangle \right) = \left\langle -\rho_r u'v' \frac{\partial U}{\partial y} \right\rangle - \left\langle g \rho' w' \right\rangle = SP - BP \\
\dot{P} &\equiv \frac{d}{dt} \left(\langle PE \rangle \right) = \left\langle g v' \rho' \left(\frac{\partial \bar{\rho}}{\partial z} \right)^{-1} \frac{\partial \bar{\rho}}{\partial y} \right\rangle + \left\langle g \rho' w' \right\rangle = TP + BP.
\end{aligned}$$
$$(3.134.a, b)$$

Earlier in the preceding section, we have identified $SP = shear\ production$ related to meridional shear, $BP = buoyancy\ production$ related to vertical exchanges and $TP = thermal\ production$ related to thermal wind.

The shear production SP of kinetic energy would be positive, if on the average, sign of horizontal turbulent correlation of momentum $-u'v'$ over selected volume does not negate that of $\partial U / \partial y$, as will be shown in later sections.

The buoyancy production term BP converts potential to kinetic energy by direct buoyant convection. However, this term cancels out in the sum $\dot{T} = \dot{K} + \dot{P}$ and therefore becomes hidden in the total. This term can actually be interpreted as follows, defining *Available Potential Energy (APE)*

$$APE = \iiint g \rho' \xi \, dx dy dz = \left\langle g \rho' \xi \right\rangle = g m_c z_c \qquad (3.135)$$

where ξ is the vertical displacement of a particle, m_c the center of gravity and z_c the mean vertical displacement of the center of mass of integrated volume, so that its time rate of change (by vertical velocity) is interpreted as buoyant production BP, related to APE as

$$BP = \frac{d}{dt} APE = \frac{d}{dt} \iiint g\rho'\xi \, dx dy dz$$

$$= \iiint g\rho' \overset{w'}{\frac{d\xi}{dt}} dx dy dz + \iiint g\xi \overset{0}{\frac{d\rho'}{dt}} dx dy dz \qquad (3.136)$$

$$= \left\langle g\rho' w' \right\rangle = gm_c \frac{dz_c}{dt}.$$

which indicates that APE is increased by raising the center of mass of the fluid against gravity.

Although the BP disappears from the total rate of energy production $\dot{T}$, it would still have a positive contribution to kinetic energy production $\dot{K} > 0$ for an unstable wave, suggesting that vertical velocity should be negatively correlated with density perturbation on the average:

$$\text{cold air} \quad \rho' > 0 \quad \text{sinking} \quad w' < 0$$
$$\text{warm air} \quad \rho' < 0 \quad \text{rising} \quad w' > 0$$

Lastly, the thermal wind production term TP creates change in potential energy, the net effects of which need to be further studied to evaluate its contribution to stable or unstable conditions

$$TP = \left\langle gv'\rho' \frac{\frac{\partial\bar{\rho}}{\partial y}}{\frac{\partial\bar{\rho}}{\partial z}} \right\rangle = g\overline{v'\rho'} \frac{\frac{\partial\bar{\rho}}{\partial y}}{\frac{\partial\bar{\rho}}{\partial z}}. \qquad (3.137)$$

Taking example from mid-latitude atmospheric disturbances in the north hemisphere, meridional and vertical gradients of mean density are typically characterized with colder air in the north and warmer air near surface,

$$\frac{\partial\bar{\rho}}{\partial y} > 0 \quad \text{and} \quad \frac{\partial\bar{\rho}}{\partial z} < 0,$$

clearly leading to a negative ratio

$$\frac{\partial\bar{\rho}}{\partial y} / \frac{\partial\bar{\rho}}{\partial z} < 0.$$

In order to have an unstable situation $\dot{T} > 0$ with conversion rate of potential to kinetic energy $\dot{P} > 0$, one must have negative correlation of meridional velocity and density $\overline{v'\rho'} < 0$ leading to $TP > 0$ on the average,

$$\text{cold air} \quad \rho' > 0 \quad \text{flowing south} \quad v' < 0$$
$$\text{warm air} \quad \rho' < 0 \quad \text{flowing north} \quad v' > 0. \qquad (3.138)$$

Fig. 3.9 Ascending and
descending motions under
unstable conditions of
slanted convection for a
zonal jet with meridional
gradients of density

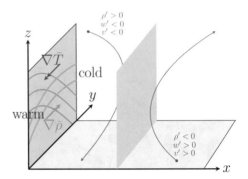

As a result of the above requirements for unstable conditions, a growing weather wave has the familiar mid-latitude cyclonic/anticyclonic pattern sketched in Fig. 3.9. It is noted however, that the energy conversions for growing patterns of unstable waves may only occur in a very special conditions that need further consideration.

First of all, BP conversion of potential to kinetic energy would not necessarily lead to unstable conditions. It seems that special conditions of slanted convection described in Fig. 3.9 is essential for net energy conversion. Furthermore, as will be shown below, the rate of change of Available Potential Energy APE conversion to Kinetic Energy occurs under special configuration of density and pressure fields, taking advantage of thermal wind regime.

In the $y - z$-plane, particle displacements (η, ζ) and velocity perturbations (v', w') are related as

$$v' = \frac{d\eta}{dt} = \frac{\partial \eta}{\partial t} + U\frac{\partial \eta}{\partial x}$$
$$w' = \frac{d\zeta}{dt} = \frac{\partial \zeta}{\partial t} + U\frac{\partial \zeta}{\partial x}. \qquad (3.139.a, b)$$

In quasi-geostrophic theory we have made use of the Boussinesq approximations leading to continuity equation (3.92.c) and thermal/buoyancy equation (3.92.c). Consistent with these approximations, it is tempting to return to the original form of the continuity equation (1.1), to obtain

$$\frac{D\rho}{Dt} = \frac{D}{Dt}(\rho' + \bar{\rho}) = \frac{\partial \rho'}{\partial t} + U\frac{\partial \rho'}{\partial x} + v'\frac{\partial \bar{\rho}}{\partial y} + w'\frac{\partial \bar{\rho}}{\partial z} = 0 \qquad (3.140)$$

with the above particle velocities integrated as

$$\left(\frac{\partial \rho'}{\partial t} + U\frac{\partial \rho'}{\partial x}\right)\left(\rho' + \eta\frac{\partial \bar{\rho}}{\partial y} + \zeta\frac{\partial \bar{\rho}}{\partial z}\right) = 0 \qquad (3.141)$$

to yield

$$\rho' + \eta\frac{\partial \bar{\rho}}{\partial y} + \zeta\frac{\partial \bar{\rho}}{\partial z} = \text{constant} = 0 \qquad (3.142)$$

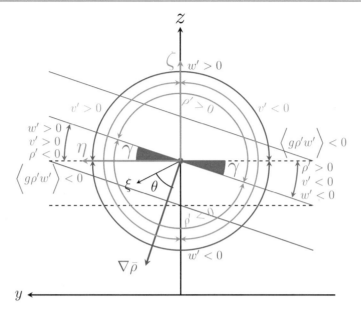

Fig. 3.10 Ranges of perturbation variables in the meridional plane, in relation to energy conversions. The blue lines show constant mean density $\bar{\rho}$ increasing near the surface and to the north, in the midlatitudes. The region marked in red is the angle γ where unstable motions make positive contribution to kinetic energy production

by setting some initial state as $\rho' = 0$ for $\eta = \zeta = 0$ to allow conservation for the perturbation density

$$\rho' = -\eta \frac{\partial \bar{\rho}}{\partial y} - \zeta \frac{\partial \bar{\rho}}{\partial z} = -\boldsymbol{\xi} \cdot \nabla \bar{\rho} \qquad (3.143)$$

where $\boldsymbol{\xi} = (\eta, \zeta)$ is the particle position vector at any point on the $y - z$ plane. Similarly the time rate of change of perturbation density is related to the planar velocity field $\mathbf{q} = (v', w')$ by

$$\frac{\partial \rho'}{\partial t} = -v' \frac{\partial \bar{\rho}}{\partial y} - w' \frac{\partial \bar{\rho}}{\partial z} = -\mathbf{q} \cdot \nabla \bar{\rho}. \qquad (3.144)$$

Finally, favorable conditions for conversion of APE, leading to increased Kinetic Energy production $\dot{K} > 0$ are revealed by analyses explained in Fig. 3.10. By setting the slope of particle trajectories α and the slope of isopycnals γ,

$$\alpha = \frac{\zeta}{\eta} \quad \text{and} \quad \gamma = -\frac{\frac{\partial \bar{\rho}}{\partial y}}{\frac{\partial \bar{\rho}}{\partial z}} \qquad (3.145.a, b)$$

Fig. 3.11 Demonstration of conditions favorable for unstable growth of waves. Particles exchanged between positions A and B along an angle $\alpha < \gamma$ lead to increased conversion of APE into KE on the average

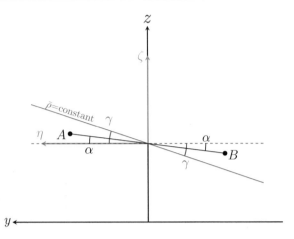

the buoyancy production term, representing rate of change of APE is re-organized as

$$
\begin{aligned}
BP = \frac{d}{dt} APE &= \iiint g\rho' w' \, dx dy dz = \left\langle g\rho' w' \right\rangle \\
&= \left\langle g \left(-\eta \frac{\partial \bar{\rho}}{\partial y} - \zeta \frac{\partial \bar{\rho}}{\partial z} \right) \frac{d\zeta}{dt} \right\rangle \\
&= \left\langle -g\zeta \frac{d\zeta}{dt} \frac{\partial \bar{\rho}}{\partial z} \left(\frac{\eta \frac{\partial \bar{\rho}}{\partial y}}{\zeta \frac{\partial \bar{\rho}}{\partial z}} + 1 \right) \right\rangle \\
&= \left\langle \underbrace{\frac{1}{2} g \frac{d\zeta^2}{dt}}_{>0} \underbrace{\left(-\frac{\partial \bar{\rho}}{\partial z} \right)}_{>0} \underbrace{\left(1 - \frac{\gamma}{\alpha} \right)}_{<0 \text{ for } \alpha < \gamma} \right\rangle .
\end{aligned}
\tag{3.146}
$$

With this view, energy conversion is found to be much constrained in geometry. Only in the narrow range of slant angles $\alpha < \gamma$, marked in red in Fig. 3.11, will the buoyancy production term BP be negative, concurrent with positive contribution to rate of change of kinetic energy $\dot{K} = d(KE)/dt > 0$. It is also observed in Fig. 3.10 that the region marked by angle γ leads to the same quantity of

$$
BP = \frac{d}{dt} APE = \left\langle g\rho' w' \right\rangle < 0 \quad \text{for slant angles} \quad \alpha < \gamma
$$

in the region marked in red as angle (γ) in Fig. 3.10. Furthermore, a particle A in the left hand side moved to point B on the right hand side will become lighter than its original position with positive change $\dot{K} = -BP > 0$. Similarly a particle B on the right hand side will become heavier than its original position at A, but still with positive change $\dot{K} = -BP > 0$. Both of these are true as long as the motion is constrained in the wedge with $\alpha < \gamma$, as shown in Fig. 3.10, consequent to the fact that ρ' and w' are negatively correlated in the wedge region.

In dimensional equivalents to the above, scaling gives

$$\alpha = \frac{\zeta}{\eta} \sim O\left(\frac{w'}{v'}\right) = O\left(\frac{U_0}{f_0 L}\frac{H}{L}\right)$$

$$\frac{\partial \bar{\rho}}{\partial y} \sim \rho_0 f_0 U_0/g, \quad \text{noting} \quad -\frac{g}{\rho_0}\frac{\partial \bar{\rho}}{\partial z} = N^2 \quad \text{and} \quad \gamma = -\frac{\partial \bar{\rho}}{\partial y}\Big/\frac{\partial \bar{\rho}}{\partial z} = O(f_0 U_0/HN^2)$$

identifying L with wavelength λ and $L_d = HN/f_0$ with the *Radius of Deformation* for a stratified fluid, and using the criteria $\alpha < \gamma$ we arrive at necessary conditions for instability indicating waves of wavelength λ greater than the given deformation scale L_d

$$\lambda > L_d.$$

3.3.7   Quasi-geostrophic Equations for Stratified Uniform Flow

The study of the proposed model of a uniform channel flow described in Fig. 3.12 makes use of the vorticity equation (3.95) is with its perturbed form (3.120.a) developed in earlier sections

$$\left(\frac{\partial}{\partial t} + \mathbf{k} \times \nabla\psi \cdot \nabla\right)\left[\nabla^2\psi + \frac{f_0^2}{\rho_r}\frac{\partial}{\partial z}\frac{\rho_r}{N^2}\frac{\partial\psi}{\partial z}\right] + f_0\beta\frac{\partial\psi}{\partial x} = 0. \qquad (3.148)$$

Defining

$$q = f_0(1 + \beta y) + \nabla^2\psi + \frac{f_0^2}{\rho_r}\frac{\partial}{\partial z}\frac{\rho_r}{N^2}\frac{\partial\psi}{\partial z}, \qquad (3.149)$$

Fig. 3.12 Uniform zonal flow with variable cross-stream velocity $U(y, z)$ with meridional thermal wind stratification

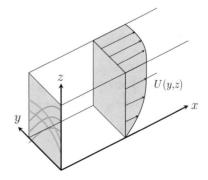

vorticity equation can alternatively be written as

$$\frac{\partial q}{\partial t} + \mathbf{u} \cdot \nabla q = 0 \tag{3.150}$$

or

$$\frac{\partial q}{\partial t} + \frac{\partial \psi}{\partial x}\frac{\partial q}{\partial y} - \frac{\partial \psi}{\partial y}\frac{\partial q}{\partial x} = 0 \tag{3.151}$$

or

$$\frac{\partial q}{\partial t} + J(\psi, q) = 0. \tag{3.152}$$

From the above vorticity equation a version is developed to allow uniform flow in the x-direction with cross-stream variations of velocity $U(y, z)$ and a perturbed state superimposed on this flow. The motion is envisioned to be in a channel of width Y in the latitudinal direction and height H in the vertical.

The motion is thus split into two parts expressed in the stream-function ψ and vorticity q components.

$$\psi = \Psi + \psi \quad \text{and} \quad q = Q + q \tag{3.153}$$

where the uniform flow component $U(y, z)$ is along the (zonal) x direction, being only a function of (meridional and vertical) y, z coordinates, prescribed in relation to the basic state stream-function

$$U(y, z) = -\frac{\partial \Psi(y, z)}{\partial y} \tag{3.154}$$

and the corresponding basic state vorticity prescribed by

$$Q(y, z) = f_0 + \hat{\beta}y + \frac{\partial^2 \Psi}{\partial y^2} + \frac{f_0^2}{\rho_r}\frac{\partial}{\partial z}\frac{\rho_r}{N^2}\frac{\partial \Psi}{\partial z}. \tag{3.155}$$

Vorticity of the superposed motion is given by

$$q = \nabla^2 \psi + \frac{f_0^2 \partial}{\rho_r \partial z}\frac{\rho_r}{N^2}\frac{\partial \psi}{\partial z} \tag{3.156}$$

and combining Eqs. (3.154)–(3.156) together gives total vorticity $q = Q + q$ with the above definitions satisfying (3.152). Analyzing the Jacobian term in Eq. (3.152) produces

$$
\begin{aligned}
J(\psi, q) &= J(\Psi + \psi, Q + q) \\
&= \frac{\partial(\Psi + \psi)}{\partial x}\frac{\partial(Q + q)}{\partial y} - \frac{\partial(\Psi + \psi)}{\partial y}\frac{\partial(Q + q)}{\partial x} \\
&= \frac{\partial\Psi}{\partial x}^{0}\frac{\partial Q}{\partial y} - \frac{\partial\Psi}{\partial y}\frac{\partial Q}{\partial x}^{0} \\
&\quad + \left(\frac{\partial\psi}{\partial x}\frac{\partial Q}{\partial y} + \frac{\partial\Psi}{\partial x}^{0}\frac{\partial q}{\partial y} - \frac{\partial\psi}{\partial y}\frac{\partial Q}{\partial x}^{0} - \frac{\partial\Psi}{\partial y}\frac{\partial q}{\partial x} \right) \\
&\quad + \left(\frac{\partial\psi}{\partial x}\frac{\partial q}{\partial y} - \frac{\partial\psi}{\partial y}\frac{\partial q}{\partial x} \right) \\
&= \frac{\partial\psi}{\partial x}\frac{\partial Q}{\partial y} - \frac{\partial\Psi}{\partial y}\frac{\partial q}{\partial x} + J(\psi, q) \\
&= \frac{\partial\psi}{\partial x}\frac{\partial Q}{\partial y} + U\frac{\partial q}{\partial x} + J(\psi, q).
\end{aligned}
\tag{3.157}
$$

With the above substitutions, vorticity equation (3.152) becomes

$$
\frac{\partial Q}{\partial t}^{0} + \frac{\partial q}{\partial t} + \left(\frac{\partial\psi}{\partial x}\frac{\partial Q}{\partial y} + U\frac{\partial q}{\partial x} \right) + J(\psi, q) = 0
\tag{3.158}
$$

where the unsteady term is canceled for basic state and nonlinear terms $J(\psi, q)$ are dropped, argued to have small contributions, yielding

$$
\left(\frac{\partial}{\partial t} + U\frac{\partial}{\partial x} \right)q + \left(\frac{\partial Q}{\partial y} \right)\frac{\partial\psi}{\partial x} = 0.
\tag{3.159}
$$

Next, primes denoting perturbed flow variables are dropped, to yield

$$
\left(\frac{\partial}{\partial t} + U\frac{\partial}{\partial x} \right)q + \left(\frac{\partial Q}{\partial y} \right)v = 0
\tag{3.160}
$$

which is equivalently written as

$$
\left(\frac{\partial}{\partial t} + U\frac{\partial}{\partial x} \right)\left[\nabla^2\psi + \frac{f_0^2}{\rho_r}\frac{\partial}{\partial z}\left(\frac{\rho_r}{N^2}\frac{\partial\psi}{\partial z} \right) \right] + R\frac{\partial\psi}{\partial x} = 0
\tag{3.161}
$$

where use has been made of Eqs. (3.155) and (3.156):

$$q = \nabla^2 \psi + \frac{f_0^2}{\rho_r} \frac{\partial}{\partial z} \left(\frac{\rho_r}{N^2} \frac{\partial \psi}{\partial z} \right)$$

$$R(y, z) \equiv \frac{\partial Q}{\partial y} = \hat{\beta} - \frac{\partial^2 U}{\partial y^2} - \frac{f_0^2}{\rho_r} \frac{\partial}{\partial z} \left(\frac{\rho_r}{N^2} \frac{\partial U}{\partial z} \right).$$

$$(3.162.a, b)$$

3.3.8 Quasi-geostrophic Motion in Stratified Uniform Flow

In the last section quasi-geostrophic vorticity equation has been developed for motions superposed on uniform zonal mean flow $U(y, z)$ with variation over meridional and vertical coordinates

$$\left(\frac{\partial}{\partial t} + U \frac{\partial}{\partial x} \right) q + R \frac{\partial \psi}{\partial x} = 0. \tag{3.163}$$

The following boundary conditions are applied at the side, surface and bottom boundaries:

$$\frac{\partial \psi}{\partial x} = 0 \quad \text{on} \quad y = 0, Y$$

$$\left(\frac{\partial}{\partial t} + U \frac{\partial}{\partial x} \right) \frac{\partial \psi}{\partial z} - \left(\frac{\partial U}{\partial z} \right) \frac{\partial \psi}{\partial x} = 0 \quad \text{on} \quad z = 0, H.$$

$$(3.164.a, b)$$

Without providing details of lengthy integration (left for Exercise 1 followed up in Chap. 7) the corresponding mechanical energy equation is provided as

$$\frac{\partial}{\partial t} \iiint \frac{1}{2} \rho_r \left[\psi_x^2 + \psi_y^2 + \frac{f_0^2}{N^2} \psi_z^2 \right] dx dy dz$$

$$= \iiint \rho_r \left[\psi_x \psi_y U_y + \frac{f_0^2}{N^2} \psi_x \psi_z U_z \right] dx dy dz. \tag{3.165}$$

Integrating assumed periodic solutions in x, within m repeating periods of $(0, X)$, denoted as

$$F(x, y, z) = F(x + mX, y, z) = \sum_{n=0}^{\infty} A_n(y.z) e^{i 2\pi n x / X}. \tag{3.166}$$

$$\overline{(\,)} = \frac{1}{X} \int_0^X (\,) \, dx \tag{3.167}$$

produces

$$
\frac{\partial}{\partial t} \int_0^H \int_0^Y \frac{1}{2}\rho_r \left[\overline{\psi_x^2} + \overline{\psi_y^2} + \frac{f_0^2}{N^2}\overline{\psi_z^2} \right] dydz
$$
$$
= \int_0^H \int_0^Y \rho_r \left[\overline{\psi_x\psi_y}U_y + \frac{f_0^2}{N^2}\overline{\psi_x\psi_z}U_z \right] dydz. \tag{3.168}
$$

Defining perturbation stream-function and by using Eq. (3.112)

$$
\psi_x = +v'
$$
$$
\psi_y = -u' \tag{3.169}
$$
$$
\psi_z = \frac{g}{f_0}\frac{\theta'}{\theta_r}. \tag{3.170}
$$

leads to the equivalent of Eq. (3.168)

$$
\frac{\partial}{\partial t} \int_0^H \int_0^Y \frac{1}{2}\rho_r \left[\overline{(u')^2} + \overline{(v')^2} + \frac{g^2}{N^2}\overline{\left(\frac{\theta'}{\theta_r}\right)^2} \right] dydz
$$
$$
= \int_0^H \int_0^Y \rho_r \left[\overline{u'v'}U_y + \frac{f_0 g}{N^2\theta_r}\overline{v'\theta'}U_z \right] dydz. \tag{3.171}
$$

First terms in Eq. (3.171) are defined as the *zonal average Kinetic Energy* $(\overline{KE})$ and *zonal average Potential Energy* $(\overline{KE})$ components, the sum of which is the *zonal average Total Energy* $\overline{TE} = \overline{KE} + \overline{PE}$ integrated over the cross-section.

$$
\overline{KE} = \int_0^H \int_0^Y \frac{1}{2}\rho_r \left[\overline{(u')^2} + \overline{(v')^2} \right] dydz = \text{kinetic energy}
$$
$$
\overline{PE} = \int_0^H \int_0^Y \frac{1}{2}\rho_r \frac{g^2}{N^2}\overline{\left(\frac{\theta'}{\theta_r}\right)^2} dydz = \text{available potential energy}
$$
$$
\tag{3.172.a, b}
$$

The energy components given in (3.172.a, b) are also often referred to as *Eddy Kinetic Energy* and *Eddy Potential Energy* because they apply to perturbations superposed on mean flow. The rate of change of total energy $\overline{TE}$ is balanced by the terms of the following equation:

$$
\frac{\partial \overline{TE}}{\partial t} = \frac{\partial}{\partial t}(\overline{KE} + \overline{PE}) = \int_0^H \int_0^Y \rho_r \left[-\overline{u'v'}\frac{\partial U}{\partial y} + \frac{f_0 g}{N^2\theta_r}\overline{v'\theta'}\frac{\partial U}{\partial z} \right] dydz. \tag{3.173}
$$

The last equation can be re-interpreted by making use of (3.114),

$$
\frac{\partial U}{\partial z} = -\frac{g}{\theta_r f_0}\frac{\partial \bar{\theta}}{\partial y} \tag{3.174}
$$

to obtain

$$\frac{\partial}{\partial t}(\overline{KE} + \overline{PE}) = -\int_0^H \int_0^Y \rho_r \left[\overline{u'v'}\frac{\partial U}{\partial y} + \frac{g}{N^2\theta_r}\overline{v'\theta'}\frac{\partial\bar{\theta}}{\partial y} \right] dydz. \qquad (3.175)$$

The last equation can be put in an alternative form using integration by parts, for a total integral yielding zero,

$$\int_0^Y A\frac{\partial B}{\partial y}dy + \int_0^Y B\frac{\partial A}{\partial y}dy = \int_0^Y \frac{\partial AB}{\partial y}dy = 0$$

permitting Eq. (3.167) to be alternatively written as

$$\frac{\partial}{\partial t}(\overline{KE} + \overline{PE}) = +\int_0^H \int_0^Y \rho_r \left[U\frac{\partial}{\partial y}\overline{u'v'} + \frac{g}{N^2\theta_r}\bar{\theta}\frac{\partial}{\partial y}\overline{v'\theta'} \right] dydz. \qquad (3.176)$$

Terms representing energy transfers between mean and perturbed flows in Eqs. (3.159) and (3.160) are very similar to *Reynolds' stress* terms that arise in turbulent fluid dynamics. Here the correlations $\overline{uv}$ and $\overline{v\theta}$ of perturbed variables multiplied with meridional horizontal shear $\partial U/\partial y$, meridional thermal gradient $\partial\bar{\theta}/\partial y$ and vertical shear $\partial U/\partial z$ create turbulent energy transfers between mean flow and perturbed fields.

Individual terms in (3.175) represent rate of change of total energy

$$\frac{\partial}{\partial t}(\overline{KE} + \overline{PE}) = \int_0^H \int_0^Y \rho_r(\dot{K} + \dot{P})\, dydz \qquad (3.177.a)$$

defining terms for individual rates of Kinetic and Potential Energy change

$$\frac{\partial}{\partial t}(\overline{KE}) = \int_0^H \int_0^Y \rho_r \dot{K}\, dydz, \qquad \frac{\partial}{\partial t}(\overline{PE}) = \int_0^H \int_0^Y \rho_r \dot{P}\, dydz \qquad (3.177.b, c)$$

with integrands (excluding the basic state density multiplier ρ_r) defined as

$$\dot{K} = -\overline{u'v'}\frac{\partial U}{\partial y} = \overline{\frac{\partial\psi}{\partial x}\frac{\partial\psi}{\partial y}}\frac{\partial U}{\partial y} \qquad (3.178)$$

$$\dot{P} = -\frac{g^2}{N^2\theta_r^2}\overline{v'\theta'}\frac{\partial\bar{\theta}}{\partial y} = \frac{f_0 g}{N^2\theta_r}\overline{v'\theta'}\frac{\partial U}{\partial z} = \frac{f_0^2}{N^2}\overline{\frac{\partial\psi}{\partial x}\frac{\partial\psi}{\partial z}}\frac{\partial U}{\partial z}. \qquad (3.179)$$

Kinetic Energy averaged over a wavelength is the product of two terms, the eddy correlation of velocities $\overline{u'v'}$ and meridional mean velocity shear $\partial U/\partial y$. By rearranging terms one obtains

$$\dot{K} = -\overline{u'v'}\frac{\partial U}{\partial y} = \overline{\frac{\partial\psi}{\partial x}\frac{\partial\psi}{\partial y}}\frac{\partial U}{\partial y} = \overline{\left(\frac{\frac{\partial\psi}{\partial x}}{\frac{\partial\psi}{\partial y}}\right)\left(\frac{\partial\psi}{\partial y}\right)^2}\frac{\partial U}{\partial y} = -\overline{\left(\frac{v'}{u'}\right)(u')^2}\frac{\partial U}{\partial y}$$

$$(3.180)$$

Fig. 3.13 (left) Streamlines ψ oriented with angle γ and its gradient $\nabla\psi$ with angle α in the $x - y$ plane, (right) relationship of the rate of Kinetic Energy change with the orientation of streamline and velocity fields

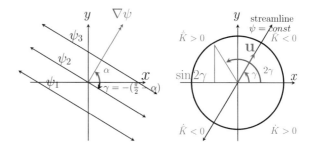

where we note

$$\left(\frac{v'}{u'}\right) = -\left(\frac{\frac{\partial\psi}{\partial x}}{\frac{\partial\psi}{\partial y}}\right) = -\frac{\mathbf{i}\cdot\nabla\psi}{\mathbf{j}\cdot\nabla\psi} = -\frac{\cos\alpha}{\cos\left(\frac{\pi}{2}-\alpha\right)} = -\frac{\cos\alpha}{\sin\alpha} = -\cot\alpha.$$

(3.181.a)

Equivalently, the slope of the streamline is evaluated as

$$\left(\frac{\partial y}{\partial x}\right)_\psi \equiv \tan\gamma = \tan\left(\alpha - \frac{\pi}{2}\right) = -\tan\left(\frac{\pi}{2}-\alpha\right) = -\cot\alpha \qquad (3.181.b)$$

and leads to

$$\dot{\overline{K}} = -\overline{\left(\frac{\partial y}{\partial x}\right)_\psi \underbrace{\left(\frac{\partial\psi}{\partial y}\right)^2}_{>0}\frac{\partial U}{\partial y}} = \overline{-(\tan\gamma)(u')^2}\frac{\partial U}{\partial y}. \qquad (3.182)$$

The angle α that a streamline makes with the horizontal appears to be important, as geometrically analyzed in Fig. 3.13. However, one should also note that this expression leads to singular cases as $\alpha \to 0$ and $\alpha \to \pi$ when the streamlines are horizontal.

Alternative interpretation for the rate of change Kinetic Energy can be given by re-organizing Eqs. (3.168)–(3.170) as

$$\dot{\overline{K}} = -\overline{(\mathbf{u}'\cdot\mathbf{i})(\mathbf{u}'\cdot\mathbf{j})}\frac{\partial U}{\partial y} = -\overline{|\mathbf{u}'|\cos\gamma.|\mathbf{u}'|\sin\gamma}\frac{\partial U}{\partial y} = -\frac{1}{2}\underbrace{\overline{|\mathbf{u}'|^2\sin 2\gamma}}_{>0}\frac{\partial U}{\partial y}.$$

(3.183)

Interpretation of this result allows to understand the rate of change of Kinetic Energy; in particular if the arising motions lead to an increase or decrease. The tendency of Kinetic Energy $\dot{\overline{K}}$ depends entirely on the sign of $\sin 2\gamma$, with twice the orientation angle γ of the velocity vector $\mathbf{u}$, with streamlines $\psi =$ constant aligned the same way over the integration period $(0, X)$ where the overbar is applied. Secondly, the tendency of $\overline{KE}$ also depends on the sign of mean flow shear $\partial U/\partial y$ multiplied with the average of the sinusoidal term applied on Kinetic Energy $\frac{1}{2}|\mathbf{u}'|^2$.

Fig. 3.14 Streamfunction and velocity fields for (upper figure) growing and (lower figure) decaying waves, for the typical case of positive mean meridional shear $\frac{\partial U}{\partial y} > 0$

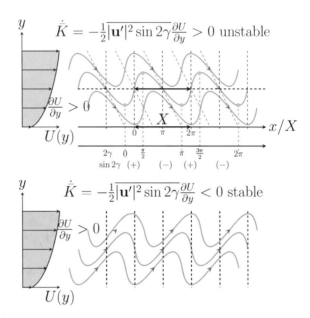

For instance, if the horizontal shear is positive $\partial U/\partial y > 0$, i.e. if the mean flow velocity increases towards the north, then the sign of $\dot{K}$ depends on the sinusoidal part. Remembering that solution is periodic with period X by Eqs. (3.165) and (3.166), the term $\sin 2\alpha$ appears doubly periodic, with zeroes at $\alpha=0, \pi/X, 2\pi/X, 3\pi/X, 4\pi/X$. Then it is obvious that the sign of $\dot{K}$ is negative in the first and third quadrants, and positive in the second and fourth quadrants, as shown in Fig. 3.13.

When, on the average, $\pi/X < \alpha < 2\pi/X$ and $3\pi/X < \alpha < 4\pi/X$ as in the upper part of Fig. 3.14, we expect $\dot{K} > 0$ indicating eddy Kinetic Energy to be increasing, meaning that energy is transferred to eddy motions at the cost of mean flow. In this case, the perturbation field appears to 'lean backward' relative to the basic current, corresponding to unstable and growing waves. In fact, this unstable state leads to rapid growth of intense storms and cyclones during periods and localities such as in the case of mid-latitude weather systems.

On the other hand, when velocity vector and streamlines are aligned in directions $0 < \alpha < \pi/X$ and $2\pi/X < \alpha < 3\pi/X$, as in the lower part of Fig. 3.14, we expect $\dot{K} < 0$ indicating eddy Kinetic Energy to be decreasing, meaning that it is transferred to the mean current. In these cases energy of perturbation decreases and helps increase the basic current, indicating decaying disturbances. In this case, waves and eddies traveling through the domain correspond to streamlines that 'lean forward' as often observed in the case of stable atmospheric waves.

Note that both the mean flow direction $U > 0$ and the horizontal shear $\partial U/\partial y > 0$ (i.e. the mean flow velocity increasing towards north in the northern hemisphere) have been assumed positive in these cases. If either the sense of the current U or the direction of the shear $\partial U/\partial y$ changes, then the sign of $\dot{K}$, i.e. the sense of energy transfers will change, multiplied with the sense of the sinusoidal part.

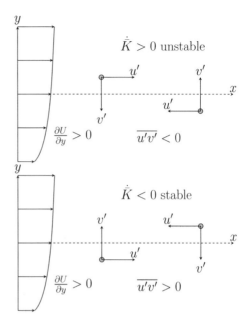

Fig. 3.15 Dependence of Kinetic Energy conversion rates on correlations of the perturbation velocity fields, (upper figure) positive energy transfer rates $\dot{K} > 0$ for negative correlations $\overline{u'v'} < 0$, (lower figure) negative energy transfer rates $\dot{K} < 0$ for positive correlations $\overline{u'v'} > 0$

An alternate interpretation of the eddy motions superposed on mean flow is based on the influence of Reynolds' stress terms specified by the product of $\overline{u'v'}$ and $\partial U/\partial y$ in Eq. (3.170), based on the above discussion on average energy transfers. The possible situations are demonstrated in Fig. 3.15.

It is clear that when either the correlation $\overline{u'v'}$ or $\partial U/\partial y$ or the direction of mean current U change sign, then the direction of energy transfers are also changed. For instance, if all conditions remain same except for $\overline{u'v'} < 0$, momentum is transferred south and away from the regions of high shear. Increasing Kinetic Energy $\dot{K} > 0$, with backward leaning waves are typical for unstable or cyclogenesis conditions.

When transfer of momentum to the north occurs in the case $\overline{u'v'} > 0$ and $\partial U/\partial y > 0$. In this case shear production of energy is negative, i.e. energy flow occurs from the eddy field to mean flow, with $\dot{K} < 0$, according to Eq. (3.178). The waves are forward leaning in decaying mode.

An additional example occurs when horizontal shear has varying character, for instance in the case of a zonal jet flow, when $U(y)$ has a maximum at some latitude, as pictured in Fig. 3.16. Then north of the jet axis where $\partial U(y)/\partial y < 0$, if either $\overline{u'v'} > 0$ or $\sin 2\alpha > 0$ (forward leaning waves) momentum and energy is transferred to eddy motions with $\dot{K} > 0$, leading to growing waves. On the other hand backward leaning waves $\overline{u'v'} > 0$ will tend to decay north of the jet.

South of the jet ($\partial U(y)/\partial y < 0$), unstable waves transferring momentum to the south ($\overline{u'v'} < 0$) will be those backward leaning waves leading to growing eddies ($\dot{K} > 0$).

This process of energy transfer between eddy and mean flow components has the same features of the barotropic instability mechanism related to horizontal shear,

Fig. 3.16 A jet flow with mean velocity $U(y)$ with eddy momentum flux moving towards the jet core results in the eddy kinetic energy to be lost and transferred to the jet, (upper figure) unstable growth by eddy transfers away from jet, (lower figure) stable decay by eddy transfers towards jet

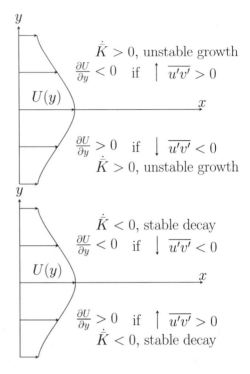

reviewed in Sect. 3.2.3. Here the only difference is that the same mechanism is applied independently at each level in three dimensional field of the stratified case.

We have so far only investigated Kinetic Energy production by interaction between eddy and mean flows, based on the first terms in Eqs. (3.175) and (3.178). The second term in Eq. (3.179) is the Potential Energy component interpreted with the use of (3.114) as

$$\dot{P} = -\frac{g^2}{N^2\theta_r^2}\overline{v'\theta'}\frac{\partial\bar{\theta}}{\partial y} = \frac{gf_0}{N^2\theta_r}\overline{v'\theta'}\frac{\partial U}{\partial z} \qquad (3.184)$$

Interpretation of Potential Energy conversions through interaction of eddy and mean currents can be studied in relation to the eddy correlation $\overline{v'\theta'}$ multiplied either by the mean horizontal potential temperature gradient $-\partial\bar{\theta}/\partial y$ or the vertical gradient of the mean current $\partial U/\partial z$, according to (3.184).

This possibility of instability, created by Potential Energy conversions is demonstrated in Fig. 3.17. By virtue of Eq. (3.173), with positive currents, mean eddy transport of heat occurs towards the north $\overline{v'\theta'} > 0$, and if at the same time mean meridional gradient of potential temperature is negative $\partial\bar{\theta}/\partial y < 0$, then the net result is that the production of Potential Energy is positive $\dot{P} > 0$, by virtue of (3.173), leading to thermal energy conversions from the mean, i.e. growing eddy motions at the expense of the meridional heat gradient. Consequently. if either the mean meridional eddy transport of heat or the mean gradient of potential temperature changes sign then we

Fig. 3.17 Creation of instability by means of positive eddy correlation $\overline{v'\theta'} > 0$ multiplied either by meridional potential temperature gradient $\partial\overline{\theta}/\partial y < 0$ or vertical gradient of the zonal current $\partial U/\partial z > 0$

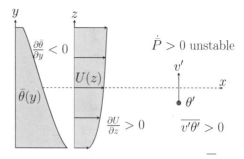

will have decaying solutions, in which case the meridional gradient will be increased at the cost of decaying eddies.

Another way of looking at Potential Energy conversion occurs in relation to the vertical shear of the mean current $\partial U/\partial z$ multiplied by the eddy correlation $\overline{v'\theta'}$.

$$
\begin{aligned}
\dot{P} &= -\frac{g^2}{N^2\theta_r^2}\,\overline{v'\theta'}\,\frac{\partial\overline{\theta}}{\partial y} = \frac{g f_0}{N^2\theta_r}\,\overline{v'\theta'}\,\frac{\partial U}{\partial z} \\
&= \frac{f_0^2}{N^2}\,\overline{\frac{\partial\psi}{\partial x}\frac{\partial\psi}{\partial z}}\,\frac{\partial U}{\partial z} = \frac{f_0^2}{N^2}\left(\frac{\frac{\partial\psi}{\partial x}}{\frac{\partial\psi}{\partial z}}\right)\left(\frac{\partial\psi}{\partial z}\right)^2\frac{\partial U}{\partial z} \\
&= -\frac{f_0^2}{N^2}\left(\frac{\partial z}{\partial x}\right)_\psi\left(\frac{\partial\psi}{\partial z}\right)^2\frac{\partial U}{\partial z}
\end{aligned}
\tag{3.185}
$$

where $(\partial z/\partial x)_\psi$ is the slope of the stream-function in $x - z$ plane evaluated for $\psi = $ constant.

From this last evaluation it can be seen that the $\partial U/\partial z > 0$, lines of constant ψ must slant westward $(\partial z/\partial x)_\psi < 0$ for growing waves $\dot{P} > 0$, indicating Potential Energy production. In the vertical, wave patterns in the upper part will be shifted west with a phase lag compared to the lower layers, as shown in Fig. 3.18.

Again, if the wave patterns in the upper part are shifted east with a phase lag compared to the lower layers, this would imply that the eddy pattern is decaying by feeding the vertical shear.

Fig. 3.18 Backward leaning streamline pattern in the case of unstable waves created by Potential Energy conversion. The upper air and surface streamline patterns are superposed to show westward lag in the upper air pattern

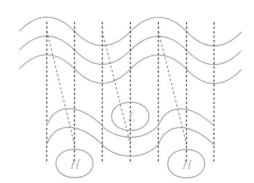

This form of Potential Energy conversion by *Baroclinic Instability* is associated with vertical shear that occurs in relation to thermal wind.

3.4   Baroclinic Instability in a Uniform Zonal Current

3.4.1   Integrated Equations

Motions created in a uniform zonal current have been studied in the above sections. Instability of these motions will be studied here. The governing vorticity equation once again is (3.160) with definitions (3.162.b, c),

$$\left(\frac{\partial}{\partial t} + U\frac{\partial}{\partial x}\right)q + R\frac{\partial \psi}{\partial x} = 0$$

$$q = \nabla^2 \psi + \frac{f_0^2}{\rho_r}\frac{\partial}{\partial z}\left(\frac{\rho_r}{N^2}\frac{\partial \psi}{\partial z}\right) \qquad (3.186.a-c)$$

$$R = \frac{\partial Q}{\partial y} = \hat{\beta} - \frac{\partial^2 U}{\partial y^2} - \frac{f_0^2}{\rho_r}\frac{\partial}{\partial z}\frac{\rho_r}{N^2}\frac{\partial U}{\partial z}$$

and boundary conditions (3.164.a, b)

$$\frac{\partial \psi}{\partial x} = 0 \text{ on } y = 0,\ Y$$

$$\left(\frac{\partial}{\partial t} + U\frac{\partial}{\partial x}\right)\frac{\partial \psi}{\partial z} - \left(\frac{\partial U}{\partial z}\right)\frac{\partial \psi}{\partial x} = 0 \text{ on } z = 0,\ H \qquad (3.187.a, b)$$

Assuming wave-solutions with amplitude $\phi(y.z)$ propagating in the x-direction,

$$\psi = Re\left\{\phi(y, z)e^{ik(x-ct)}\right\} \qquad (3.188)$$

and substituting in Eqs. (3.186.a–c) and (3.187.a, b) gives

$$(U - c)\left[-k^2\phi + \frac{\partial^2 \phi}{\partial y^2} + \frac{f_0^2}{\rho_r}\frac{\partial}{\partial z}\left(\frac{\rho_r}{N^2}\frac{\partial \phi}{\partial z}\right)\right] + R\phi = 0 \qquad (3.189)$$

with boundary conditions

$$\phi = 0 \text{ on } y = 0,\ Y$$

$$(U - c)\frac{\partial \phi}{\partial z} - \left(\frac{\partial U}{\partial z}\right)\phi = 0 \text{ on } z = 0, H. \qquad (3.190.a, b)$$

Denoting complex wave speed c and its conjugate c^* by

$$c = c_r + ic_i, \qquad c^* = c_r - ic_i$$

and noting

$$(U - c)(U - c^*) = |U - c|^2 = [(u - c_r) - ic_i][(u - c_r) + ic_i] = (u - c_r)^2 + c_i^2,$$

Equation (3.189), multiplied by $\rho_r \phi^*(U - c^*)$ is integrated across a meridional section:

$$\int_0^H \int_0^Y \rho_r |U - c|^2 \left[-k^2 \phi^* \phi + \underbrace{\phi^* \frac{\partial^2 \phi}{\partial y^2}}_{(1)} + \underbrace{\frac{f_0^2}{\rho_r} \phi^* \frac{\partial}{\partial z} \left(\frac{\rho_r}{N^2} \frac{\partial \phi}{\partial z} \right)}_{(2)} \right] dy dz$$

$$+ \int_0^H \int_0^Y \rho_r (U - c^*) R \phi^* \phi \, dy dz = 0.$$

(3.191)

The term marked as (1) in the above integral is simplified by integration by parts and by making use of the boundary conditions (3.190.a)

$$\int_0^Y \rho_r |U - c|^2 \left(\phi^* \frac{\partial^2 \phi}{\partial y^2} \right) dy = \int_0^Y \rho_r |U - c|^2 \left(\frac{\partial}{\partial y} \phi^* \frac{\partial \phi}{\partial y} - \frac{\partial \phi^*}{\partial y} \frac{\partial \phi}{\partial y} \right) dy$$

$$= \rho_r |U - c|^2 \left[\phi^* \frac{\partial \phi}{\partial y} \right]_0^{Y \to 0} - \int_0^Y \rho_r |U - c|^2 \left| \frac{\partial \phi}{\partial y} \right|^2 dy$$

(3.192.a)

while the term marked as (2) is similarly reduced to by making use of boundary conditions (3.190.b)

$$\int_0^H |U - c|^2 f_0^2 \phi^* \frac{\partial}{\partial z} \left(\frac{\rho_r}{N^2} \frac{\partial \phi}{\partial z} \right) dz = \int_0^H |U - c|^2 f_0^2 \left[\frac{\partial}{\partial z} \frac{\rho_r}{N^2} \phi^* \frac{\partial \phi}{\partial z} dz - \frac{\rho_r}{N^2} \frac{\partial \phi^*}{\partial z} \frac{\partial \phi}{\partial z} \right]$$

$$= \left[\rho_r |U - c|^2 \frac{f_0^2}{N^2} \phi^* \frac{\partial \phi}{\partial z} \right]_0^H - \int_0^H \rho_r |U - c|^2 \frac{f_0^2}{N^2} \left| \frac{\partial \phi}{\partial z} \right|^2 dz$$

$$= \left[\rho_r \frac{f_0^2}{N^2} \left(\frac{\partial U}{\partial z} \right) \phi^* \phi \right]_0^H - \int_0^H \rho_r |U - c|^2 \frac{f_0^2}{N^2} \left| \frac{\partial \phi}{\partial z} \right|^2 dz,$$

(3.192.b)

with these simplifications inserted back in (3.191). The sum of planetary and zonal mean current vorticity term R in (3.193) is modified so as to replace it with $\hat{R}$

$$\hat{R} = R + \left[\frac{f_0^2}{N^2} \frac{\partial U}{\partial z} \right] (\delta(z - H) - \delta(z))$$

(3.193)

where the *delta function* $\delta(z)$ has been used to incorporate boundary conditions at $z = 0$ and $z = H$,

$$\int_0^Y \int_0^H \left\{ \rho_r |U - c|^2 \left[k^2 |\phi|^2 + \left| \frac{\partial \phi}{\partial y} \right|^2 + \frac{f_0^2}{N^2} \left| \frac{\partial \phi}{\partial z} \right|^2 \right] - \rho_r (U - c^*) \hat{R} |\phi|^2 \right\} dA = 0,$$

(3.194)

re-organized by denoting positive-definite quantities in brackets as

$$P(\phi) = \left[k^2 |\phi|^2 + \left| \frac{\partial \phi}{\partial y} \right|^2 + \frac{f_0^2}{N^2} \left| \frac{\partial \phi}{\partial z} \right|^2 \right]$$

$$S(\phi) = \hat{R} |\phi|^2$$

(3.195.a, b)

further by changing notation to simplify multiple integrals evaluated at cross-sectional area $A = YH$ of the model channel

$$\int_A (\,) dA = \int_0^Y \int_0^H (\,) dy dz.$$

The complex valued Eq. (3.194) has real and imaginary parts

$$\int_A \left\{ \rho_r |U - c|^2 P(\phi) - \rho_r (U - c_r) S(\phi) \right\} dA - i c_i \int_A \rho_r S(\phi) dA = 0. \quad (3.196.a)$$

The imaginary part of (3.196.a) helps to question the possibility to have unstable solutions:

$$c_i \int_A \rho_r S(\phi) dA = 0.$$

(3.196.b)

For unstable wave solutions to occur, it is necessary that $c_i \neq 0$, so as to make the integral zero, which means that $\hat{R}$ has to change sign at least once in the meridional cross-section of the model channel.

It is further noted that the equality

$$\underbrace{\int_A \rho_r |U - c|^2 P(\phi) \, dA}_{>0} = \int_A \rho_r \underbrace{(U - c_r)}_{>0} S(\phi) \, dA$$

(3.197)

limits the real part of phase speed to $c_r < U$ everywhere, by virtue of (3.196.a). Based on this result, and by taking note of equation (3.196.b), it can be proved that the following integral is positive definite:

$$\int_A \rho_r U S(\phi) \, dA > 0.$$

(3.198)

3.4.2   Howard Semicircle Theorem

Limits for real and imaginary parts of wave speeds $c_i \neq 0$ and $c_r < U$ determine the stability of solutions. Alternatively, the following form of the Eq. (3.197) using property (3.198), being multiplied by $(U - c)$ instead of $(U - c^*)$ gives

$$\int_A \rho_r (U - c)^2 P(\phi)\, dA = \int_A \rho_r (U - c) S(\phi) dA. \tag{3.199}$$

which is then expanded to yield two equations

$$(U - c) = (U - c_r - ic_i)$$
$$(U - c)^2 = [(U - c_r) - ic_i]^2 = (U - c_r)^2 - c_i^2 - 2ic_i (U - c_r) \tag{3.200.a, b}$$
$$= [U^2 + c_r^2 - 2Uc_r - c_i^2] - i[2c_i (U - c_r)]$$

providing the real and imaginary parts of (3.199). The real part gives

$$\int_A \rho_r \left\{ \left[U^2 + c_r^2 - 2Uc_r - c_i^2 \right] P(\phi) - (U - c_r) S(\phi) \right\}\, dA \tag{3.201.a}$$

while the imaginary part gives

$$-2ic_i \times \left\{ \int_A \rho_r (U - c_r) P(\phi)\, dA = \frac{1}{2} \int_A \rho_r S(\phi)\, dA \right\} = 0 \tag{3.201.b}$$

Substituting (3.201.b) in (3.201.a) gives

$$\int_A \rho_r U^2 P(\phi)\, dxdy = \int_A \rho_r (c_r^2 + c_i^2) P(\phi) dA + \int_A \rho_r U S(\phi)\, dA. \tag{3.202}$$

Defining U_n as the minimum and U_x as the maximum current values and making use of the property

$$\int_A \rho_r (U - U_n)(U - U_x) P\, dA \leq 0, \tag{3.203}$$

leads to

$$\int_A \rho_r (U - U_n)(U - U_x) P\, dA$$
$$= \int_A \rho_r U^2 P\, dA - (U_n + U_x) \int_A \rho_r U P\, dA + U_n U_x \int_A \rho_r P\, dA$$
$$= \left\{ c_r^2 + c_i^2 - (U_n + U_x) c_r + U_n U_x \right\} \int_A \rho_r P\, dA + \int_A \rho_r \left(U - \frac{U_n + U_x}{2} \right) S(\phi)\, dA$$
$$= \left\{ [c_r - \frac{U_x + U_n}{2}]^2 + c_i^2 - [\frac{U_x - U_n}{2}]^2 \right\} \int_A \rho_r P\, dA + \int_A \rho_r \left(U - \frac{U_n + U_x}{2} \right) S(\phi)\, dA$$
$$= \left\{ [c_r - \bar{U}]^2 + c_i^2 - U_r^2 \right\} \int_A \rho_r P\, dA + \int_A \rho_r (U - \bar{U}) S\, dA \leq 0. \tag{3.204}$$

where $\bar{U} = (U_n + U_x)/2$ is the mean velocity and $U_r = (U_x - U_n)/2$ the difference between maximum and minimum velocity. As noted by (3.195.b) and (3.196.a), the term $\bar{U} \int_A \hat{R}|\phi|^2 dA = 0$ for unstable waves, so that by dividing (3.204)

$$\left\{ [c_r - \bar{U}]^2 + c_i^2 - U_r^2 \right\} + \frac{\int_A \rho_r U S(\phi) dA}{\int_A \rho_r P \, dA} \leq 0. \tag{3.205}$$

and noticing both the numerator and denominator of the added ratio term are positive numbers, we stipulate that (c_r, c_i) lies in the semicircle as shown. The above inequality is the *Howard Semicircle Theorem* which is used to set bounds on phase speed and growth rate of unstable baroclinic motions. The additive term in (3.205) is defined as

$$\Delta^2 \equiv \frac{\int_A \rho_r U S(\phi) dA}{\int_A \rho_r P(\phi) \, dA} = \frac{\int_A \rho_r U \hat{R}|\phi|^2 dA}{\int_A \rho_r \left[k^2|\phi|^2 + \left| \frac{\partial \phi}{\partial y} \right|^2 + \frac{f_0^2}{N^2} \left| \frac{\partial \phi}{\partial z} \right|^2 \right] dA} \tag{3.206}$$

with $\hat{R}$ defined by Eqs. (3.186) and (3.193). Re-setting

$$c_0^2 = U_r^2 + \Delta^2. \tag{3.207}$$

gives

$$\left(\frac{c_0}{U_r} \right)^2 = 1 + \left(\frac{\Delta}{U_r} \right)^2 \tag{3.208}$$

so that (3.206) takes the form

$$(c_r - \bar{U})^2 + c_i^2 \leq c_0^2. \tag{3.209}$$

The instability conditions are therefore found for all real wave speed (c_r) and imaginary growth rate (c_i) components inside the semicircle defined by (3.191). The limits of these characteristic numbers are related to the velocity minimum (U_n) and maximum (U_x) values found in the mean zonal velocity profile $U(y, z)$ across any meridional section A.

The semicircle is centered at a value equivalent to the average velocity $c_r = \bar{U} = (U_x + U_n)/2$. The semicircle has radius c_0 modified by an additive positive constant approximately prescribed by (3.209) from the value $U_r = (U_x - U_n)/2$ (shown by the dashed line in Fig. 3.21). Because it is proved that $c_r \leq U(y, z)$ and essentially $U_n \leq U(y, z) \leq U_x$ in general, instability must be limited for wave speeds $c_r < U_x$, as shown, with real wave speed bounded within $c_{rn} \leq c_r \leq U_x$. Wave speed and growth rate should lie in the shaded area in Fig. 3.19 for unstable motions.

At this stage a definite value has not been assigned for the correction term Δ^2 in (3.206) only giving an educated guess, with further details can be found in Pedlosky (1994). Estimates can be made by considering weighted terms in the integrals

Fig. 3.19 Howard semi-circle limits for unstable waves are shown by the shaded region

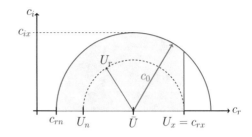

$$\Delta^2 = \frac{\int_A U \hat{\beta} |\phi|^2 + \cdots dA}{\int_A k^2 |\phi|^2 + \cdots dA} \simeq \frac{\bar{U} \hat{\beta}}{k^2}.$$

Some typical numerical values can be provided as follows:

$$U_n = 5 \text{ m/s}, \quad U_x = 25 \text{ m/s}, \quad U_r = 10 \text{ m/s}$$

$$\text{latitude } \phi_0 = 40°, \text{ earth's radius } R_e = 6356 \text{ km}$$

$$f_0 = 2\Omega \sin \phi_0 = 2\frac{2\pi}{86400} \sin 40° = 92 \times 10^{-6} \text{ s}^{-1}$$

$$\beta = \frac{\cot \phi_0}{R_e} = \frac{\cot 40°}{6.36 \times 10^6} = 0.188 \times 10^{-6} \text{ s}^{-1}$$

$$\hat{\beta} = f_0 \beta = 1.76 \times 10^{-11} \text{ s}^{-1} \text{ m}^{-1}$$

$$\lambda = 3000 \text{ km} = 3 \times 10^6 \text{ m}$$

$$k = \frac{2\pi}{\lambda} = \frac{2\pi}{3 \times 10^6} \simeq 2 \times 10^{-6} \text{ m}^{-1}$$

$$\Delta^2 = \frac{\beta U_r}{k^2} \simeq \frac{10 \times 17.6 \times 10^{-12}}{(2 \times 10^{-6})^2} \simeq 44 \text{ m}^2 \text{s}^{-2}$$

$$(\Delta/U_r)^2 = 0.44$$

so that

$$\frac{c_0}{U_r} = \sqrt{1 + \left(\frac{\Delta}{U_r}\right)^2} \simeq 1.20$$

which shows a 20% increase in the radius enclosing the instability region. Based on these numbers, the expected range of the phase speed c_r could be

$$U_n = \bar{U} - U_r\sqrt{1 + (\Delta/U_r)^2} \leq c_r \leq U_x$$

$$-2 \, m/s \leq c_r < 25 \text{ m/s}.$$

The maximum growth rate kc_{ix} can be found as

$$kc_{ix} = kU_r\sqrt{1 + (\Delta/U_r)^2} = 2 \times 10^{-6} \times 20 \times 1.20 = 5 \times 10^{-5} \text{ s}^{-1}$$

which gives a maximum e-folding time of

$$(kc_{ix})^{-1} = 0.23 \ d.$$

Exercises

Exercise 1
This exercise aims to develop energy conservation of baroclinic motions in the presence of a basic zonal flows, based on developments considered in Chap. 3. In particular, the quasi-geostrophic vorticity equation developed in Sect. 3.3.7 will be applied to motions superposed on a uniform zonal flow $U(y, z)$ in the x-direction

$$\left(\frac{\partial}{\partial t} + U\frac{\partial}{\partial x}\right)q + R\frac{\partial \psi}{\partial x} = 0$$

where use has been made of Eqs. (3.161) and (3.162.a, b), allowing conservation of relative vorticity q along the flow and a source term involving components of planetary $\hat{\beta}$ and mean flow $U(y.z)$ vorticity components, given as follows:

$$q = \nabla^2 \psi + \frac{f_0^2}{\rho_r}\frac{\partial}{\partial z}\left(\frac{\rho_r}{N^2}\frac{\partial \psi}{\partial z}\right)$$

$$R(y, z) = \hat{\beta} - \frac{\partial^2 U}{\partial y^2} - \frac{f_0^2}{\rho_r}\frac{\partial}{\partial z}\left(\frac{\rho_r}{N^2}\frac{\partial U}{\partial z}\right).$$

The motion is envisioned to be confined in a zonal channel with width of Y in the latitudinal direction, with height H in the vertical, as shown in Fig. 3.11. This would be a model for a bounded fluid along a uniform zonal channel of infinite length that could mimic a slice of the atmosphere on a β-plane, later to be wrapped around the globe with some extension of our imagination, to allow periodic wave solutions.

Although the confined geometry perhaps does not fit within bounds of our imagination, we allow the following boundary conditions to be applied at the sides, surface and bottom:

$$\frac{\partial \psi}{\partial x} = 0 \quad \text{on} \quad y = 0, Y$$

$$\left(\frac{\partial}{\partial t} + U\frac{\partial}{\partial x}\right)\frac{\partial \psi}{\partial z} - \left(\frac{\partial U}{\partial z}\right)\frac{\partial \psi}{\partial x} = 0 \quad \text{on} \quad z = 0, H$$

Can you derive total energy conservation for the flow confined by the channel geometry, starting from the above closed form vorticity equation and specified boundary conditions, assuming periodic solutions (wave motion) along the zonal direction?

Exercise 2

In Sect. 3.3.7 and Exercise 1, quasi-geostrophic vorticity equations (3.161) and (3.162.a, b) are developed for a baroclinic perturbation superposed on uniform zonal flow $U(y, z)$, with a term R representing planetary and zonal flow components contributing to quasi-geostrophic relative vorticity q. Once again the flow contained in a zonal channel is described, with side, bottom and top boundary conditions provided in Exercise 1, which will now be specialized for the following problem.

Because it is in general difficult to obtain closed form solutions for the above, we will seek physical guidance based on a simplified model, by setting $R = 0$. With this choice our attention is limited to smaller scales of motion, by first having set planetary vorticity effects to zero $\hat{\beta} = 0$, also ignoring vorticity generated by horizontal and vertical second order derivatives of zonal velocity U in the original equations. Basic state density variations are ignored by taking $\rho_r \approx \rho_0 = $ constant, with a constant stratification parameter $N^2 = N_0^2$.

In agreement with the above simplifications, it is tempting to specify a simple zonal current $U = U(z)$ not allowing any meridional variation, instead only specifying a linear dependence of current only in the vertical

$$U(z) = U_0 \left(\frac{z}{H} \right).$$

Based on the above, a simplified vorticity equation is obtained=

$$\left(\frac{\partial}{\partial t} + U_0 \frac{z}{H} \frac{\partial}{\partial x} \right) \left[\nabla^2 \psi + \frac{f_0^2}{N_0^2} \frac{\partial^2 \psi}{\partial z^2} \right] = 0$$

satisfying the following boundary conditions at the side walls, top and bottom boundaries:

$$\frac{\partial \psi}{\partial x} = 0 \quad \text{on} \quad y = 0, Y$$

$$\frac{\partial}{\partial t} \frac{\partial \psi}{\partial z} - \frac{U_0}{H} \frac{\partial \psi}{\partial x} = 0 \text{ on } z = 0$$

$$\left(\frac{\partial}{\partial t} + U_0 \frac{\partial}{\partial x} \right) \frac{\partial \psi}{\partial z} - \frac{U_0}{H} \frac{\partial \psi}{\partial x} = 0 \text{ on } z = H.$$

Solution to the above well defined problem will lead to perfect analysis of mean and eddy motions pertaining to common situations in the atmosphere and ocean. Perhaps the best application is the mid-latitude cyclones and atmospheric jet flows.

Internal Waves

4

4.1 Basic Equations

4.1.1 Re-Scaled Equations

Having developed basic theory of stratified geophysical flows in Chap. 1, perturbations around basic states of the atmosphere and ocean have been considered in Chap. 2. Quasi-geostrophic equations of motion have been developed and applied to the ocean and atmosphere in Chap. 3.

Time scales of quasi-geostrophic flows have been characterized by time required for a fluid particle to traverse the basin, i.e. $t_0 \sim L/U$, where L is the basin size and U is velocity. Small variations in bottom topography were allowed with $\alpha \equiv h_b/H \ll 1$, where h_b is topographic amplitude and H is total depth. The shallow water approximation $\lambda = H/L \ll 1$ were also justified for motions of the earth's atmosphere and ocean, with induced vertical velocity w being sufficiently small, $w = O(Uh_b/L) = O(\alpha\lambda U) \ll U$. Rossby number $\delta \equiv U/(f_0 L)$ and vertical velocity $w \sim \delta\lambda U$ are the same order as $\alpha \ll 1$. The divergence parameter $\mu = f_0^2 L^2/(gH)$ has been found to be a small non-dimensional measure. The resulting Eqs. (2.56a–f) were the basis from which quasi-geostrophic equations were developed by perturbation analysis in terms of the small parameter δ.

In the present section, some of the earlier assumptions of quasi-geostrophic theory are relaxed, to enable studies of flows with balances far from the geostrophic basic state. In essence, we will be considering i.e. ageostrophic motions with smaller horizontal scale, possibly allowing non-hydro-static pressure distribution. In essence, time is scaled with respect to the inertial period, $t \sim f_0^{-1}$, also relaxing the conditions on vertical velocity and allowing greater changes in bottom topography $\alpha = h_b/H = O(1)$, leading to scaling vertical velocity as $w \sim \lambda U$. Yet, Rossby number δ and divergence parameter μ continue to be assumed as small parameters, while the aspect ratio λ need not be so small, $\lambda < O(1)$.

© Springer Nature Switzerland AG 2021

101

E. Özsoy, *Geophysical Fluid Dynamics II*, Springer Textbooks in Earth Sciences, Geography and Environment,

https://doi.org/10.1007/978-3-030-74934-7_4

New scales introduced to modify Eqs. (2.56.a–f) and versions of the thermodynamic equation developed for atmosphere and ocean perturbations (2.56.d,f) are now combined to give

$$(\rho_r + \delta\mu\tilde{\rho})\left\{\frac{\partial \mathbf{u}}{\partial t} + \delta\mathbf{u} \cdot \nabla\mathbf{u} + \delta w\frac{\partial \mathbf{u}}{\partial z}\right\} + (\rho_r + \delta\mu\tilde{\rho})(1 + \delta\beta y)\mathbf{k} \times \mathbf{u} = -\nabla\tilde{p}$$

$$(\rho_r + \delta\mu\tilde{\rho})\lambda^2\left\{\frac{\partial w}{\partial t} + \delta(\mathbf{u} \cdot \nabla)w + \delta w\frac{\partial w}{\partial z}\right\} = -\frac{\partial \tilde{p}}{\partial z} - \tilde{\rho}$$

$$\mu\left\{\frac{\partial \tilde{\rho}}{\partial t} + \delta\mathbf{u} \cdot \nabla\tilde{\rho} + \delta\frac{\partial}{\partial z}\tilde{\rho}w\right\} + (\rho_r + \delta\mu\tilde{\rho})\,\nabla \cdot \mathbf{u} + \frac{\partial}{\partial z}\rho_r w = 0$$

$$\frac{\partial s}{\partial t} + \delta\mathbf{u} \cdot \nabla s + \delta w\frac{\partial s}{\partial z} + \frac{1}{\mu}w\frac{\partial \rho_r}{\partial z} = 0$$

$$(4.1.a - d)$$

Keeping terms of $O(\delta)$ and alike, these equations are further simplified

$$\rho_r\left\{\frac{\partial \mathbf{u}}{\partial t} + \delta\mathbf{u} \cdot \nabla\mathbf{u} + \delta w\frac{\partial \mathbf{u}}{\partial z}\right\} + \rho_r\mathbf{k} \times \mathbf{u} = -\nabla\tilde{p}$$

$$\rho_r\lambda^2\left\{\frac{\partial w}{\partial t} + \delta(\mathbf{u} \cdot \nabla)w + \delta w\frac{\partial w}{\partial z}\right\} = -\frac{\partial \tilde{p}}{\partial z} - \tilde{\rho}$$

$$(4.2.a - d)$$

$$\mu\left\{\frac{\partial \tilde{\rho}}{\partial t} + \delta\mathbf{u} \cdot \nabla\tilde{\rho} + \delta\frac{\partial}{\partial z}\tilde{\rho}w\right\} + \rho_r\nabla \cdot \mathbf{u} + \frac{\partial}{\partial z}\rho_r w = 0$$

$$\frac{\partial s}{\partial t} + \delta\mathbf{u} \cdot \nabla s + \delta w\frac{\partial s}{\partial z} + \frac{1}{\mu}w\frac{\partial \rho_r}{\partial z} = 0.$$

Only keeping terms of $O(1)$ and dropping nonlinear terms of smaller order, linear forms of the equations are obtained

$$\frac{\partial \mathbf{u}}{\partial t} + \mathbf{k} \times \mathbf{u} = -\nabla\phi$$

$$\lambda^2\frac{\partial w}{\partial t} + \frac{\partial \phi}{\partial z} = s$$

$$(4.3.a - d)$$

$$\nabla \cdot \mathbf{u} + \frac{1}{\rho_r}\frac{\partial}{\partial z}(\rho_r w) = 0$$

$$\frac{\partial s}{\partial t} + S_0 N^2(z)w = 0$$

where perturbation variables $\phi = \tilde{p}/\rho_r$ and $s = \tilde{\theta}/\theta_r$ for the atmosphere, and $s = -\tilde{\rho}/\rho_r$ for the ocean are applied. A new variable σ is introduced representing *reduced gravity*, a restoring force proportional to the ratio of perturbation and basic state densities,

$$\phi = \frac{\tilde{p}}{\rho_r}$$

$$\sigma = gs = \begin{cases} g\tilde{\theta}/\theta_r & \text{for the atmosphere} \\ -g\tilde{\rho}/\rho_r & \text{for the ocean.} \end{cases}$$

$$(4.4.a - c)$$

The equations in dimensional form are

$$\frac{\partial \mathbf{u}}{\partial t} + f_0 \mathbf{k} \times \mathbf{u} = -\nabla\phi$$

$$\frac{\partial w}{\partial t} + \frac{\partial \phi}{\partial z} = \sigma$$

$$\nabla \cdot \mathbf{u} + \frac{1}{\rho_r}\frac{\partial}{\partial z}(\rho_r w) = 0$$

$$\frac{\partial \sigma}{\partial t} + N^2(z)w = 0.$$

$$(4.5.a - d)$$

These simplifications, often referred to as *Boussinesq approximations*, neglect compressible effects in the atmosphere and ocean. Further simplification is justified by expanding the second terms on the right hand sides of Eqs. (4.5.b, c),

$$\frac{\partial \phi}{\partial z} = \frac{\partial(\tilde{p}/\rho_r)}{\partial z} = \frac{1}{\rho_r}\left(\frac{\partial \tilde{p}}{\partial z} - \tilde{p}\frac{1}{\rho_r}\frac{\partial \rho_r}{\partial z}\right)$$

$$\frac{1}{\rho_r}\frac{\partial}{\partial z}(\rho_r w) = \frac{\partial w}{\partial z} + w\frac{1}{\rho_r}\frac{\partial \rho_r}{\partial z}.$$

The first terms on the right hand sides consist of a vertical gradient scaled with inverse of depth $1/H_0$, while latter terms in the above expansions involve *scale height* $H_s^{-1} = -\frac{1}{\rho_r}\frac{d\rho_r}{dz}$ defined in relation to basic state density gradient in Eq. (2.18).

By comparing orders of magnitude in second and last terms,

$$\frac{-\tilde{p}\frac{1}{\rho_r}\frac{\partial \rho_r}{\partial z}}{\frac{\partial \tilde{p}}{\partial z}} \simeq \frac{w\frac{1}{\rho_r}\frac{\partial \rho_r}{\partial z}}{\frac{\partial w}{\partial z}} \simeq O\left(\frac{H_s^{-1}}{H_0^{-1}}\right) = O\left(\frac{H_0}{H_s}\right).$$

we can arrive at the conclusion that the latter of the terms in the above expansions can be dropped, since for most small scale internal processes, depth is much smaller than the scale height. In Sect. 2 the scale height for the atmosphere was estimated to be about $10\,km$, close to the thickness of troposphere. For the ocean it is estimated to be about $200\,km$, which is much larger than the maximum depth of the ocean. In either case, the ratio $H_0/H_s \leq 1$, and therefore vertical gradients of basic state density can be ignored (except in the definition of $N^2(z)$ in Eq. 4.4.c). In fact, it can be noted that both basic state density gradients and compressible effects parameterized in $N^2(z)$ can be neglected elsewhere in the simplified equations. As ρ_r does not seem to have a relevant contribution in dynamics (except in the stratification parameter), basic state density is set to $\rho_r = \rho_0 =$ constant in the revised Boussinesq equations:

$$\frac{\partial \mathbf{u}}{\partial t} + f_0 \mathbf{k} \times \mathbf{u} = -\frac{1}{\rho_0}\nabla p$$

$$\frac{\partial w}{\partial t} = -\frac{1}{\rho_0}\frac{\partial p}{\partial z} + \sigma$$

$$\nabla \cdot \mathbf{u} + \frac{\partial w}{\partial z} = 0 \qquad\qquad (4.6.a - d)$$

$$\frac{\partial \sigma}{\partial t} + N^2(z)w = 0.$$

Linear Eqs. (4.6.a–d) can be written in component form as follows

$$\frac{\partial u}{\partial x} + \frac{\partial v}{\partial y} + \frac{\partial w}{\partial z} = 0$$

$$\frac{\partial u}{\partial t} - f_0 v = -\frac{1}{\rho_0}\frac{\partial p}{\partial x}$$

$$\frac{\partial v}{\partial t} + f_0 u = -\frac{1}{\rho_0}\frac{\partial p}{\partial y} \qquad\qquad (4.7.a - d)$$

$$\frac{\partial w}{\partial t} = -\frac{1}{\rho_0}\frac{\partial p}{\partial z} + \sigma$$

$$\frac{\partial \sigma}{\partial t} + N^2 w = 0.$$

4.1.2   Oscillatory Motion in Stratified Environment (N^2)

Information on basic state of a stratified fluid is parameterized by $N^2(z)$, while density is set to $\rho_r(z) = \rho_0 =$ constant in all other terms of Boussinesq equations (4.7.a–d). The parameter N is referred to as *Brunt-Väisälä* or *stability frequency*, defined in equations as defined in (2.74) and (2.81)

$$N^2(z) = \begin{cases} \frac{g}{\theta_r}\frac{d\theta_r}{dz} = -\frac{g}{\rho_r}\frac{d\rho_r}{dz} - \frac{g^2\rho_r}{\gamma p_r} & \text{(atmosphere)} \\ -\frac{g}{\rho_r}\frac{d\Delta_r}{dz} = -\frac{g}{\rho_r}\frac{d\rho_r}{dz} - \frac{g^2}{c_s^2} & \text{(ocean).} \end{cases} \qquad (4.8.a, b)$$

The last terms on the right hand side of (4.8.a, b) represent *compressibility* effects in basic state, although perturbation equations (4.7.a–d) formally do not include compressible effects. Basic state properties are only included in total vertical derivative of density in N^2, considering only small scale perturbations are in the present chapter.

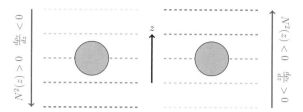

Fig. 4.1 Brunt-Väisälä parameter representing (left hand figure) stable case $N^2 > 0$ with increasing density, (right hand figure) unstable case $N^2 < 0$ with decreasing density with depth

The role of basic state stratification parameter N^2 is revealed by Eqs. (4.7.d, e), with stable and unstable stratification cases in Fig. 4.1.

Neglecting vertical gradient of pressure, i.e. the first term on the right hand side of Eq. (4.7.d), solely considers vertical acceleration due to buoyancy forces. While at the same time, Eq. (4.7.e) indicates perturbation density (buoyancy) is related to vertical motion. Differentiating (4.7.d) and using (4.7.e) an approximate equation of motion can be obtained

$$\frac{\partial^2 w}{\partial t^2} + N^2 w \simeq 0, \tag{4.9}$$

with typical oscillatory solutions for positive values of stratification parameter $N^2 > 0$,

$$w = A e^{iNt} + B e^{-iNt} \tag{4.10}$$

where N is the frequency of oscillation. In this case ($N^2 > 0$), the motion is said to be stable with free oscillations created, owing to restoring forces of *stable stratification*.

On the other hand, if $N^2 \equiv -\hat{N}^2 < 0$, the motion would be described by

$$w = \hat{A} e^{\hat{N}t} + \hat{B} e^{-\hat{N}t} \tag{4.11}$$

where $\hat{N}$ represents either the growth or decay rates of the motion. If the first term dominates, the motion will soon become unstable and in fact unbounded, or else motion may die out in the second case, depending on boundary and initial conditions. The above simple set of considerations demonstrates the role of basic state stability parameter N^2.

In the atmosphere, definition of potential temperature in (1.72.a, b)

$$N^2 = \frac{g}{\theta_r} \frac{\partial \theta_r}{\partial z} = \frac{g}{T_r} \left(\frac{p_r}{p*} \right)^{\frac{R}{c_p}} \left\{ \left(\frac{P_*}{P_r} \right)^{\frac{R}{c_p}} \frac{dT_r}{dz} - T_r \frac{R}{c_p} \left(\frac{p_*}{p_r} \right)^{\frac{R}{c_p} - 1} \frac{p_* \, dp_r}{p_r^2 \, dz} \right\}$$

$$= \frac{g}{T_r} \left\{ \frac{dT_r}{dz} - T_r \frac{R}{c_p} \frac{1}{p_r} \frac{dp_r}{dz} \right\}$$

which by making use of thermodynamic and hydrostatic equations (1.64) and (2.13) gives

Fig. 4.2 Thermal stratification corresponding to (left side figure) stable $N^2 > 0$ and (right side figure) unstable $N^2 < 0$ cases determined by lapse rate

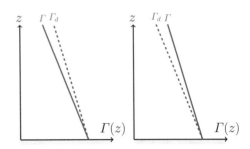

$$N^2 = \frac{g}{T_r} \left\{ \frac{dT_r}{dz} + \frac{g}{c_p} \right\}.$$ (4.12)

Definitions of *dry adiabatic lapse rate*

$$\Gamma_d = \frac{g}{c_p}$$ (4.13)

and *lapse rate*

$$\Gamma = -\frac{dT_r}{dz}$$ (4.14)

are part of meteorological terminology characterizing basic state atmosphere. The dry adiabatic lapse rate is the rate of change of temperature with elevation that would occur in adiabatic (isentropic) atmosphere without humidity (since we have assumed constant proportions in Sect. 2.1.2). This vertical temperature variation is a result of adiabatic compression, with typical values of the rate calculated as

$$\Gamma_d = \frac{g}{c_p} = \frac{9.81 \, m \, s^{-2}}{1.005 \times 10^3 \, m^2 \, s^{-2} \circ K^{-1}} \simeq 10^\circ C/km.$$

Equation (4.12)

$$N^2 = \frac{g}{T_r} \{ \Gamma_d - \Gamma \}$$ (4.15)

can be used to diagnose atmospheric static stability as shown in Fig. 4.2. If $\Gamma < \Gamma_d$ ($N^2 > 0$), the basic state atmosphere is expected to be stable, and if $\Gamma > \Gamma_d$ ($N^2 < 0$) it would have a tendency to become statically unstable.

The physical interpretation of static stability in Fig. 4.2 can be explained as follows:

$\Gamma < \Gamma_d$ *(stable)*: A fluid parcel (initially at equilibrium) moved upwards will cool at a rate given by the dry adiabatic lapse rate Γ_d while the surrounding basic state fluid will cool at a rate of $\Gamma = -\frac{dT_r}{dz} < \Gamma_d$, so that the cooling rate of the fluid parcel is faster to become heavier than the surrounding fluid. In this case, negative buoyancy restoring force will try to return the parcel to its initial position. Similarly, if the parcel is moved downward from its initial position it will warm more rapidly to become lighter than the surrounding fluid, again encountering a restoring buoyancy force.

$\Gamma < \Gamma_d$ *(unstable)*: A fluid parcel moved upwards will cool less rapidly than the surrounding fluid and will therefore become lighter compared to the surroundings. It will thus continue rising indefinitely. The same parcel moved downward will become heavier than the surrounding fluid and continue sinking.

In the ocean, the stratification parameter is expressed by Eq. (2.81)

$$N^2 = -\frac{g}{\rho_r}\frac{d\rho_r}{dz} - \frac{g^2}{c_s^2}. \qquad (4.16)$$

The first term represents density stratification, while the second term expresses changes that would occur if compressibility of sea-water is accounted for. The latter effect is often ignored in the ocean, except for the deep ocean. A typical estimate for density changes due to the second term can be given as $g^2/c_s^2 \simeq 4 \times 10^{-7}s^{-2}$. Density stratification will become very small in the deep ocean, and the two terms will tend to cancel each other, approaching isentropic conditions $\frac{\partial \Delta_r}{\partial z} \to 0$ or $N^2 \to 0$.

Ignoring compressibility effects in the upper ocean, Eq. (4.16) can be approximated by the vertical gradient of density

$$N^2 \simeq -\frac{g}{\rho_r}\frac{d\rho_r}{dz} \simeq -\frac{g}{\rho_0}\frac{d\rho_r}{dz}$$

4.1.3 Total Energy in a Boussinesq Fluid

An equation for the conservation of mechanical energy can easily be derived, based on the linearized form of the Boussinsq equations in (4.6.d).

In the present section some simplifications are made to set both total and basic state density $\rho = \rho_r \simeq \rho_0 =$ constant everywhere, except in the stratification parameter $N^2(z)$ as imposed in the last section. With some foresight, however, nonlinear terms can be reinstated with a small parameter of $O(\delta)$ in Eqs. (4.2.a–d), terms of higher order formerly excluded in Eqs. (4.6.d). The nonlinear advection terms will be needed to generalize energy conservation statement to material volumes of a Boussinesq fluid, following motion. Re-inserting nonlinear terms of $O(\delta)$ in Eq. (4.6a–d) gives

$$\frac{\partial \mathbf{u}_h}{\partial t} + \mathbf{u}_h \cdot \nabla_h \mathbf{u}_h + f_0 \mathbf{k} \times \mathbf{u}_h = -\frac{1}{\rho_0} \nabla_h p$$

$$\frac{\partial w}{\partial t} + \mathbf{u}_h \cdot \nabla w = -\frac{1}{\rho_0} \frac{\partial p}{\partial z} + \sigma$$

$$(4.17.a - d)$$

$$\nabla_h \cdot \mathbf{u}_h + \frac{\partial w}{\partial z} = 0$$

$$\frac{\partial \sigma}{\partial t} + \mathbf{u}_h \cdot \nabla_h \sigma + w \frac{\partial \sigma}{\partial z} + N^2(z)w = 0$$

Alternatively denoting 3-D velocity field as $\mathbf{u}_3$, the 3-D divergence operator is obtained reverting from the 2-D notation used for horizontal fields in earlier sections

$$\mathbf{u}_3 = \mathbf{u} + w\mathbf{k}, \qquad \nabla_3 = \nabla + \mathbf{k}\frac{\partial}{\partial z}, \qquad \nabla = \nabla_h = \mathbf{i}\frac{\partial}{\partial x} + \mathbf{j}\frac{\partial}{\partial y}.$$

Three dimensional representation of the equations are

$$\frac{\partial \mathbf{u}_3}{\partial t} + \mathbf{u}_3 \cdot \nabla_3 \mathbf{u}_3 + f_0 \mathbf{k} \times \mathbf{u}_h = -\frac{1}{\rho_0} \nabla_3 p + \sigma \mathbf{k}$$

$$\nabla_3 \cdot \mathbf{u}_3 = 0 \qquad\qquad (4.18.a - c)$$

$$\frac{\partial \sigma}{\partial t} + \mathbf{u}_3 \cdot \nabla_3 \sigma + N^2(z)w = 0.$$

Taking dot product of the momentum equation (4.19.a) with velocity $\mathbf{u}_3$ and multiplying Eq. (4.19.c) with $\sigma/N^2($ yields

$$\frac{\partial}{\partial t}\left(\frac{1}{2}\mathbf{u}_3 \cdot \mathbf{u}_3\right) + \mathbf{u}_3 \cdot (\mathbf{u}_3 \cdot \nabla_3 \mathbf{u}_3) + f_0 \underbrace{\mathbf{u}_3 \cdot (\mathbf{k} \times \mathbf{u}_h)}_{\,0} = -\frac{1}{\rho_0}\mathbf{u}_3 \cdot \nabla_3 p + \sigma w$$

$$\frac{\partial}{\partial t}\left(\frac{1}{2}\frac{\sigma^2}{N^2}\right) + \frac{\sigma}{N^2}(\mathbf{u}_3 \cdot \nabla_3 \sigma) + \sigma w = 0.$$

$$(4.19.a, b)$$

With following simplifications and the continuity equation (4.18.b)

$$\mathbf{u}_3 \cdot (\mathbf{u}_3 \cdot \nabla \mathbf{u}_3) = \mathbf{u}_3 \cdot \nabla_3 \left(\frac{1}{2}\mathbf{u}_3 \cdot \mathbf{u}_3\right) - \underbrace{\mathbf{u}_3 \cdot (\mathbf{u}_3 \times \nabla_3 \times \mathbf{u}_3)}_{\,0}$$

$$\frac{\sigma}{N^2}\mathbf{u}_3 \cdot \nabla_3 \sigma = \mathbf{u}_3 \cdot \nabla_3 \left(\frac{1}{2}\frac{\sigma^2}{N^2}\right)$$

$$\mathbf{u}_3 \cdot \nabla_3 p = \nabla_3 \cdot p\mathbf{u}_3 - p\underbrace{\nabla_3 \cdot \mathbf{u}_3}_{\,0} = \nabla_3 \cdot p\mathbf{u}_3$$

then yields

$$\frac{\partial}{\partial t}\left(\frac{1}{2}\mathbf{u}_3 \cdot \mathbf{u}_3\right) + \mathbf{u}_3 \cdot \nabla_3\left(\frac{1}{2}\mathbf{u}_3 \cdot \mathbf{u}_3\right) = -\frac{1}{\rho_0}\nabla_3 \cdot p\mathbf{u}_3 + \sigma w$$

$$\frac{\partial}{\partial t}\left(\frac{1}{2}\frac{\sigma^2}{N^2}\right) + \mathbf{u}_3 \cdot \nabla_3\left(\frac{1}{2}\frac{\sigma^2}{N^2}\right) + \sigma w = 0$$

$$(4.20.a, b)$$

so that by adding these together, yields

$$\int_V \left\{\frac{\partial}{\partial t}\left(\frac{1}{2}\mathbf{u}_3 \cdot \mathbf{u}_3 + \frac{1}{2}\frac{\sigma^2}{N^2}\right) + \mathbf{u}_3 \cdot \nabla_3\left(\frac{1}{2}\mathbf{u}_3 \cdot \mathbf{u}_3 + \frac{1}{2}\frac{\sigma^2}{N^2}\right)\right\} dV = -\frac{1}{\rho_0}\int_V \nabla_3 \cdot p\mathbf{u}_3 dV.$$

$$(4.21)$$

Letting T denote total energy for a material volume

$$T = \frac{1}{2}\mathbf{u}_3 \cdot \mathbf{u}_3 + \frac{\sigma^2}{2N^2}$$

and noting

$$\mathbf{u}_3 \cdot \nabla_3 T = \nabla_3 \cdot T\mathbf{u}_3 - T\nabla_3 \cdot \mathbf{u}_3 {}^{\nearrow 0}$$

using continuity equation (4.18.b) and divergence theorem (GFD-I 1.44.c),

$$\int_V \frac{\partial T}{\partial t} dV + \int_V \nabla_3 \cdot T\mathbf{u}_3 \, dV = -\frac{1}{\rho_0}\int_S p\mathbf{u}_3 \cdot \mathbf{n} \, dS \qquad (4.22)$$

then making use of divergence theorem (GFD-I 1.29), and Leibnitz rule of integration (GFD-I 1.44.c), this leads to cancellation of some terms

$$\frac{d}{dt}\int_V T \, dV - \int_S T\mathbf{u}_3 \cdot \mathbf{n} \, dS + \int_S T\mathbf{u}_3 \cdot \mathbf{n} \, dS = -\frac{1}{\rho_0}\int_S p\mathbf{u}_3 \cdot \mathbf{n} \, dV S$$

leading to the conservation statement for total energy $\rho_0 T$ for a material volume V enclosed by surface S

$$\frac{d}{dt}\int_V \rho_0 \left\{\frac{1}{2}\mathbf{u}_3 \cdot \mathbf{u}_3 + \frac{\sigma^2}{2N^2}\right\} dV = -\int_S p\mathbf{u}_3 \cdot \mathbf{n} \, dS. \qquad (4.23)$$

This equation indicates that the material derivative (rate of change) of total kinetic and thermal (potential) energy integrated over a fixed volume of Boussinesq fluid is conserved, with energy transmitted between the parcel and environment, through work done by pressure applied at the bounding surface.

4.1.4   Further Manipulation of Boussinesq Equations

Based on selected scales and simplifications imposed in section (4.1), governing equations applicable to small-scale motions of a stratified fluid can be developed, neglecting nonlinear terms of small order as in a Boussinesq fluid. However $\rho_r(z)$ will be used for basic state density instead of the typical use of $\rho_0 = $ constant consistently adopted in earlier Eq. (4.5.a–d). This simplification appears to be more relevant in the case of atmosphere rather than ocean. In addition, a fixed value of the Coriolis parameter will be assumed with $f = f_0 =$ constant, considering the small scales appropriate for internal motions.

Taking horizontal divergence of the horizontal momentum equation (4.5.a),

$$\nabla \cdot \frac{\partial \mathbf{u}}{\partial t} + f\nabla \cdot \mathbf{k} \times \mathbf{u} = -\nabla^2 \phi. \tag{4.24}$$

where the second term by vector identity becomes

$$f\nabla \cdot \mathbf{k} \times \mathbf{u} = -f\mathbf{k} \cdot \nabla \times \mathbf{u} \tag{4.25}$$

and inserting in Eq. (4.24), the equality takes the form

$$\frac{\partial \nabla \cdot \mathbf{u}}{\partial t} - f\mathbf{k} \cdot \nabla \times \mathbf{u} + \nabla^2 \phi = 0. \tag{4.26}$$

Next, taking curl of (4.5.a), and taking dot product with $f\mathbf{k}$ gives

$$f\mathbf{k} \cdot \nabla \times \frac{\partial \mathbf{u}}{\partial t} + f^2 \mathbf{k} \cdot \nabla \times (\mathbf{k} \times \mathbf{u}) = f\underbrace{\nabla \times \nabla \phi}_{0}$$

where the second term is reduced by vector identity

$$f^2 \mathbf{k} \cdot \nabla \times \mathbf{k} \times \mathbf{u} = f^2 \nabla \cdot \mathbf{u}. \tag{4.27}$$

and inserting in (4.1.4) yields

$$f\mathbf{k} \cdot \left(\nabla \times \frac{\partial \mathbf{u}}{\partial t} \right) + f^2 \nabla \cdot \mathbf{u} = 0. \tag{4.28}$$

Differentiating (4.26) with respect to time and replacing the second term from (4.28) yields

$$\frac{\partial^2 \nabla \cdot \mathbf{u}}{\partial t^2} + f^2 \nabla \cdot \mathbf{u} + \nabla^2 \frac{\partial \phi}{\partial t} = 0. \tag{4.29}$$

Finally, by inserting continuity equation (4.3.c), Eq. (4.29) becomes

$$\frac{\partial^2}{\partial t^2} \frac{\partial \rho_r w}{\partial z} + f^2 \frac{\partial \rho_r w}{\partial z} - \rho_r \nabla^2 \frac{\partial \phi}{\partial t} = 0. \tag{4.30}$$

Next, differentiating the vertical momentum equation (4.5.b) with time and inserting from buoyancy equation (4.5.d) gives

$$\frac{\partial^2 \phi}{\partial z \partial t} = -\frac{\partial^2 w}{\partial t^2} - N^2 w. \tag{4.31}$$

Differentiating (4.29) and substituting (4.31) results

$$\frac{\partial^2}{\partial z^2}\left(\frac{\partial^2 \rho_r w}{\partial t^2} + f^2 \rho_r w\right) + \nabla^2 \frac{\partial^2 \rho_r w}{\partial t^2} + N^2 \nabla^2 \rho_r w = 0 \tag{4.32}$$

Alternatively by letting $W = \rho_r w$, (4.32) is written as

$$\left(\frac{\partial^2}{\partial t^2} + f^2\right)\frac{\partial^2 W}{\partial z^2} + \left(\frac{\partial^2}{\partial t^2} + N^2\right)\nabla^2 W = 0. \tag{4.33}$$

or

$$\frac{\partial^2}{\partial t^2}\left(\nabla^2 W + \frac{\partial^2 W}{\partial z^2}\right) + f^2 \frac{\partial^2 W}{\partial z^2} + N^2 \nabla^2 W = 0. \tag{4.34}$$

Note that the new variable $W = \rho_r w$ contains the basic state density information, but if constant density $\rho_r = \rho_0$ is assumed as in the Boussinesq sense of dumping basic state density variations in N^2, then w becomes the state variable in Eq. (4.34). Assuming $\left(\frac{H_0}{H_s}\right) << 1$ and simplifying density as $\rho_r = \rho_0 =$ constant, Eq. (4.31) is ultimately reduced to

$$\frac{\partial^2}{\partial t^2}\left[\frac{\partial^2 w}{\partial x^2} + \frac{\partial^2 w}{\partial y^2} + \frac{\partial^2 w}{\partial z^2}\right] + f^2 \frac{\partial^2 w}{\partial z^2} + N^2\left(\frac{\partial^2 w}{\partial x^2} + \frac{\partial^2 w}{\partial y^2}\right) = 0. \tag{4.35}$$

To solve internal wave problems with the governing Eq. (4.34), boundary conditions must be applied at the surface and the bottom. At the bottom, $z = -h(x, y)$, an inviscid boundary condition can be used

$$W = \rho_r w = -\rho_r \mathbf{u} \cdot \nabla h \quad \text{on } z = -h. \tag{4.36}$$

At the surface, rigid boundary conditions can be set as $W = \rho_r w = 0$, but if realistic boundary conditions at the ocean surface ($z = \eta(x, y, t)$) are desired, inviscid boundary conditions could be applied, with friction neglected at the outset. Total pressure $p = p_r + \tilde{p}$ (basic plus perturbation pressure) needs to be specified continuously at the surface, and equal to atmospheric pressure p_a

$$\frac{Dp}{Dt} = \frac{Dp_a}{Dt} \quad \text{on } z = \eta \tag{4.37}$$

with nonlinear forcing applied as

$$\frac{Dp_r}{Dt} + \frac{D\tilde{p}}{Dt} = w\frac{\partial p_r}{\partial z} + \frac{\partial \tilde{p}}{\partial t} + \mathbf{u} \cdot \nabla\tilde{p} = \frac{Dp_a}{Dt} \quad \text{on} \ z = \eta. \tag{4.38}$$

Making use of hydrostatic approximation $\tilde{p} = -\rho_r g z$, neglecting nonlinear terms and accounting for temporal variations in atmospheric pressure, one could impose

$$\frac{\partial \tilde{p}}{\partial t} - g\rho_r w = \frac{\partial p_a}{\partial t} \quad \text{on} \quad z = 0. \tag{4.39}$$

Substitution into (4.30) to obtain the proper dynamic surface boundary conditions for pressure, the operator in curly brackets is applied to W

$$\left\{ \left(\frac{\partial^2}{\partial t^2} + f^2 \right) \frac{\partial}{\partial z} + g\nabla^2 \frac{\partial}{\partial t} \right\} W = \frac{1}{\rho_r} \frac{\partial p_a}{\partial t} \quad \text{on} \ z = 0. \tag{4.40}$$

4.2   Internal Gravity Waves Without the Effects of Earth's Rotation

4.2.1   A Note on 3-D Internal Waves

In general, Eqs. (4.34) and simplified form (4.35) allow *internal wave* solutions in a Boussinesq stratified fluid. As reviewed in the last section, these equations have three dimensional character, with comparable velocities in horizontal and vertical directions to first order. The investigated internal modes are thus scaled differently from motions at geophysical scales, e.g. quasi-geostrophic theory reviewed in Chap. 3, where vertical velocity had been assumed much smaller than the horizontal.

Internal wave motion in 3-D has some peculiar characteristics which are not fully explored in early literature. Reduction of these problems to 2-D versions of the same enables simplified understanding of the basic nature of these motions.

For instance, expounding on the problem of a small object oscillated in a stratified environment leads to puzzling behavior, especially when trying to solve the problem in bounded domain. This is because of the hyperbolic nature of the governing equations especially in the case of time dependent or oscillatory motions. For instance, inserting $w \sim e^{i\omega t}$ in the 3-D Eq. (4.34) gives

$$\frac{\partial^2 w}{\partial z^2} - \mu^2 \nabla_h^2 w = 0 \tag{4.41}$$

where

$$\mu^2 = \frac{N^2 - \omega^2}{\omega^2 - f^2}. \tag{4.42}$$

for constant values of the Coriolis and stratification parameters f and N. Positive values of $\mu^2 > 0$ for a typical range $f \leq \omega \leq N$ ensures that the above system is hyperbolic. This 3-D wave equation has wave speed given as μ and the usual time variable t replaced by the vertical coordinate z. On the other hand, 3-D solutions of the above hyperbolic equation requires waves, say from an oscillating source, to propagate along characteristic surfaces of a double cone $z = \pm\sqrt{x^2 + y^2}/\mu$, with a shape referred to as "St. Andrews cross" in the literature. Observations and theory suggests that energy is beamed along these surfaces. A consequence of this rather peculiar geometry of wave spreading is the difficulty to have basin boundary conditions satisfied when the domain is limited, leading to possible multiple reflections, in need of viscous forces, so far been neglected.

4.2.2   Internal Waves in the 2-D Plane

In order to demonstrate some of the peculiar characteristics of internal waves in 3-D, we make the choice of developing equivalent 2-D systems, more instructive to demonstrate basic mechanics. Orientation of the axes in a 2-D slice can be selected along the $x - z$ plane without loss of generality

$$\frac{\partial^2}{\partial t^2}\left(\frac{\partial^2 w}{\partial x^2} + \frac{\partial^2 w}{\partial z^2}\right) + f^2\frac{\partial^2 w}{\partial z^2} + N^2\frac{\partial^2 w}{\partial x^2} = 0. \tag{4.43}$$

Although earth's rotation is important at geophysical scales, momentarily they will be ignored at local scale, setting the Coriolis parameter to zero ($f = 0$) simplifying (4.43) into

$$\frac{\partial^2}{\partial t^2}\left(\frac{\partial^2 w}{\partial x^2} + \frac{\partial^2 w}{\partial z^2}\right) + N^2\frac{\partial^2 w}{\partial x^2} = 0. \tag{4.44}$$

Proposing plane-wave solutions to (4.44) in an unbounded fluid

$$w = \hat{w}(z)e^{i(kx-\omega t)}, \tag{4.45}$$

reduces the problem to solving for amplitude $\hat{w}(z)$

$$\frac{d^2\hat{w}}{dz^2} + \left(\frac{N^2}{\omega^2} - 1\right)k^2\hat{w} = 0. \tag{4.46}$$

Here $k = 2\pi/\lambda_x$ represents wave-number of waves propagating in the x-direction, with λ_x being wavelength, $\omega = 2\pi/T$ angular frequency, and T period of oscillation.

The variable coefficient $N^2 = N^2(z)$ in Eq. (4.44) defines vertical structure of the solution. Solutions to (4.46) are oscillatory in z-direction if $\omega < N$, or evanescent if $\omega > N$. Therefore, Brunt-Väisälä frequency N is an upper cut-off for the vertical component of internal gravity wave propagation. Waves can propagate away from a local disturbance only if $\omega < N$.

Fig. 4.3 Waves spreading
from a point source
oscillated with fixed
frequency ω_0 in a stratified
fluid with given $N^2(z)$
profile

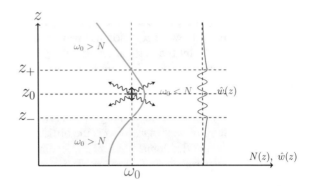

For example, envision the situation in Fig. 4.3 where a point source is oscillated
at level $z = z_0$ with fixed frequency, $w \sim e^{i\omega_0 t}$ in an infinite stratified fluid, where
the basic state density $\rho_r(z)$ is prescribed to be stable

$$N^2 = -\frac{g}{\rho_r} \frac{\partial \rho_r}{\partial z} > 0 \qquad (4.47)$$

Solutions to (4.44) would indicate wave motion between levels $z_- < z < z_+$,
disseminated from the source at z_0 in the same depth range satisfying the criteria
$\omega_0 < N(z)$, as shown in Fig. 4.3. Waves propagating away from the source would
turn around and only be able to propagate in the horizontal direction when they
encounter limits $\omega_0 = N$ at $z = z_-$ and $z = z_+$, essentially being reflected back at
these levels. In the shaded regions $z < z_-$ and $z > z_+$ where the stratification only
allows frequencies $\omega_0 > N$, there would be motion, but the character of this motion
would only be decaying or evanescent, away from the source region.

Internal waves generated in the ocean tend to be trapped near the *pycnocline*
region of maximum density gradient, for suitable frequencies $\omega_0 < N(z)$. Waves
generated near the pycnocline are expected to spread in both the vertical and horizontal regions in the suitable depth range, while they are expected to be evanescent,
with only horizontal components of wave motion outside the given depth range.
The wave amplitude created by an arbitrary source in the pycnocline region is only
schematically illustrated in Fig. 4.3.

Considering a sharp density stratification profile ρ_r going through a step-wise
jump $N^2 \rightarrow \infty$ at an interface implies that the cutoff for frequency ω will be lowered,
leading to oscillations being localized and intensified at the interface. Local analyses
for an interfacial layer are left for Exercise 4.2, reviewed in Chap. 7.

Next, consider the case of uniform stratification with $N^2 = N_0^2$=constant, resulting from density profiles given as

$$\rho_r = \rho_0 e^{-\alpha z} \quad \text{or} \quad \rho_r = \rho_0(1 - \alpha z). \qquad (4.48)$$

with corresponding constant coefficient $N_0^2 = \alpha g$, noting also the relation to scale
height $\alpha = H_s^{-1}$ in the case of exponential stratification. A plane-wave solution of
Eq. (4.46) for frequencies $\omega < N_0$ would give

Fig. 4.4 Plane internal waves for constant stratification $N^2 = N_0^2$

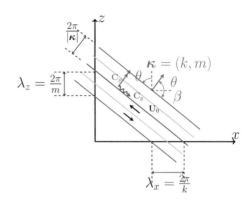

$$\hat{w}(z) = A_0 e^{imz} \tag{4.49}$$

or

$$w = A_0 e^{i(kx+mz-\omega t)} = A_0 e^{i(\boldsymbol{\kappa}\cdot\mathbf{r}-\omega t)} \tag{4.50}$$

where m satisfies

$$m^2 = \left(\frac{N_0^2}{\omega^2} - 1\right) k^2 \quad \text{or} \quad \omega^2 = \frac{N_0^2 k^2}{k^2 + m^2}. \tag{4.51}$$

Defining wave number vector $\boldsymbol{\kappa} = (k, m)$ and coordinates $\mathbf{r} = (x, z)$, the *dispersion equation* (4.51) relates frequency $\omega = 2\pi/T$ to horizontal and vertical wavenumber components $k = 2\pi/\lambda_x$ and $m = 2\pi/\lambda_z$ with λ_x and λ_z wavelengths in the respective directions in Fig. 4.4.

The angle θ in Fig. 4.4 points in direction of propagation, with horizontal and vertical components in the $x - z$ plane

$$\tan\theta = \frac{\lambda_x}{\lambda_z} = \frac{m}{k} \tag{4.52}$$

which by virtue of (4.51) turns out to be

$$\tan^2\theta = \frac{N_0^2}{\omega^2} - 1 \tag{4.53}$$

and therefore

$$\frac{N_0^2}{\omega^2} = 1 + \tan^2\theta = \frac{1}{\cos^2\theta} \tag{4.54}$$

which gives

$$\omega = N_0 \cos \theta. \tag{4.55}$$

This form of the dispersion equation states that the angle θ of propagation is determined uniquely by the ratio of wave frequency to the stability frequency. For θ to correspond to a realistic solution, we must have $\omega < N_0$ based on Eq. (4.55). Note that waves have *fixed* direction θ *independent of the wavelength*, for any given ratio of frequency ω/N_0.

Furthermore, let the velocity field be expressed as

$$\mathbf{u} = \mathbf{U}_0 e^{i(kx+mz-\omega t)} = \mathbf{U}_0 e^{i(\mathbf{k}\cdot\mathbf{r}-\omega t)} \tag{4.56}$$

where $\kappa = (k, m)$ and $\mathbf{r} = (x, z)$ and $\mathbf{U}_0$ is a constant. Substituting (4.56) into the continuity equation (4.23.a) results in

$$\begin{aligned}
\nabla \cdot \mathbf{u} &= \mathbf{U}_0 . \nabla e^{i(\kappa\cdot\mathbf{r}-\omega t)} \\
&= i(\mathbf{U}_0 \cdot \kappa) e^{i(\kappa\cdot\mathbf{r}-\omega t)} \\
&= 0,
\end{aligned} \tag{4.57}$$

which requires that

$$\mathbf{U}_0 \cdot \kappa = 0. \tag{4.58}$$

This result proves that the direction of *particle motion* is always perpendicular to the wave-number vector and therefore to the direction of propagation. It is concluded that internal gravity waves are essentially *transverse waves*, with velocity of motion perpendicular to wave propagation, parallel to the wave phase lines, as shown in Fig. 4.4.

Spreading properties of waves are defined in Fig. 4.4 by *wave-number vector κ* oriented along the direction of propagation,

$$\kappa = (k, m) \tag{4.59}$$

and by *phase velocity* $\mathbf{C}_p$ in the direction of propagation, perpendicular to lines of constant phase

$$\mathbf{C}_p = \frac{\omega}{|\kappa|} \frac{\kappa}{|\kappa|}. \tag{4.60}$$

The *group velocity* of energy propagation is defined as

$$\mathbf{C}_g = \left(\frac{\partial \omega}{\partial k}, \frac{\partial \omega}{\partial m} \right). \tag{4.61}$$

Phase velocity is obtained from Eqs. (4.55) and (4.60)

$$\mathbf{C}_p = \frac{N_0 \cos \theta}{|\boldsymbol{\kappa}|} \frac{\boldsymbol{\kappa}}{|\boldsymbol{\kappa}|} \tag{4.62}$$

with vector components

$$\mathbf{C}_p = \frac{N_0}{|\boldsymbol{\kappa}|} \cos \theta \, (\cos \theta, \sin \theta). \tag{4.63}$$

To calculate group velocity (4.49) is re-written as

$$\omega = N_0 \left(\frac{k^2}{k^2 + m^2} \right)^{1/2} = N_0 \frac{k}{|\boldsymbol{\kappa}|} == N_0 \frac{k}{\kappa}, \tag{4.64}$$

letting $\kappa = |\boldsymbol{\kappa}|$ and differentiating according to (4.61) to obtain

$$\mathbf{C}_g = \frac{N_0}{|\boldsymbol{\kappa}|} \sin \theta (\sin \theta, -\cos \theta) \tag{4.65}$$

which shows that the group velocity vector is at right angles to the phase velocity vector. The magnitudes of the phase and group speeds are respectively

$$C_p = |\mathbf{C}_p| = \frac{N_0}{\kappa} \cos \theta = \frac{\omega}{\kappa} \tag{4.66}$$

and

$$C_g = |\mathbf{C}_g| = \frac{N_0}{\kappa} \sin \theta = \frac{\left(N_0^2 - \omega^2 \right)^{1/2}}{\kappa}. \tag{4.67}$$

It becomes clear that squared phase and group speeds sum up to a constant

$$C_p^2 + C_g^2 = \left(\frac{N_0}{\kappa} \right)^2. \tag{4.68}$$

When either one of these speeds increase, the other decreases depending on angle θ, as indicated by Eq. (4.68). Comparing (4.63) and (4.65), vectors $\mathbf{C}_g$ and $\mathbf{C}_p$, indicate that phase and energy is propagation directions are at right angles to each other

$$\mathbf{C}_p \perp \mathbf{C}_g. \tag{4.69}$$

The angles θ and $\beta = \pi/2 - \theta$ are given as (4.55)

$$\theta = \cos^{-1} \left(\frac{\omega}{N_0} \right) \quad \text{and} \quad \beta = \sin^{-1} \left(\frac{\omega}{N_0} \right). \tag{4.70}$$

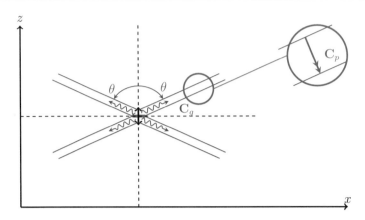

Fig. 4.5 Beams of wave energy from an oscillatory source spreading at angle θ from the vertical (blue), with wave rays (phases) proceeding in perpendicular direction within the beam (red)

Fig. 4.6 The relationship between phase velocity $\mathbf{C}_p$, group velocity $\mathbf{C}_g$ and frequency ω for waves spreading from an oscillatory source in a stratified fluid with constant N_0^2

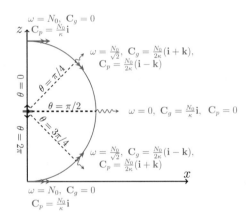

An experiment envisaged in Fig. 4.5 could verify characteristics of wave motion in a uniformly stratified fluid ($N = N_0 =$ constant) by oscillations of a small disc. Because of the peculiar pattern of transverse waves, beams of energy (blue wavy patterns) will spread at angle θ from the vertical, while wave rays or phases move within these beams perpendicular to the beam (red vector) as shown in Fig. 4.5.

Under stationary conditions, propagation away from the disc occurs along beams carrying energy with group velocity $\mathbf{C}_g$. Although wave rays propagating are difficult to be directly observed, taking a cross section along the beam at micro-scale, packets of waves with phase velocity $\mathbf{C}_p$ appear to propagate perpendicular to the beam, with phase and group velocity vectors at right angles to each other as shown in Fig. 4.5.

Investigation of phase and group velocity as a function of oscillation frequency ω is made in Fig. 4.6, with vector components of phase and group velocities obtained from (4.63) and (4.65)

Fig. 4.7 Rays and beams, respectively of phase and energy propagation can be limited in range $\omega < N_*$

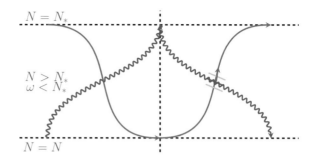

$$\mathbf{C}_p = \frac{N_0}{\kappa}(\mathbf{i}\cos^2\theta + \mathbf{k}\cos\theta\sin\theta) = \frac{N_0}{2\kappa}[\mathbf{i}(1 + \cos 2\theta) + \mathbf{k}\sin 2\theta]$$

$$\mathbf{C}_g = \frac{N_0}{\kappa}(\mathbf{i}\sin^2\theta - \mathbf{k}\sin\theta\cos\theta) = \frac{N_0}{2\kappa}[\mathbf{i}(1 - \cos 2\theta) - \mathbf{k}\sin 2\theta].$$

$$(4.72.a, b)$$

When beams are oriented vertically either at $\theta=0$ or $\theta = \pi$, the frequency reaches its maximum possible value $\omega = N_0$, group velocity vanishes and propagation is aligned along x with phase velocity $C_p = N_0/\kappa$, according to (4.72.a, b) and Fig. 4.6. The above situation is relevant to the reflection of an internal wave at level $\omega = N_0$ of a fluid with variable stratification $N(z)$, where group velocity vanishes while the ray is vertical.

On the other hand, the beams being horizontally oriented with $\theta = \pi/2$ ($\beta = 0$) corresponds to steady state conditions with zero frequency $\omega/N = 0$, with increased horizontal component of group velocity $C_g = N/\kappa$, but without any phase propagation $C_p = 0$, by virtue of (4.72.a, b) and as shown in Fig. 4.6. The situation may have relevance if either $\omega \to 0$ (steady flow) or near a density interface $N \to \infty$, where the ray is horizontal.

As rays and beams are refracted subject to density stratification, their orientations change as shown in Fig. 4.6. When the frequency of oscillation is limited in the range $\omega < N_* < N$, internal waves can become trapped within a layer of maximum stratification as in Fig. 4.7, for instance in the case of a thermocline.

4.2.3 Refraction of Internal Gravity Waves

Consider two layers of fluid with different (but constant values of N_1 and N_2 respectively. While a wave beam passes from one layer to the other frequency

$$\omega = N_1 \sin \beta_1 = N_2 \sin \beta_2 \tag{4.73}$$

should be conserved, yielding Snell's law of refraction (originally applicable to acoustics or light rays)

Fig. 4.8 Refraction of a
beam in a stratified medium
with step-wise change from
N_1^2 to N_2^2, reproducing an
analogue to Snell's law with
$N_2^2 > N_1^2$, $\beta_1 > \beta_2$

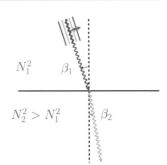

$$\frac{\sin \beta_1}{\sin \beta_2} = \frac{N_2}{N_1} \tag{4.74}$$

If $N_2^2 > N_1^2$, $\beta_2 < \beta_1$ and the ray would look like as in Fig. 4.8.

In a non-uniformly stratified fluid with variable $N^2 = N^2(z)$, the rays should be become closer to vertical as N decreases, and would finally be reflected at the level $N = N_0 = \omega$.

4.2.4 Reflection of Internal Gravity Waves on a Slope

In the earlier sections it has been shown that for a fixed frequency ω and constant stratification $N = N$, internal waves can propagate energy with fixed angle β (or phase propagation with angle θ), prescribed by Eq. (4.70). When internal waves are reflected from a side boundary, the angle β is to be conserved.

Consider a boundary slanted at an angle α with the horizon, angle of the reflected ray has no relation to orientation α of the boundary, only requiring that the angle β of reflected wave is conserved.

If the side boundary has steeper slope than the incidence angle of the internal wave, $\alpha > \beta$, then the reflection occurs directly in reverse direction of the wave approach, as shown in Fig. 4.9.

The distance between individual rays is proportional to the wave length $2\pi/\kappa$ and the phase speed $\mathbf{C}_p$. We thus observe that the wavelength and phase speed of a wave changes upon reflection.

Considering a wedge with milder bottom slope in comparison to the incident wave, $\alpha < \beta$, the wave reverses its direction in the vertical, conserving the same angle. In this case wave energy can only travel towards the tip of the wedge, as shown in Fig. 4.10. Since no energy is reflected back in the incoming direction, the mild slope configuration effectively absorbs energy of internal gravity waves. Since energy flux will be increased towards the tip, waves should eventually break, which means that nonlinear and diffusive processes come into play.

Fig. 4.9 Reflection of internal waves from a coast with topographic slope $\alpha > \beta$

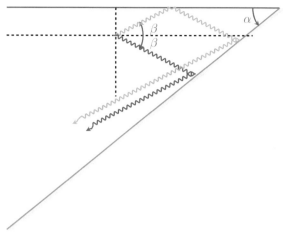

Fig. 4.10 Trapping of internal waves at a coast with topographic slope $\alpha < \beta$

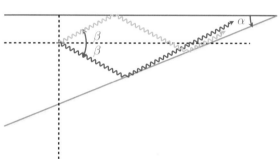

4.3   Internal Waves Under Joint Effects of Earth's Rotation and Stratification

In earlier sections, internal gravity waves were studied at local scales, by setting the Coriolis parameter to zero, $f = 0$. Investigating motions at slightly larger scales, internal waves start become influenced by the effects of earth's rotation, based on Eqs. (4.34) and (4.35). Assuming periodic solutions

$$w = \hat{w}(x, y, z)\, e^{-i\omega t}, \tag{4.75}$$

Equation (4.35) is reduced to

$$\left(f^2 - \omega^2\right) \frac{\partial^2 \hat{w}}{\partial z^2} + \left(N^2(z) - \omega^2\right) \nabla_h^2 \hat{w} = 0. \tag{4.76}$$

Employing separation of variables

$$\hat{w} = W(z) F(x, y) \tag{4.77.a}$$

leads to

$$\frac{\nabla_h^2 F}{F} = \frac{\omega^2 - f^2}{N^2(z) - \omega^2} \frac{1}{W} \frac{\partial^2 W}{\partial z^2} = -\kappa_h^2 \qquad (4.77.b)$$

where κ_h is the separation constant, yielding

$$\nabla_h^2 F + \kappa_h^2 F = 0 \qquad (4.78.a)$$

and

$$\frac{d^2 W}{dz^2} + \left(\frac{N^2 - \omega^2}{\omega^2 - f^2}\right) \kappa_h^2 W = 0. \qquad (4.78.b)$$

These equations determine horizontal and vertical components of the motion. We assume the fluid is bounded by rigid horizontal surfaces at the bottom and top, with boundary conditions

$$W(0) = 0 \qquad (4.79.a)$$

$$W(-h) = 0. \qquad (4.79.b)$$

For given values of $N(z)$ and f, for a fixed frequency ω, Eqs. (4.78.a, b), (4.79.a, b) constitute an eigenvalue problem which depends on the constant κ_h and its multiplier in terms of frequencies.

It is noted that for $\kappa_h^2 > 0$, Eq. (4.78.a) allows horizontally propagating wave solutions where κ_h is the horizontal wave-number. Multiplying (4.78.b) by W and integrating vertically we have

$$\int_{-h}^{0} \left[W \frac{d^2 W}{dz^2} dz + \kappa_h^2 \left(\frac{N^2 - \omega^2}{\omega^2 - f^2}\right) W^2 \right] dz = 0. \qquad (4.80)$$

Upon integrating the first term by parts and utilizing (4.79.a, b), it can be shown that

$$\kappa_h^2 \int_{-h}^{0} q(z) W^2 dz = \int_{-h}^{0} \left(\frac{dW}{dz}\right)^2 dz > 0 \qquad (4.81)$$

where

$$q(z) = \frac{N^2(z) - \omega^2}{\omega^2 - f^2}. \qquad (4.82)$$

As already noted, horizontally propagating wave solutions are possible only for positive numbers ($\kappa_h^2 > 0$), as the right hand side of (4.81) and W^2 are also positive valued,

$$\int_{-h}^{0} q\,(z)\,W^2 dz > 0 \tag{4.83}$$

which then allows evaluation of possible solutions. Types of solution therefore depend on behavior of the function $q(z)$ contributing to the integral in (4.83):

(i) *internal waves in whole domain*: Provided that $\kappa_h^2 > 0$ and the function $q(z) > 0$ everywhere in the domain $-h \leq z \leq 0$, then (4.83) is satisfied and internal waves can exist in all the domain, with horizontal and vertical oscillatory solutions of (4.78.a, b).

(ii) *internal waves in part of the domain*: If $q(z)$ has positive values $q(z) > 0$ in some part of the domain, and negative values $q(z) < 0$ elsewhere, propagating internal waves can still exist provided that $\int_{-h}^{z} q(z)W^2 dz > 0$ is satisfied. Solutions are oscillatory in the vertical direction only in the region where $q(z) > 0$, and evanescent in the regions where $q(z) < 0$ according to (4.78.b).

(iii) *no internal waves*: If $q(z) < 0$ in the interval $-h \leq z \leq 0$, (4.81) will not be satisfied and therefore internal wave solutions are not possible.

4.3.1   Internal Inertia-Gravity Waves

The first one of the above cases is investigated

$$q(z) = \frac{N^2 - \omega^2}{\omega^2 - f^2} \geq 0 \quad \text{and} \quad \kappa_h^2 > 0 \tag{4.84}$$

ensuring *existence* of proper internal waves in the whole domain. Two cases are evident.

4.3.1.1 Case(1)—Super-inertial Motions ($\omega > f$)
If $\omega > f$, one should also have $\omega < N$ to satisfy (4.84),

$$f < \omega < N. \tag{4.85}$$

According to (4.85) this case would require $N > f$. This is the common case in most part of the ocean. Earth's rotation represented by f has a maximum at the North Pole

$$f = 2\Omega \sin \phi \leq 2\Omega \tag{4.86}$$

and therefore typically

$$f^2 \leq 2\Omega^2 = (2 \times 2\pi \; rad/day \;)^2 = 2.1 \times 10^{-8} \; s^{-2}, .$$

Fig. 4.11 Super-inertial internal waves with fixed frequency $f_0 < \omega < N_0$ at latitudes away from the equator

On the other hand, for a weakly stratified ocean with density change on the order of 10^{-3} in a depth of $100\,m$ in the upper ocean, the stratification parameter can be estimated as

$$N^2 = -\frac{g}{\rho_r}\frac{d\rho_r}{dz} \simeq \frac{10ms^{-2} \times 0.001\,g\,cm^{-3}}{1\,g\,cm^{-3} \times 100\,m} \simeq 10^{-4}s^{-2}.$$

These above order of magnitude estimates already confirm validity of cutoff frequencies in the range $f < \omega < N$ as specified by (4.85). For these oscillatory solutions, f is the lower cutoff and N is the upper cut-off. Waves with fixed frequency ω satisfying (4.85) are therefore called *internal inertia-gravity waves*, which are subject to joint effects of earth's rotation and density stratification.

Waves within the lower part of the range $f < \omega < N$, closer to the inertial frequency can reach the cutoff $\omega = f_0$ at some latitude ϕ_0 where $f_0 = 2\Omega\sin\phi_0$. These inertia-gravity internal waves would be reflected at latitude ϕ_0 and wave motions would not be allowed at southerly latitudes. The propagation of inertial waves with fixed frequency $\omega > f_0$ closer to the pole is sketched in Fig. 4.11.

4.3.1.2 Case(2)—Sub-inertial Motions ($\omega < f$)

Sub-inertial wave motions with frequency $\omega < f$ could exist, provided that condition (4.84) is satisfied. The only requirement for wave motion at these sub-inertial frequencies would have to be $\omega > N$, resulting in the frequency range

$$N < \omega < f. \tag{4.87}$$

This essentially requires $N < f$, in reversed order of cutoff frequencies in most atmosphere and ocean domains. Yet, rather exceptional situations could occur in the deep ocean, where static stability tends to vanish $N \to 0$. Sub-inertial waves exceptionally could appear within given frequency and mid-latitude intervals shown in Fig. 4.12.

Fig. 4.12 Sub-inertial internal waves with fixed frequency $N < \omega < f_0$ possibly at deep waters of mid-latitudes away from the equator and the pole

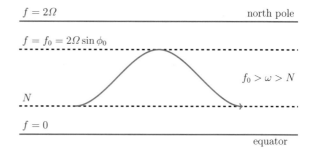

$f = 2\Omega$ north pole

$f = f_0 = 2\Omega \sin \phi_0$

$f_0 > \omega > N$

N

$f = 0$

equator

4.3.2 Horizontal Modes of Motion

Horizontal modes of motion determined by Eq. (4.81.a) can be investigated using

$$\frac{\partial^2 F}{\partial x^2} + \frac{\partial^2 F}{\partial y^2} + \kappa_h^2 F = 0. \tag{4.88}$$

where κ_h represents the horizontal wave number. Applying separation of variables once again

$$F(x, y) = G(x) H(y) \tag{4.89}$$

gives

$$-\frac{1}{G}\frac{d^2 G}{dx^2} = \frac{1}{H}\frac{d^2 H}{dy^2} + \kappa_h^2 = k^2 \tag{4.90}$$

where k is the new separation constant, and noting that $l^2 = \kappa_h^2 - k^2$,

$$\frac{d^2 G}{dx^2} + k^2 G = 0 \tag{4.91.a}$$

$$\frac{d^2 H}{dy^2} + l^2 H = 0. \tag{4.91.b}$$

The horizontal wave number is composed of

$$\kappa_h^2 = k^2 + l^2. \tag{4.92}$$

with components k and l respectively in the x and y direction.

Horizontal wave propagation is allowed for $\kappa_h^2 > 0$, as specified in (4.88). Wave numbers k, l could individually contribute to the sum of squares, either being real or imaginary numbers. Choosing k to be real, while l either real or imaginary, allows to classify these waves.

Since Boussinesq horizontal momentum equations are the same as for homogeneous fluids studied earlier, internal horizontal wave modes of current interest are similar to those investigated in Chap. 6, Sect. 6.6.5 of GFD-I.

4.3.3   Case 1: Poincaré Waves, $\kappa_h^2 > k^2$

In this case, both k and l are real, so that (4.92) requires $l^2 > 0$; accordingly (4.91.a, b) can be used to construct solutions classified as *Internal Poincaré Waves*

$$F(x, y) = Ae^{\pm ikx}e^{\pm ily}. \tag{4.93}$$

Wave propagation is in the direction of the horizontal wave number vector according to (4.92)

$$\kappa_h = (k, l). \tag{4.94}$$

These are typical sinusoidal waves propagating along the horizontal wave number vector κ_h. For waves reflected near a straight coast, cellular patterns of standing Poincaré waves in the homogeneous case have been demonstrated In Sect. 6.6.5, GFD-I. Similar results can be obtained for internal waves under appropriate boundary conditions.

4.3.4   Case 2: Kelvin Waves, $\kappa_h^2 < k^2$

Alternative solutions to (4.88) known as *Internal Kelvin Waves* are developed by letting $l^2 = \kappa_h^2 - k^2 = -\gamma^2 < 0$ in (4.92)

$$F(x, y) = Ae^{\pm ikx}e^{\pm \gamma y}. \tag{4.95}$$

For example, in a semi-infinite fluid bounded by a coast aligned with the x-axis, such solutions for the homogeneous case have been demonstrated in Sect. 6.6.5 GFD-I.

Horizontal motion represented by Boussinesq equations is the same as homogeneous fluids in GFD-I. In particular, if a solution form (4.92) is applied in Eq. (4.28), one obtains

$$-i\omega\mathbf{k} \cdot \nabla \times \mathbf{u} + f\nabla \cdot \mathbf{u} = 0.$$

Solutions can be generated for a straight coast aligned along the x-direction neighboring a semi-infinite ocean domain $y > 0$, similar to that in GFD-I, Sect. 6.6.5. By letting $\partial/\partial x = ik$ and $\partial/\partial y = -\gamma$ one obtains

$$-i(\gamma\omega - kf)u + (ik\omega - \gamma f)\not{v} = 0 \quad \text{at} \quad y = 0$$

and application of no-flux boundary condition $v = 0$ at the coast implies

$$\gamma = kf/\omega$$

which then gives the typical Kelvin wave solution $\sim e^{-\gamma y} e^{i(kx-\omega t)}$ riding along the coast in x direction, subject to exponential decay in y with maximum amplitude at the coast. Details of dispersion relations for Poincaré and Kelvin waves will be given in the next section for the specific case of constant stratification.

4.4   Inertia-Gravity Waves: Constant Stratification N^2

Constant stratification is considered in this section with

$$N^2 = -\frac{g}{\rho_0}\frac{d\rho_r}{dz} = N_0^2$$

in Eqs. (4.78.a, b), letting

$$m^2 = \frac{N^2 - \omega^2}{\omega^2 - f^2}\kappa_h^2 \tag{4.96}$$

be the vertical wave-number. Equation (4.96) provides the dispersion relation establishing the relationship between frequency ω versus horizontal and vertical wave numbers κ_h and m

$$\omega^2 = \frac{N^2\kappa_h^2 + f^2 m^2}{\kappa_h^2 + m^2} = \frac{N^2\left(\kappa^2 + l^2\right) + f^2 m^2}{k^2 + l^2 + m^2} \tag{4.97}$$

where κ_h is the horizontal wave-number described by (4.92) and (4.94). Defining a wave number vector

$$\kappa = (k, l, m) = (\kappa_h, m) \tag{4.98}$$

with magnitude

$$|\kappa|^2 \equiv \kappa^2 = k^2 + l^2 + m^2 = \kappa_h^2 + m^2, \tag{4.99}$$

Dispersion relation is alternatively written as

$$\omega^2 = N^2 \left(\frac{\kappa_h}{\kappa}\right)^2 + f^2 \left(\frac{m}{\kappa}\right)^2, \tag{4.100}$$

where it is noted that

$$\frac{\kappa_h}{\kappa} = \cos\theta \le 1 \quad \text{and} \quad \frac{m}{\kappa} = \sin\theta \le 1 \tag{4.101.a, b}$$

by virtue of (4.92), (4.94), (4.97) and (4.99). θ is the angle that wave number vector makes with the horizontal plane. Then (4.98) can be written as

$$
\begin{aligned}
\omega^2 &= N^2 \cos^2 \theta + f^2 \sin^2 \theta \\
&= \left(N^2 - f^2\right) \cos^2 \theta + f^2 \\
&= N^2 + \left(f^2 - N^2\right) \sin^2 \theta
\end{aligned}
\tag{4.102}
$$

and therefore

$$
\begin{aligned}
\cos^2 \theta &= \frac{\omega^2 - f^2}{N^2 - f^2} = \left(\frac{\kappa_h}{\kappa}\right)^2, \\
\sin^2 \theta &= \frac{N^2 - \omega^2}{N^2 - f^2} = \left(\frac{m}{\kappa}\right)^2, \\
\tan^2 \theta &= \frac{N^2 - \omega^2}{\omega^2 - f^2} = \left(\frac{m}{\kappa_h}\right)^2.
\end{aligned}
\tag{4.103.$a-c$}
$$

Equations (4.103.a-c) define phase propagation direction for given frequency ω, f and N. This equation reduces to (4.55) for the non-rotating case.

It is seen from (4.100) that as

$$
\kappa_h \to 0, \quad \kappa \to m, \quad \theta \to \frac{\pi}{2} \quad \text{then} \quad \omega \to f,
\tag{4.104}
$$

so that *inertial waves* with inertial frequency $\omega = f$ propagate only vertically. On the other hand, as

$$
m \to 0, \quad \kappa \to \kappa_h, \quad \theta \to 0 \quad \text{then} \quad \omega \to N,
\tag{4.105}
$$

waves with buoyancy frequency $\omega = N$ can only propagate horizontally.

Frequency ranges of possible wave motions, by virtue of (4.104) and (4.105) are given as

$$
f < \omega < N \quad \text{if} \quad f < N
\tag{4.106}
$$

and

$$
N < \omega < f \quad \text{if} \quad N < f.
\tag{4.107}
$$

The first regime (4.106) rather than the second one (4.107) is the most likely one of the two regimes in most part of the atmosphere and ocean, as often stratification overcomes inertial effects.

Frequencies N and f represent upper and lower cutoff values respectively in the case of *internal inertia-gravity* waves. Most other results obtained in Sect. 4.2 apply to the present case with rotation. For instance, it can be shown that the phase velocity $\mathbf{C}_p$ and wave number κ are in the same direction, whereas $\mathbf{C}_p$ and $\mathbf{C}_g$ are perpendicular.

The particle motions are transverse to the direction of phase propagation, as discussed earlier in Sect. 4.2.2.

Results are summarized for the common inertia-gravity wave regime (4.106) in the following. Although internal waves can truly be 3-D in nature as discussed in Sect. 4.2.1, the present theory is extended to plane waves truly of 2-D, for demonstration of basic results. In particular, plane wave characteristics are considered to take place in the same plane.

The phase velocity vector $\mathbf{C}_p$ will be separated into horizontal and vertical components

$$\mathbf{C}_p = (\frac{\omega}{\kappa}\frac{\kappa_h}{\kappa}, \frac{\omega}{\kappa}\frac{m}{\kappa}) = C_{p_h}\frac{\kappa_h}{\kappa_h} + C_{p_v}\mathbf{k} \tag{4.108}$$

Horizontal and vertical components of the phase velocity can be calculated from (4.97), making use of (4.101) and (4.103.a-c):

$$C_{p_h} = \frac{\omega}{\kappa}\frac{\kappa_h}{\kappa} = \frac{\omega}{\kappa}\left(\frac{\omega^2 - f^2}{N^2 - f^2}\right)^{1/2} = \frac{\omega}{\kappa}\cos\theta \tag{4.109.a}$$

$$C_{p_v} = \frac{\omega}{\kappa}\frac{m}{\kappa} = \frac{\omega}{\kappa}\left(\frac{N^2 - \omega^2}{N^2 - f^2}\right)^{1/2} = \frac{\omega}{\kappa}\sin\theta. \tag{4.109.b}$$

Corresponding components of group velocity vector, direction not ascertained as of yet are

$$\mathbf{C}_g = \mathbf{C}_{g_h} + \mathbf{C}_{g_v} = (\frac{\partial\omega}{\partial\kappa_h}, \frac{\partial\omega}{\partial m}). \tag{4.110}$$

Differentiating (4.103.a) with respect to horizontal wave number κ_h, (4.103.b) with respect to the vertical wave number m, then making use of

$$\frac{\partial}{\partial\kappa_h}\frac{\omega^2 - f^2}{N^2 - f^2} = \frac{2\omega}{N^2 - f^2}\frac{\partial\omega}{\partial\kappa_h} = \frac{\partial}{\partial\kappa_h}\left(\frac{\kappa_h^2}{\kappa_h^2 + m^2}\right) = \frac{2\kappa_h m^2}{\kappa^4}$$

$$\frac{\partial}{\partial m}\frac{N^2 - \omega^2}{N^2 - f^2} = \frac{-2\omega}{N^2 - f^2}\frac{\partial\omega}{\partial m} = \frac{\partial}{\partial m}\left(\frac{m^2}{\kappa^2 + m^2}\right) = \frac{2\kappa_h^2 m}{\kappa^4} \tag{4.111, a, b}$$

once again inserting from equations (4.103.a, b), horizontal and vertical components of the group velocity are calculated as

$$C_{g_h} = \frac{\partial\omega}{\partial\kappa_h} = \frac{(\omega^2 - f^2)\kappa_h m^2}{\kappa^4\omega} = \frac{N^2 - \omega^2}{\omega\kappa}\left(\frac{\omega^2 - f^2}{N^2 - f^2}\right)^{1/2} \tag{4.112.a}$$

$$C_{g_v} = \frac{\partial\omega}{\partial m} = -\frac{(N^2 - f^2)\kappa_h^2 m}{\kappa^4\omega} = -\frac{\omega^2 - f^2}{\omega\kappa}\left(\frac{N^2 - \omega^2}{N^2 - f^2}\right)^{1/2}. \tag{4.112.b}$$

Magnitudes of phase and group velocities can be calculated from the above as

$$C_p = \sqrt{C_{p_h}^2 + C_{p_v}^2} = \frac{\omega}{\kappa} \tag{4.113.a}$$

$$C_g = \sqrt{C_{g_h}^2 + C_{g_v}^2} = \frac{\omega}{\kappa} \left(\frac{N^2}{\omega^2} - 1 \right)^{1/2} \left(1 - \frac{f^2}{\omega^2} \right)^{1/2} \tag{4.113.b}$$

which are then used to obtain

$$C_p^2 + C_g^2 = \frac{\omega^2}{\kappa^2} \left[\frac{N^2}{\omega^2} + \frac{f^2}{\omega^2} - \frac{f^2}{\omega^2} \frac{N^2}{\omega^2} \right]. \tag{4.114}$$

Important changes occur at both cutoff frequencies (4.104) and (4.105). According to (4.109.a, b) and (4.112.a, b), expressing horizontal and vertical components of phase and group velocities respectively vanish at frequencies $\omega = f$ ($\kappa_h = 0$) and $\omega = N$ ($m = 0$). The vanishing of phase and group velocities at both cutoff frequencies indicates possible resonant conditions under forcing, when the frequency of the oscillation matches either of these natural frequencies.

The relations between various variables can be obtained for by substitution into (4.7) of the wave solution

$$e^{i(kx+ly+mz-\omega t)}$$

for each of the variables, to yield

$$ku + lu + mw = 0 \tag{4.115.a}$$

$$-i\omega u - fv = -ik\frac{p}{\rho_0} \tag{4.115.b}$$

$$-i\omega v + fu = -il\frac{p}{\rho_0} \tag{4.115.c}$$

$$-i\omega w = -im\frac{p}{\rho_0} + \sigma \tag{4.115.d}$$

$$-i\omega\sigma + N^2 w = 0. \tag{4.115.e}$$

From (4.115.e) we have

$$\sigma = -i\frac{N^2}{\omega} w, \tag{4.116.a}$$

and substituting this one into (4.115.d) gives

Fig. 4.13 3-D rotation of the velocity vector u on a slanted surface

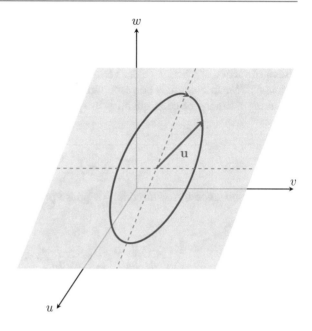

$$w = -\frac{m\omega}{N^2 - \omega^2} \frac{p}{\rho_0}. \tag{4.116.b}$$

From (4.115.b, c) and (4.116.b) one obtains

$$u = \frac{k\omega + ilf}{\omega^2 - f^2} \frac{p}{\rho_0} = -\left(\frac{k\omega + ilf}{m\omega}\right)\left(\frac{N^2 - \omega^2}{\omega^2 - f^2}\right) w$$

$$v = \frac{l\omega - ikf}{\omega^2 - f^2} \frac{p}{\rho_0} = -\left(\frac{l\omega - ikf}{m\omega}\right)\left(\frac{N^2 - \omega^2}{\omega^2 - f^2}\right) w. \tag{4.116.c, d}$$

Considering horizontal propagation only in x-direction ($l = 0$) without loss of generality, the 2-D version of (4.115.a) becomes

$$ku + mw = 0 \tag{4.117}$$

which then gives

$$u = -\frac{m}{k}w = -(\tan\theta)\,w. \tag{4.118}$$

Accordingly (4.115.c) and (4.118) yields

$$v = -\frac{if}{\omega}u = \frac{if}{\omega}(\tan\theta)\,w \tag{4.119}$$

Equations (4.118) and (4.119) show that the current vector rotates as an ellipse on an inclined plane, as shown in Fig. 4.13. The sense of rotation is clockwise (anti-cyclonic) due to Coriolis forces.

4.4.1   Energetics of Internal Waves in a Uniformly Stratified Fluid

The mechanical energy Eq. (4.23) developed in Sect. 4.1.3 can be utilized to recon-
struct elements of energy conservation for waves in a uniformly stratified fluid
$N^2 = N_0^2$. Mean quantities corresponding to the perturbation field can be obtained
by taking averages of each quantity in space over a wavelength, denoted by over
bars:

$$\overline{()} = \frac{\kappa}{2\pi} \int_0^{2\pi/\kappa} () \, ds \tag{4.120}$$

where s is distance in the direction of phase propagation. Equation (4.23) can be
written as

$$\frac{d}{dt} \int_V \overline{T} \, dv = - \int_S \overline{\mathbf{F}_e} \cdot \mathbf{n} \, ds \tag{4.121}$$

transformed by making use of the divergence theorem (GFD-I, 1.34) and Leibnitz
rule (GFD-I, 1.44.c), into

$$\frac{\partial \overline{T}}{\partial t} + \nabla \cdot \overline{\mathbf{F}_e} = 0 \tag{4.122}$$

where V is a fixed volume enclosed by surface S, with definitions of average energy
and fluxes obtained from (4.23)

$$\overline{T} = \overline{KE} + \overline{PE} \tag{4.123.a}$$

$$\overline{KE} = \frac{1}{2}\rho_0 \overline{\mathbf{u} \cdot \mathbf{u}} = \frac{1}{2}\rho_0 (\overline{\mathbf{u}_h \cdot \mathbf{u}_h} + \overline{w^2}) \tag{4.123.b}$$

$$\overline{PE} = \frac{1}{2}\rho_0 \frac{\overline{\sigma^2}}{N^2} \tag{4.123.c}$$

$$\overline{\mathbf{F}_e} = \overline{p\mathbf{u}} = \overline{p\mathbf{u}_h} + \overline{pw} \tag{4.123.d}$$

where the 3-D velocity is denoted as $\mathbf{u} = \mathbf{u}_3$. Considering, without loss of generality,
waves described in the $x - y$ plane with horizontal propagation along x direction,

$$w = Re \left\{ w_0 e^{i(\kappa s - \omega t)} \right\} \tag{4.124}$$

Equations (4.118) and (4.119) give

$$u = Re \left\{ -w_0 \tan \theta \, e^{i(\kappa s - \omega t)} \right\} \tag{4.125}$$

and

$$v = Re\left\{\frac{if}{\omega}w_0\tan\theta\,e^{i(\kappa s-\omega t)}\right\}. \tag{4.126}$$

Alternatively, the velocity vector can be written as

$$\mathbf{u} = Re\left\{\hat{\mathbf{u}}e^{i(\kappa s-\omega t)}\right\} \tag{4.127}$$

with amplitude

$$\hat{\mathbf{u}} = \left\{\begin{matrix}\hat{u}\\\hat{v}\\\hat{w}\end{matrix}\right\} = \left\{\begin{matrix}-\tan\theta\\i\frac{f}{\omega}\tan\theta\\1\end{matrix}\right\}w_0. \tag{4.128}$$

Pressure field is obtained from (4.116.b)

$$p = Re\left\{-\frac{\rho_0}{m\omega}\left(N^2-\omega^2\right)w_0e^{i(\kappa s-\omega t)}\right\} \tag{4.129}$$

and buoyancy is given by (4.116.a):

$$\sigma = Re\left\{-i\frac{N^2}{\omega}w_0e^{i(\kappa s-\omega t)}\right\} \tag{4.130}$$

Next, we will make use of averaging rules for two complex periodic functions specified as

$$\begin{aligned}A &= Re\left\{A_0e^{i\kappa s}\right\} = Re\,(\alpha)\\B &= Re\left\{B_0e^{i\kappa s}\right\} = Re\,(\beta)\,.\end{aligned} \tag{4.131}$$

The average product of two periodic functions are obtained as follows, denoting the complex conjugate by an asterix (*) sign

$$\begin{aligned}\overline{AB} = \overline{Re(\alpha)Re(\beta)} &= \frac{1}{4}\overline{(\alpha+\alpha^*)(\beta+\beta^*)} = \frac{1}{4}\overline{(\alpha\beta+\alpha^*\beta^*+\alpha^*\beta+\alpha\beta^*)}\\&= \frac{1}{2}[\overline{Re(\alpha\beta)}+\overline{Re(\alpha^*\beta)}] = \frac{1}{2}[Re(\overline{\alpha\beta})+Re(\overline{\alpha^*\beta})] = \frac{1}{2}Re\overline{(\alpha^*\beta)}\end{aligned} \tag{4.132}$$

It remains to be verified from (4.120) that the first term averaged over half the wavelength vanishes and therefore the second term gives

$$\overline{AB} = \frac{1}{2}Re\overline{(\alpha^*\beta)} = \frac{1}{2}Re(A_0^*B_0) = \frac{1}{2}Re(A_0B_0^*) \tag{4.133}$$

By making use of this result, the average kinetic energy (4.123.b) is calculated as

$$\overline{KE} = \frac{\rho_0}{2}\overline{\mathbf{u}\cdot\mathbf{u}} = \frac{\rho_0}{2}\frac{1}{2}Re(\mathbf{u}_h^* \cdot \mathbf{u}_h + w^2)$$

$$= \frac{\rho_0}{4}\left[1 + \left(1 + \frac{f^2}{\omega^2}\right)\tan^2\theta\right]w_0^2 \tag{4.134}$$

$$= \frac{\rho_0 w_0^2}{4}\left(\frac{1 + \frac{f^2}{\omega^2}\sin^2\theta}{\cos^2\theta}\right)$$

The average potential energy is then calculated as

$$\overline{PE} = \frac{\rho_0}{2N^2}\overline{\sigma^2} = \frac{\rho_0}{2N^2}\frac{1}{2}\frac{N^4 w_0^2}{\omega^2} = \frac{\rho_0 w_0^2}{4}\frac{N^2}{\omega^2}$$

$$= \frac{\rho_0 w_0^2}{4}\left(\frac{1 - \frac{f^2}{\omega^2}\sin^2\theta}{\cos^2\theta}\right), \tag{4.135}$$

where use has been made of (4.102) or (4.103). The average total energy is obtained as

$$\overline{T} = \overline{KE} + \overline{PE} = \frac{\rho_0 w_0^2}{2\cos^2\theta} = \frac{\rho_0 w_0^2}{2}\left(1 + \frac{m^2}{\kappa_h^2}\right). \tag{4.136}$$

Since f does not enter (4.136), the total energy given above is independent of rotation. However, it is important to note that energy is not equally partitioned in the presence of rotation. The ratio between mean kinetic and potential energy is calculated by making use of (4.102) and (4.103a-c) and some trigonometric identities in (4.134) and (4.135),

$$\frac{\overline{KE}}{\overline{PE}} = \frac{\left[1 + \left(1 + \frac{f^2}{\omega^2}\right)\tan^2\theta\right]}{\frac{N^2}{\omega^2}} = \frac{1 + \tan^2\theta + \frac{f^2}{\omega^2}\tan^2\theta}{\frac{N^2}{\omega^2}}$$

$$= \frac{\omega^2}{N^2\cos^2\theta} + \frac{f^2}{N^2}\tan^2\theta = 1 + \frac{2f^2}{N^2}\tan^2\theta = 1 + 2\left[\frac{1 - \left(\frac{\omega}{N}\right)^2}{\left(\frac{\omega}{f}\right)^2 - 1}\right].$$

$$\tag{4.137}$$

The mean energy flux can be calculated as

$$\overline{\mathbf{F}_e} = \overline{p\mathbf{u}} = \frac{1}{2} Re(\overline{p^*\mathbf{u}})$$

$$= \frac{\rho_0 w_0^2 (N^2 - \omega^2)}{2m\omega} \left\{ \begin{array}{c} \tan\theta \\ 0 \\ -1 \end{array} \right\}$$

$$= \left\{ \begin{array}{c} \sin\theta \\ 0 \\ -\cos\theta \end{array} \right\} \frac{\rho_0 w_0^2 (N^2 - \omega^2)}{2\omega\kappa \sin\theta \cos\theta} \tag{4.138}$$

replacing $m/\kappa = \sin^2\theta$ from (4.103.b). Now dividing by $\overline{T}$ gives (using Eqs. (4.136) and 4.103):

$$\frac{\overline{\mathbf{F}_e}}{\overline{T}} = \left\{ \begin{array}{c} \sin\theta \\ 0 \\ -\cos\theta \end{array} \right\} \frac{(N^2 - \omega^2)}{\kappa\omega} \frac{\cos\theta}{\sin\theta}$$

$$= \left\{ \begin{array}{c} \sin\theta \\ 0 \\ -\cos\theta \end{array} \right\} \frac{\omega}{\kappa} \left(\frac{N^2}{\omega^2} - 1 \right)^{1/2} \left(1 - \frac{f^2}{\omega^2} \right)^{1/2} \tag{4.139}$$

$$= \mathbf{C}_g = C_g \mathbf{e}$$

where the magnitude of $\mathbf{C}_g$ is given in the same way as (4.113.b), where the unit vector $\mathbf{e}$ is directed along group velocity, perpendicular to wave-number and the phase velocity vector $\mathbf{C}_p$ at angle θ.

The resulting balance (4.139) proves that the mean flux of energy is equivalent to group velocity multiplied by total mean energy

$$\overline{\mathbf{F}_e} = \mathbf{C}_g \overline{T}. \tag{4.140}$$

The above considerations expressed in (4.137) indicate that mean energy is not equipartioned in a rotating stratified fluid. On the other hand, since $N \gg f$ in typical atmosphere and ocean environments, at intermediate range of frequencies $f \ll \omega \ll N$ in the range $0 \ll \theta \ll \pi/2$, energy is closer to being equally partitioned at the higher frequencies. As $\omega \to N$, $\theta \to 0$ we find $\overline{KE} \simeq \overline{PE}$. On the other hand, as $\omega \to f$, $\theta \to \frac{\pi}{2}$ there will be a shift in energy transfers favoring kinetic energy being trapped in inertial motions $\overline{KE} \gg \overline{PE}$.

In general, the energy flux and group velocity is perpendicular to phase propagation. The vertical components of phase and group velocities (or phase propagation and energy flux) are always in opposite senses.

4.4.2   Momentum Flux in a Stratified Fluid

Consider a material surface S enclosing a material volume V in a stratified fluid. The surface force in the absence of friction is

$$\mathbf{F}_s = \int_S p\mathbf{n}dS \tag{4.141}$$

and is balanced by (cf. GFD-I Eqs. 1.16, 1.18 and 1.19):

$$\mathbf{F}_s = \int_S p\mathbf{n}ds = \int_V \frac{\partial}{\partial t}(\rho\mathbf{u})dV + \int_S \rho\mathbf{u}(\mathbf{u}\cdot\mathbf{n})dS \tag{4.142}$$

The first term on the right hand side is the body force $\mathbf{F}_b$ due to the rate of change of momentum of the fluid enclosed in V, and the second term $\mathbf{F}_f$ is a surface force due to the momentum flux through the surface S:

$$\mathbf{F}_s = \mathbf{F}_b + \mathbf{F}_f \tag{4.143}$$

Only the second term representing flux of momentum in a stratified fluid is evaluated in Eq. (4.142), where the pressure and density are the sum of basic state and perturbation fields $p = p_r + p', \rho = \rho_r + \rho'$ while velocity $\mathbf{u}$ is due to perturbations. Neglecting small perturbations in p and ρ, the momentum flux term becomes

$$\mathbf{F}_f = \int_S \rho_r\mathbf{u}(\mathbf{u}\cdot\mathbf{n})dS. \tag{4.144}$$

Here S is a closed material surface. Similarly, let us investigate the contribution of momentum flux through a portion of a *fixed* surface S_0, i.e. integral taken over S_0. Furthermore let us make the approximation $\rho_r = \rho_0$=constant and consider the uniformly stratified case of the last section.

It is clear from (4.144) that $\mathbf{F}_f$ is in the direction of the particle velocity $\mathbf{u}$ which is perpendicular to $\boldsymbol{\kappa}$. Let S_0 be a rectangle with unit width and a length equal to the wavelength, such that $\boldsymbol{\kappa}$ is in the plane of S_0. Consider the average force (or momentum flux) across the surface S_0

$$\overline{\mathbf{f}_f} = \frac{1}{S_0}\overline{F_f} = \frac{\kappa}{2\pi}\int_0^{\frac{2\pi}{\kappa}} \rho_0\overline{\mathbf{u}(\mathbf{u}\cdot\mathbf{n})}ds, \tag{4.145}$$

which is the average momentum flux per unit wavelength.

The velocity vector is given in (4.127) and (4.128). The normal vector $\mathbf{n}$ can be written as

$$\mathbf{n} = -\sin\theta\mathbf{i} + \cos\theta\mathbf{k} \tag{4.146}$$

such that

$$\mathbf{u} \cdot \mathbf{n} = -\frac{w_0}{\cos \theta} e^{i(\kappa s - \omega t)}.$$ (4.147)

The mean momentum flux is than calculated as (cf. 4.145)

$$\begin{aligned}
\overline{\mathbf{f}_f} &= \frac{1}{2} Re \left\{ \rho_0 \overline{\mathbf{u}(\mathbf{u} \cdot \mathbf{n})} \right\} \\
&= -\frac{1}{2} \rho_0 \begin{pmatrix} \sin \theta \\ 0 \\ -\cos \theta \end{pmatrix} \frac{w_0^2}{\cos^2 \theta} \\
&= \left(\frac{\rho_0 w_0^2}{2 \cos^2 \theta} \right) \mathbf{n}.
\end{aligned}$$ (4.148)

Therefore a mean force is exerted in the direction normal to the phase propagation direction. Comparing with (4.138) and (4.139) it can be observed that this mean force (or momentum flux) is opposite to the direction of the energy flux.

Now consider a moving object in a stratified fluid. since there is an energy input into the fluid by the object, waves may be created and the mean energy flux of these waves would then be directed away from the object by radiation of energy. However there will be a mean momentum flux opposite to the energy flux, hence towards the object, which means that a force will be exerted by the fluid on the object. This force is called "wave resistance", or "radiation resistance", or "wave drag".

4.4.3 Semi-Infinite Boussinesq Fluid with Moving Sinusoidal Boundary

Consider a sinusoidal boundary with wavelength $\lambda = 2\pi/k$ forced to move with speed c in the x-direction on the surface of a uniformly stratified semi-infinite fluid with N, as represented in Fig. 4.14.

Let the equation for the surface be

$$\eta = A e^{ik(x - ct)}.$$ (4.149)

If the surface boundary conditions are linearized for a sufficiently small value of A, the moving boundary will be expected to induce a vertical velocity of

$$w = w_0 e^{ik(x - ct)} \quad \text{on} \ z = 0,$$ (4.150)

at the surface. Since vertical velocity satisfies (4.49)–(4.51) in interior of the fluid and linearized surface boundary condition (4.150), the solution should have the same x, t dependence, and with the replacements

Fig. 4.14 Effects of a sinusoidal boundary moved with wavelength $\lambda = 2\pi/k$, with speed c in the x-direction on the surface of a uniformly stratified semi-infinite fluid with N

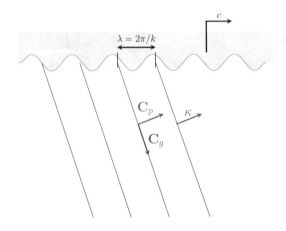

$$\omega = kc \quad \text{and} \quad \kappa_h = k \tag{4.151}$$

where ω is called the encounter frequency, since the fluid encounters kc undulations during the corresponding period of time at a given point, while vertical velocity amplitude of the motion $W(z)$ should satisfy (4.151)

$$\frac{d^2W}{dz^2} + m^2 W = 0 \tag{4.152}$$

where

$$m^2 = \frac{N^2 - k^2 c^2}{k^2 c^2 - f^2} k^2, \tag{4.153}$$

with boundary condition (4.152) are applied at the surface

$$W(0) = w_0. \tag{4.154}$$

A *radiation boundary condition* is needed to have bounded solutions far away from the boundary, $z \to -\infty$, specifically requiring that waves must propagate energy in the negative z-direction. The solution to (4.152) depends on wave number k, which is now *specified* by configuration of the solid boundary. It can be noted that if either the wavelength k or the speed c of the moving boundary are selected to be outside of the range in (4.153), $m^2 = -\gamma^2 < 0$, then the solutions to (4.152) could have the evanescent forms in the vertical, with decreasing amplitude in the $-z$ direction, while still generating horizontal wave motions with

$$W(z) = w_0 e^{\pm \gamma z} \tag{4.155}$$

whereas, if selected speed and wave numbers of the boundary motion yield $m^2 > 0$, then it is possible have internal waves propagating in the vertical and horizontal components traveling far away from the boundary with

$$W(z) = w_0 e^{\pm imz}. \tag{4.156}$$

The radiation boundary condition requires that as $z \to -\infty$ the vertical component of group velocity $C_{g_v} > 0$ yields downward energy flux. On the other hand it can be verified from (4.153) that evanescent solutions $m = \pm i\gamma$, the group velocity becomes pure imaginary. Since evanescent modes do not propagate energy in the vertical, they are not acceptable solutions. On the other hand, for given $m^2 > 0$ in (4.156), only positive value can be accepted with $m = m_1 = (N^2 - k^2 c^2)k/(k^2 c^2 - f^2)^{1/2} > 0$, since only this solution would give $C_{g_v} < 0$ in (4.133), allowing energy to travel away from the boundary. The solution in this case would be

$$w = w_0 e^{i(kx+mz-kct)}. \tag{4.157}$$

Since only $m > 0$ can be accepted with $m^2 > 0$, and for $N > f$, Eq. (4.153) implies that

$$f < kc < N \tag{4.158}$$

or

$$\frac{2\pi c}{N} < 2\pi k^{-1} < \frac{2\pi c}{f_0} \tag{4.159}$$

as wavelength limits for waves propagating away from the sinusoidal boundary, carrying with them a downward energy flux. On the other hand, the solution (4.157) would still imply phase propagation into the boundary.

The selected configuration is represented in Fig. 4.14. The group velocity and energy flux (4.161) are perpendicular to phase propagation (κ) and have a downward component. The momentum flux per unit wavelength $2\pi/\kappa$ is opposite to the energy flux and is given by (4.148). To balance the horizontal component

$$\overline{f_f} = |\overline{\mathbf{f}_f}| \sin\theta = \frac{\rho_0 w_0^2}{2\cos\theta} \tan\theta \tag{4.160}$$

a horizontal force *per unit wavelength* $2\pi/k$ of the sinusoidal boundary

$$\overline{f_x} = \frac{k}{\kappa}\overline{f_f} = \overline{f_f}\cos\theta = \frac{\rho_0 w_0^2}{2}\tan\theta = \frac{\rho_0 w_0^2}{2}\left(\frac{k^2 c^2 - f^2}{N^2 - f^2}\right)^{\frac{1}{2}} \tag{4.161}$$

must be applied at the moving boundary.

4.5   Internal Waves in Mean Flow

4.5.1   3D Waves in Mean Flow

The existence of a basic state of rest had been questioned in an absolute sense in Sect. 3.1. In reality, ideal static conditions are often violated in geophysical systems. For instance, the mean circulation of the atmosphere is driven by differential heating of the equatorial and polar regions; similarly the mean surface circulation is generated by similar mechanisms in the ocean, leading to several scales of motion.

In Sect. 4.1, Boussinesq equations were developed for fluid motion expressed as small perturbations from a basic state of rest, following a series of simplifying approximations, based on linear versions of the basic equations. Momentarily reverting to non-linear forms of the Boussinesq equations yield

$$\frac{D\mathbf{u}}{Dt} + f\mathbf{k} \times \mathbf{u}_h = -\nabla\phi + \sigma\mathbf{k}$$
$$\nabla \cdot \mathbf{u} = 0 \qquad\qquad (4.162.a-c)$$
$$\frac{D\sigma}{Dt} + N^2 w = 0.$$

The 3D perturbation velocity $\mathbf{u}'$, pressure $\phi = p/\rho_0$, buoyancy $\sigma = -g\theta/\theta_0$ for the atmosphere and $\sigma = -g\rho/\rho_0$ for the ocean ($\rho_r \simeq \rho_0$ (cf. 4.4.a-c) are given in Boussinesq manner and the horizontal component of the Coriolis term given as $f\mathbf{k} \times \mathbf{u}_h$.

Motions superposed on mean flow are considered, albeit in simplest form. Total velocity field is comprised of simple depth dependent zonal mean shear flow velocity $\mathbf{U}(z)$ assumed uniform in zonal direction x,

$$\mathbf{u} = \mathbf{U}(z) + \mathbf{u}' = U(z)\mathbf{i} + \mathbf{u}' \qquad\qquad (4.163)$$

expanding material derivative terms in (4.162.a, c) and keeping only the linear, first order terms

$$\frac{D\mathbf{u}}{Dt} = \frac{\partial\mathbf{u}'}{\partial t} + \mathbf{u}' \cdot \nabla\mathbf{U} + \mathbf{U} \cdot \nabla\mathbf{u}' + \cancel{\mathbf{u}' \cdot \nabla\mathbf{u}'} + \cancel{\mathbf{U} \cdot \nabla\mathbf{U}}$$
$$= \frac{\partial\mathbf{u}}{\partial t} + U\frac{\partial\mathbf{u}}{\partial x} + w\frac{dU}{dz}\mathbf{i} + U\frac{\partial w}{\partial x}\mathbf{k} \qquad\qquad (4.164.a,b)$$
$$\frac{D\sigma}{Dt} = \frac{\partial\sigma}{\partial t} + \mathbf{U} \cdot \nabla\sigma + \cancel{\mathbf{u}' \cdot \nabla\sigma} = \frac{\partial\sigma}{\partial t} + U\frac{\partial\sigma}{\partial x}.$$

Linear Boussinesq equations accounting for the effects of zonal mean shear flow can therefore be written as

$$\left(\frac{\partial}{\partial t} + U\frac{\partial}{\partial x}\right)\mathbf{u}_h + w\frac{dU}{dz}\mathbf{i} + f\mathbf{k} \times \mathbf{u}_h = -\nabla_h\phi$$

$$\left(\frac{\partial}{\partial t} + U\frac{\partial}{\partial x}\right)w = -\frac{\partial\phi}{\partial z} + \sigma$$

$$\nabla_h \cdot \mathbf{u}_h + \frac{\partial w}{\partial z} = 0 \qquad (4.165.a-d)$$

$$\left(\frac{\partial}{\partial t} + U\frac{\partial}{\partial x}\right)\sigma + N^2 w = 0$$

where the rate of change operator accounts for temporal and spatial derivatives in relation to current U

$$\mathfrak{L} = \left(\frac{\partial}{\partial t} + U\frac{\partial}{\partial x}\right). \qquad (4.166)$$

Eliminating σ between (4.165.b) and (4.165.d) gives

$$(\mathfrak{L}^2 + N^2)w = -\mathfrak{L}\frac{\partial\phi}{\partial z} \qquad (4.167)$$

Now, taking the divergence and curl of horizontal momentum equation (4.165.a) and using vector identity $\nabla_h \cdot (\mathbf{k} \times \mathbf{u}_h) = -\mathbf{k} \cdot (\nabla_h \times \mathbf{u}_h)$ for the Coriolis term

$$\mathfrak{L}(\nabla_h \cdot \mathbf{u}_h) + \nabla_h \cdot \left(w\frac{dU}{dz}\mathbf{i}\right) - f\mathbf{k} \cdot (\nabla_h \times \mathbf{u}_h) = -\nabla_h^2\phi \qquad (4.168)$$

and similarly taking curl of (4.165.a) and using vector identity $\nabla_h \times \mathbf{k} \times \mathbf{u}_h = \nabla_h \cdot \mathbf{u}_h$ for the Coriolis term gives

$$\mathfrak{L}(\nabla_h \times \mathbf{u}_h) + \nabla_h \times \left(w\frac{dU}{dz}\mathbf{i}\right) + f\nabla_h \cdot \mathbf{u}_h = 0. \qquad (4.169)$$

Operating on first equation with $\mathfrak{L}$ and taking dot product of the second equation with $f\mathbf{k}$ and subtracting gives

$$\mathfrak{L}^2(\nabla_h \cdot \mathbf{u}_h) + \mathfrak{L}\nabla_h \cdot \left(w\frac{dU}{dz}\mathbf{i}\right) + f\mathbf{k} \cdot \nabla_h \times \left(w\frac{dU}{dz}\mathbf{i}\right) + f^2(\nabla_h \cdot \mathbf{u}_h) = -\mathfrak{L}(\nabla_h^2\phi).$$
$$(4.170)$$

Substituting continuity equation (4.165.c)

$$-(\mathfrak{L}^2 + f^2)(\frac{\partial w}{\partial z}) + \mathfrak{L}\nabla_h \cdot \left(w\frac{dU}{dz}\mathbf{i}\right) + f\mathbf{k} \cdot \nabla_h \times \left(w\frac{dU}{dz}\mathbf{i}\right) = -\mathfrak{L}(\nabla_h^2\phi)$$
$$(4.171)$$

noting that

$$\nabla_h \cdot \left(w \frac{dU}{dz} \mathbf{i} \right) = \frac{dU}{dz} \left(\frac{\partial w}{\partial x} \right) \tag{4.172.a}$$

$$\mathbf{k} \cdot \nabla_h \times \left(w \frac{dU}{dz} \mathbf{i} \right) = -\frac{dU}{dz} \left(\frac{\partial w}{\partial y} \right) \tag{4.172.b}$$

and differentiating with z, and inserting

$$-(\mathfrak{L}^2 + f^2) \left(\frac{\partial w}{\partial z} \right) + \frac{dU}{dz} \left(\mathfrak{L} \frac{\partial w}{\partial x} - f \frac{\partial w}{\partial y} \right) = -\mathfrak{L}(\nabla_h^2 \phi), \tag{4.173}$$

The problem suddenly becomes complicated as we aim to reach a simple form. The final step, making use of (4.167), with differentiation in the vertical produces additional terms which are far more complicated in comparison to closed form Eqs. (4.34) and (4.35) obtained in Sect. 4.1.4, eliminating the possibility to reach governing equations of high order for 3D motions superposed on a uniform flow.

The main difficulty comes from the need to take z derivatives of terms involving the shear velocity $U(z)$

$$\frac{\partial}{\partial z} \mathfrak{L} = \frac{\partial}{\partial z} \left(\frac{\partial}{\partial t} + U(z) \frac{\partial}{\partial x} \right) = \mathfrak{L} \frac{\partial}{\partial z} + \frac{dU}{dz} \frac{\partial}{\partial x}$$

$$\frac{\partial}{\partial z} \mathfrak{L}^2 = \frac{\partial}{\partial z} \left(\frac{\partial}{\partial t} + U(z) \frac{\partial}{\partial x} \right)^2 = \mathfrak{L}^2 \frac{\partial}{\partial z} + 2\mathfrak{L} \frac{dU}{dz} \frac{\partial}{\partial x}$$

and following these steps produces a complicated set with terms on the left hand side being much similar to (4.34), with a number of extra terms that are moved to the right hand side:

$$(\mathfrak{L}^2 + N^2) \nabla_h^2 w + (\mathfrak{L}^2 + f^2) \frac{\partial^2 w}{\partial z^2} = f \left(w, \phi, \frac{dU}{dz}, \frac{\partial}{\partial x}, \frac{\partial}{\partial y} \right) \tag{4.174}$$

where f is a complicated function of the basic variables and their derivatives.

4.5.2   2D Waves in Shear Flow

In the last section, it has been demonstrated that further investigation of 3D waves riding on uniform, unidirectional shear flow is not straightforward, at least through (4.174). Reverting is now in order to consider 2D problem aligned with the plane of mean flow variables and coordinates (x, z). 2D Boussinesq Eqs. (4.162a–c) were already linearized and adapted to (4.165.a–d) in the earlier sections. Further, asymmetric Coriolis effects will be ignored in local scale motions, by setting $f = 0$.

Re-writing these equations in shorthand, putting $U' = dU/dz$ and letting subscripts denote partial differentiation, and with a definition of advective derivative given in (4.166)

$$\mathcal{L} = \left(\frac{\partial}{\partial t} + U(z) \frac{\partial}{\partial x} \right) = ()_t + U(z)()_x$$

Equations (4.165.a–d) take the form

$$\mathcal{L}u + wU' = -\phi_x$$
$$\mathcal{L}w = -\phi_z + \sigma$$
$$u_x + w_z = 0 \qquad (4.175.a - d)$$
$$\mathcal{L}\sigma + N^2 w = 0.$$

A stream function ψ is defined to reduce these equations

$$u = \psi_z \ \text{ and } \ w = -\psi_x \qquad (4.176)$$

which readily satisfy the continuity equation (4.165.c). Combining equations (4.175.b, d) gives

$$(\mathcal{L}^2 + N^2)\psi_x = \mathcal{L}\phi_z. \qquad (4.177)$$

By differentiating (4.175.a) with respect to z and expanding, one obtains

$$(\mathcal{L}u)_z + (wU')_z = \mathcal{L}u_z + U'u_x + wU'' + w_z U' = -\phi_{xz}$$

with the cancellations based on (4.175.c). Combining (4.177) and (4.5.2) and using (4.176) gives the closed form equation which together with appropriate boundary and initial conditions can be used to obtain solutions.

$$\mathcal{L}^2 \psi_{zz} - \mathcal{L}U'' \psi_x - (\mathcal{L}^2 + N^2)\psi_{xx} = 0. \qquad (4.178)$$

4.5.3   Energy Conservation in Uniform Shear Flow

Consider now 3D non-linear Boussinesq Eqs. (4.162.a–c) applied to total velocity composed of mean shear flow and perturbations $\mathbf{u} = U(z)\mathbf{i} + \mathbf{u}'$ as specified in (4.163). Keeping only linear terms as in (4.164.a, b) and dropping primes resulted in Eqs. (4.165.a–d).

The appropriate *Energy Conservation Equations* for motions superposed on mean zonal shear flows are obtained from the above equations. By taking dot product of equation (4.165.a) with $\mathbf{u}_h$ and (4.165.b) with w and adding these together,

$$\frac{\partial}{\partial t}\left(\frac{1}{2}\mathbf{u}\cdot\mathbf{u}\right) + U\frac{\partial}{\partial x}\left(\frac{1}{2}\mathbf{u}\cdot\mathbf{u}\right) + uw\frac{dU}{dz} + \underbrace{f\mathbf{u}_h \cdot \mathbf{k}\times\mathbf{u}_h}_{} \overset{0}{} = -\frac{1}{\rho_0}\mathbf{u}\cdot\nabla p + \underbrace{\mathbf{u}\cdot\sigma\mathbf{k}}_{w\sigma}$$

one obtains the rate of change of Kinetic Energy deriving from 3D velocity $\mathbf{u}$ and pressure p, also by making use of the 3D continuity equation (4.162.b)

$$\mathfrak{L}\,(KE) = \left(\frac{\partial}{\partial t} + U\frac{\partial}{\partial x}\right)\left(\frac{1}{2}\mathbf{u}\cdot\mathbf{u}\right) + uw\frac{dU}{dz} = -\frac{1}{\rho_0}\nabla\cdot p\mathbf{u} + w\sigma \qquad (4.179)$$

Then, multiplying Eq. (4.165.d) with σ/N^2, the rate of change of Potential Energy is obtained as

$$\mathfrak{L}\,(PE) = \left(\frac{\partial}{\partial t} + U\frac{\partial}{\partial x}\right)\left(\frac{\sigma^2}{2N^2}\right) = -w\sigma. \qquad (4.180)$$

Adding components (4.179) and (4.180) together gives the rate of change of Total Energy

$$\mathfrak{L}\,(TE) = \left(\frac{\partial}{\partial t} + U\frac{\partial}{\partial x}\right)\left(\frac{1}{2}\mathbf{u}\cdot\mathbf{u} + \frac{\sigma^2}{2N^2}\right) = -\frac{1}{\rho_r}\nabla\cdot p\mathbf{u} - uw\frac{dU}{dz}. \qquad (4.181)$$

These energy rates of change can now be integrated over the fluid volume V, enclosed by surface S

$$\int_V dV = \int_0^X \int_0^Y \int_0^Z dx\,dy\,dz. \qquad (4.182)$$

The discussion based on energy conservation arguments is much similar to what has been considered for geophysical scales of motion in Sect. 3.3.5. In order to derive average rates of change of energy, variables $\mathbf{u}$, p, σ, are assumed to be periodic in direction x, with period X. Averaging is denoted with overbar and integrating over fixed volume

$$\overline{(\,)} = \frac{1}{X}\int_0^X (\,)dx, \quad \text{and} \quad \int_0^X \frac{\partial}{\partial x}(\,)dx = 0. \qquad (4.183)$$

expressing total energy conservation as

$$\frac{d}{dt}\int \rho_r \left[\frac{1}{2}\overline{\mathbf{u}\cdot\mathbf{u}} + \frac{\overline{\sigma^2}}{2N^2}\right]dV = -\int_S \overline{p\mathbf{u}\cdot\mathbf{n}}\,dS - \int_V \rho_r\overline{uw}\frac{dU}{dz}dV \qquad (4.184)$$

The terms on the right hand side represent energy sources and sinks. The first term represents energy flux or work done by external pressure, positive for unit vector $\mathbf{n}$ pointing out from the surface S. The second term is positive only if $\overline{uw}$ and dU/dz, have opposite signs, i.e. if the disturbance field gains energy from the mean *shear* flow; negative when the sign is reversed. Next, to explain the correlation term $\overline{uw}$, imagine a material surface of the fluid at some level $z = h(x, t)$ defined by

$$\phi = z - h(x, t) = 0. \tag{4.185}$$

The position and vertical velocity of the surface (4.185) is given by the kinematic condition for a material surface, considering a progressive wave disturbance preserving its initial form $u = u(x - ct)$, so that

$$w = \mathfrak{L}h = \frac{\partial h}{\partial t} + U\frac{\partial h}{\partial x} = (U - c)\frac{\partial h}{\partial x}. \tag{4.186}$$

The horizontal force exerted on the material surface is $p\mathbf{n}$, where $\mathbf{n}$ is the unit normal vector to the surface

$$\mathbf{n} = \frac{\nabla\phi}{|\nabla\phi|} = \frac{-\mathbf{i}\frac{dh}{dx} + \mathbf{k}}{\sqrt{1 + (\frac{dh}{dx})^2}} \tag{4.187}$$

The horizontal component of the pressure force is therefore found as

$$\overline{F_x} = \frac{1}{X}\int_0^X p\mathbf{n}\cdot\mathbf{i}\,ds = \frac{1}{X}\int_0^X \frac{-p\frac{dh}{dx}}{\sqrt{1 + (\frac{dh}{dx})^2}}ds. \tag{4.188}$$

with the length ds of the curve given by

$$ds \equiv \sqrt{1 + \left(\frac{dh}{dx}\right)^2}\,dx, \tag{4.189}$$

the integral is reduced to

$$\overline{F_x} = \frac{1}{X}\int_0^X -p\frac{dh}{dx}dx \tag{4.190}$$

Integration by parts then yields

$$\int_0^X -p\frac{dh}{dx}dx = -ph\big|_0^X + \int_0^X h\frac{dp}{dx}dx$$

where the first term is evaluated to vanish as a result of periodicity. Making use of the horizontal momentum equation in the direction of a shear current,

$$\left(\frac{\partial}{\partial t} + U\frac{\partial}{\partial x}\right)u + w\frac{dU}{dz} = -\frac{1}{\rho_r}\frac{\partial p}{\partial x} \tag{4.191}$$

so that

$$(U - c)\frac{\partial u}{\partial x} + w\frac{dU}{dz} = -\frac{1}{\rho_r}\frac{\partial p}{\partial x}. \tag{4.192}$$

Substituting Eq. (4.192) in (4.5.3) gives

$$\int_0^X h\frac{dp}{dx}dx = -\rho_r(U - c)\int_0^X h\frac{\partial u}{\partial x}dx - \rho_r\frac{dU}{dz}\int_0^X hw\,dx \tag{4.193}$$

and integration by parts of the first term in (4.193) and inserting from (4.186) yields

$$\int_0^X h\frac{\partial u}{\partial x}dx = uh|_0^{X} - \int_0^X u\frac{\partial h}{\partial x}dx = -\frac{1}{U-c}\int_0^X uw\,dx.$$

By inserting the kinematic condition (4.186), the second term in (4.193) vanishes

$$\int_0^X hw\,dx = (U - c)\int_0^X \frac{\partial}{\partial x}\left(\frac{h^2}{2}\right)dx = 0. \tag{4.194}$$

Finally, the force F_x is calculated as

$$\overline{F_x} = \frac{1}{X}\int_0^X h\frac{dp}{dx}dx = \rho_r\frac{1}{X}\int_0^X uw\,dx = \rho_r\overline{uw} \tag{4.195}$$

This is related to the vertical flux of horizontal momentum due to fluctuations, i.e. Reynolds' stresses. The average force over one wavelength can be calculated from

$$\overline{F_x} = \rho_r\overline{uw} = \rho_r\frac{k}{2\pi}\int_0^{\frac{2\pi}{k}} uw\,dx \tag{4.196}$$

But for internal waves (or other transverse waves):

$$\kappa \cdot \mathbf{U} = ku + lw = 0 \text{ or } u = -\frac{l}{k}w \tag{4.197}$$

and the flux of momentum in the vertical direction is therefore calculated to be

$$\overline{F_x} = \rho_r\overline{uw} = -\rho_r\frac{l}{k}\overline{w^2} \neq 0. \tag{4.198}$$

If energy equation is averaged over horizontal wavelength of $2\pi/k$

$$\frac{d\overline{E}}{dt} = -\rho_r\frac{dU}{dz}\overline{uw} - \frac{\partial}{\partial z}\overline{pw} \tag{4.199}$$

these terms can be interpreted as the source / sink of energy, accounting for energy exchange between the shear flow and the disturbance.

Theorem If there is no z-dependence of the vertical flux of horizontal momentum, then there is no exchange between the basic flow and disturbance.

If $F_x = \rho_r \int_0^{2\pi/k} uw\,dx = \rho_r \overline{uw}$ =constant, then

$$
\begin{aligned}
-\rho_r \frac{dU}{dz}\overline{uw} &= -\rho_r \frac{k}{2\pi}\int_0^{2\pi k} \frac{dU}{dz} uw\,dx \\
&= -\rho_r \frac{\partial}{\partial z}(\overline{uw}U) + \rho_r U \frac{\partial}{\partial z}\overline{uw} \qquad (4.200)\\
&= -\rho_r \frac{\partial}{\partial z}(U\overline{uw}).
\end{aligned}
$$

Then we can define the energy flux (by letting $\rho_r \simeq \rho_0$) as

$$
\overline{F} = \overline{pw} + \rho_0 U\overline{uw} \qquad (4.201)
$$

and therefore the energy conservation is expressed as

$$
\frac{d\overline{E}}{dt} = -\frac{\partial \overline{F}}{\partial z} \qquad (4.202)
$$

i.e. the rate of change of mean energy is balanced by the divergence of the energy flux. Energy can be transmitted from one place to another by the mean flow but no energy exchange between the shear flow and disturbance can take place.

4.5.4 Taylor-Goldstein Eigenvalue Equation

The 2D wave Eq. (4.178) was developed in Sect. 4.5.3:

$$
\mathfrak{L}^2\psi_{zz} - \mathfrak{L}U''\psi_x - (\mathfrak{L}^2 + N^2)\psi_{xx} = 0.
$$

Consider a progressive wave disturbance that preserves its initial form

$$
\psi = \psi(x - ct) = \hat{\psi}e^{ik(x-ct)}. \qquad (4.203)
$$

Remembering the corresponding form of the advective derivative

$$
\mathfrak{L} = \left(\frac{\partial}{\partial t} + U(z)\frac{\partial}{\partial x}\right) = ik(U - c)
$$

and using in Eq. (4.178) gives

$$
-k^2(U - c)^2\hat{\psi}_{zz} + k^2(U - c)U''\hat{\psi} = (-k^2(U - c)^2 + N^2)(-k^2)\hat{\psi}.
$$

Simplifying, this leads to the *Taylor-Goldstein eigenvalue equation*

Fig. 4.15 A velocity profile with shear U' changing from zero to a constant value $U' = -\alpha$ in the upper half

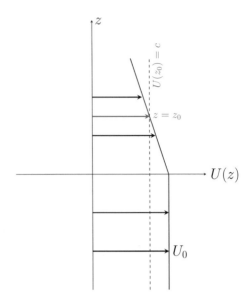

$$\left[\frac{d^2}{dz^2} - k^2 + \frac{N^2}{(U - c)^2} - \frac{U''}{(U - c)}\right]\hat{\psi} = 0. \qquad (4.204)$$

With appropriate homogeneous boundary conditions this equation forms the basis for eigenvalue problems to investigate stratified shear flows.

4.5.5   Critical Layer in a Stratified Shear Flow

The important concept of *Critical Layer* arising in stratified shear flow is demonstrated through a simple example. Higher order terms in x direction are momentarily ignored, by approximating $\frac{\partial^2 w}{\partial x^2} \ll \frac{\partial^2 w}{\partial z^2}$ in Eq. (4.178). This simplification essentially amounts to set $k^2 = 0$, so as to leave out the second term in (4.204). Further simplification is justified by selecting a linear shear profile at most, so that $U'' = 0$, leaving out the fourth term in (4.204). With these simplifications the eigenvalue equation becomes:

$$(U(z) - c)^2 \frac{\partial^2 \hat{\psi}}{\partial z^2} + N^2 \hat{\psi} = 0 \ \text{ for } \ z > 0 \qquad (4.205)$$

A discontinuous velocity profile shown in Fig. 4.15 is selected as

$$U(z) = \begin{cases} U_0, & z < 0 \\ U_0 - \alpha z, & z > 0 \end{cases}. \qquad (4.206)$$

Let some wave motion with horizontal propagation speed c be set up in the lower part of the domain $z < 0$ where the current speed is constant $U = U_0$, in the range $0 < c < U_0$.

The solution in lower part of the domain, $z < 0$, will be given as

$$\hat{\psi} = Ae^{+imz} \text{ for } z < 0$$

$$m^2 = \frac{N^2}{(U_0 - c)^2} \tag{4.210a, b}$$

where m is the vertical wave number for upward propagating waves, assuming the wave motion originates from the lower part of the domain, according to (4.203)

$$\psi = \psi(x - ct) = \hat{\psi}e^{ik(x-ct)} = Ae^{i(kx+mz-kct)}. \tag{4.211}$$

On the other hand, with velocity profile $U(z) = U_0 - \alpha z$ imposed in Eq. (4.205) in the upper domain $z > 0$,

$$(U_0 - \alpha z - c)^2 \frac{\partial^2 \hat{\psi}}{\partial z^2} + N^2 \hat{\psi} = 0, \tag{4.212}$$

a solution is proposed as

$$\hat{\psi} = (U(z) - c)^q. \tag{4.213}$$

Calculating first and second derivatives from (4.213),

$$\hat{\psi}_z = q(U - c)^{q-1} \left(\frac{\partial U}{\partial z} \right) = -\alpha q(U - c)^{q-1} \tag{4.214}$$

$$\hat{\psi}_{zz} = -\alpha q(q - 1)(U - c)^{q-2} \left(\frac{\partial U}{\partial z} \right)$$
$$= \alpha^2 q(q - 1)(U - c)^{q-2}, \tag{4.215}$$

and substituting into Eq. (4.212) gives

$$q(q - 1) + \frac{N^2}{\alpha^2} = 0. \tag{4.216}$$

Two roots can be obtained from the above equation:

$$q_\oplus = \frac{1}{2} + \left(\frac{1}{4} - R \right)^{\frac{1}{2}} = \frac{1}{2} + i \left(R - \frac{1}{4} \right)^{\frac{1}{2}}$$
$$q_\ominus = \frac{1}{2} - \left(\frac{1}{4} - R \right)^{\frac{1}{2}} = \frac{1}{2} - i \left(R - \frac{1}{4} \right)^{\frac{1}{2}}. \tag{4.217.a, b}$$

where

$$R = \frac{N^2}{\alpha^2} = \frac{N^2}{\left(\frac{\partial U}{\partial z}\right)^2}. \tag{4.218}$$

Emerging from the solution is R, which is identified as *Richardson number*, a ratio measuring the opposing effects of stratification and shear in the studied flow.

It can be noted that the exponents $q_\oplus$ and $q_\ominus$. change from real numbers to complex values, depending on the sign of $(R - 1/4)$. When Richardson Number R is smaller than the critical value $R_c = 1/4$, the exponents in (4.217) turn out to be real numbers, which otherwise should be complex. The importance of this physical parameter is already apparent in determining the behavior of the solution.

To start constructing a solution in the upper domain $z > 0$, the following positive quantities are defined

$$U_\delta = U_0 - c > 0 \quad \text{and} \quad \gamma = \sqrt{R - \frac{1}{4}} > 0 \tag{4.219}$$

$$
\begin{aligned}
B_\oplus &= |B_\oplus| e^{i\delta_\oplus}, \quad B_\ominus = |B_\ominus| e^{i\delta_\ominus} \\
(U_\delta - \alpha z)^{i\gamma} &= e^{i\gamma \ln (U_\delta - \alpha z)}, \quad (U_\delta - \alpha z)^{i\gamma} = e^{-i\gamma \ln (U_\delta - \alpha z)}.
\end{aligned} \tag{4.220.a, b}
$$

The roots of (4.217.a, b) is then inserted in the solution (4.213) to give

$$
\begin{aligned}
\hat{\psi}(z) &= B_\oplus (U(z) - c)^{q_\oplus} + B_\ominus (U(z) - c)^{q_\ominus} \\
&= B_\oplus (U_0 - c - \alpha z)^{q_\oplus} + B_\ominus (U_0 - c - \alpha z)^{q_\ominus} \\
&= \sqrt{U_0 - c - \alpha z} \left\{ B_\ominus (U_0 - c - \alpha z)^{i\gamma} + B_\ominus (U_0 - c - \alpha z)^{-i\gamma} \right\} \\
&= \sqrt{U_\delta - \alpha z} \left\{ |B_\oplus| e^{i[\gamma \ln (U_\delta - \alpha z) + \delta_\oplus]} + |B_\ominus| e^{-i[\gamma \ln (U_\delta - \alpha z) + \delta_\ominus]} \right\}.
\end{aligned} \tag{4.221}
$$

By matching the solutions (4.210) and (4.221) at $z = 0$, the constant A is obtained as

$$A = B_\oplus U_\delta^{q_\oplus} + B_\ominus U_\delta^{q_\ominus} = \sqrt{U_\delta} \left[B_\oplus U_\delta^{i\gamma} + B_\ominus U_\delta^{-i\gamma} \right] \tag{4.222}$$

The solution is then converted to trigonometric form

$$\hat{\psi}(z) = \sqrt{U_\delta - \alpha z} \left\{ S_1 \cos [\gamma \ln (U_\delta - \alpha z) + \delta_1] + i S_2 \sin [\gamma \ln (U_\delta - \alpha z) + \delta_2] \right\} \tag{4.223}$$

with constants (4.220) converted to appropriate ones. Arbitrarily setting the amplitude to $A = 1$ in principle would establish the relations between various constants.

The most essential parameter in the solution is the Richardson number R, which establishes the nature of the solution, depending on its value being greater or smaller than the critical value R_c. The condition

$$R > R_c = 1/4 \quad \text{or} \quad \gamma = \sqrt{R - \frac{1}{4}} > 0 \tag{4.224}$$

turns out to be the condition for waves to exist according to (4.219), with exponents $\pm i\gamma$ imply stable wave solutions.

On the other hand, if the exponent was to be replaced by a real constant μ according to alternatives in (4.217.a, b)

$$R < R_c = 1/4 \quad \text{or} \quad \mu \equiv i\gamma = \sqrt{\frac{1}{4} - R} > 0 \tag{4.225}$$

would result in evanescent solutions, instead of wave motion. Therefore it can be concluded that Richardson Number being higher than the critical value is an indicator of stable motions.

Further elaboration below indicates that the flow will not support wave solutions as the Richardson Number approaches critical value. A shift can be made in vertical coordinates with origin moved to level $z = z_0$

$$z_0 = (U_0 - c)/\alpha. \tag{4.226}$$

Taking real part with appropriate real constants S_0 and δ_0, the solution in region $z > 0$ becomes

$$Re[\hat{\psi}(z)] = S_0\sqrt{U_\delta - \alpha z} \, \cos\left[\gamma \ln (U_\delta - \alpha z) + \delta_0\right]$$
$$= S_0\sqrt{\alpha(z_0 - z)} \, \cos\left[\sqrt{R - 1/4} \, \ln \alpha(z_0 - z) + \delta_0\right], \quad in \ \ z > 0 \tag{4.227}$$

Remembering the solution (4.210a, b) in the lower half domain

$$Re[\hat{\psi}(z)] = A \cos mz, \quad in \ \ z < 0$$

and matching the two solutions at $z = 0$, constants can be related to each other

$$S_0 = A/\sqrt{U_0 - c}, \quad \delta_0 = -[\sqrt{R - 1/4} \, \ln (U_0 - c)]. \tag{4.228}$$

Other details of the solution are skipped to show important physical characteristics for the regime $R > 1/4$ and $U_0 - c > 0$. As $U \to c$ near $z \to z_0$, amplitude $\hat{\psi} \to 0$ vanishes at that level, while on the other hand, wavelength of the oscillation (argument of the cosine term) varies much too rapidly, the local vertical wave number approaching $\sqrt{R - 1/4} \, \ln (z_0 - z) \to -\infty$ and the solution blows up at the *critical layer* $z = z_0$. This strange behavior near the critical layer is indicative of the fact that solution breaks down at level $z = z_0$! Calculating velocity

$$
\begin{aligned}
\hat{u} = \frac{\partial \hat{\psi}}{\partial z} &= U' \left\{ q_\oplus B_\oplus (U(z) - c)^{q_\oplus - 1} + q_\ominus B_\ominus (U(z) - c)^{q_\ominus - 1} \right\} \\
&= -\frac{\alpha}{\sqrt{\alpha(z_0 - z)}} \left\{ q_\oplus B_\oplus [\alpha(z_0 - z)]^{i\gamma} + q_\ominus B_\ominus [\alpha(z_0 - z)]^{-i\gamma} \right\}
\end{aligned}
\tag{4.229}
$$

from which it is easy to see that $u \to \infty$, with all related variables becoming unbounded at critical level $z = z_0$. Waves can not propagate above this level and shear becomes large at $z = z_0$. Then, to dissipate and absorb this energy, viscous effects so far not taken into account in the present inviscid theory have to come into play at the critical level $z = z_0$, where wave speed actually reaches its maximum $c = U(z_0)$.

Above the critical level where waves cannot penetrate, solutions of evanescent nature are possible

$$
c - U_0 > 0 \quad \text{and} \quad \mu = \sqrt{\frac{1}{4} - R} > 0
\tag{4.230}
$$

letting wave speed c exceed local mean current speed U, imposing $\mu^2 = -\gamma^2$. In this case, only decaying solutions with height if $\mu > 1/2$ can be accepted

$$
\hat{\psi}(z) \sim B[\alpha(z - z_0)]^{(\frac{1}{2} - \mu z)} \quad z > 0.
\tag{4.231}
$$

4.5.6 Stability of Simple Stratified Shear Flows

Repeating Taylor-Goldstein eigenvalue Eq. (4.204), obtained in earlier sections

$$
\left[\frac{d^2}{dz^2} - k^2 + \frac{N^2}{(U - c)^2} - \frac{U''}{(U - c)} \right] \hat{\psi} = 0
$$

is re-written in the following form, with primes denoting differentiation with respect to z

$$
(U - c)^2 \hat{\psi}'' - (U - c)U'' \hat{\psi} + \left[N^2 - k^2 (U - c)^2 \right] \hat{\psi} = 0.
\tag{4.232}
$$

A solution to this equation with propagating waves in x direction can be sought for a fluid confined between solid boundaries at $z = 0$ and $z = H$, with boundary conditions

$$
\hat{\psi}_x(0) = ik\hat{\psi}(0) = 0, \quad \text{and} \quad \hat{\psi}_x(H) = ik\hat{\psi}(H) = 0.
\tag{4.233}
$$

With transformed variables

$$\hat{\psi}(z) = F(z)(U(z) - c)$$

$$F = \frac{\hat{\psi}}{(U - c)} \qquad \qquad (4.234.a - c)$$

$$F' = \frac{\hat{\psi}'(U - c) - U'\hat{\psi}}{(U - c)^2}$$

and

$$\left[(U - c)^2 F'\right]' = (U - c)\hat{\psi}'' + \cancel{\hat{\psi}'U'} - \cancel{U'\hat{\psi}'} - U''\hat{\psi}$$

replacing (4.234.a-c) and (4.5.6) in (4.133), with some cancellations, Taylor-Goldstein equation takes the following form

$$\left[(U - c)^2 F'\right]' + \left[N^2 - k^2(U - c)^2\right] F = 0 \qquad (4.235)$$

with boundary conditions

$$F(0) = 0, \quad \text{and} \quad F(H) = 0. \qquad (4.237.a, b)$$

4.5.7 Stability Analyses and the Howard Semi-Circle Theorem

Taylor (1931) and Goldstein (1931) considered special shear profiles $U(z)$ to obtain solutions of the eigenvalue problem stated in (4.235) and (4.237.a, b). Stability would be ensured if the solution does not have growing tendency with time, i.e. if c does not have imaginary positive component. Howard (1961) proved Miles (1961) theorem in a more elegant way, by integrating the set of equations multiplied with complex conjugate F^*,

$$\int_0^H \left\{ F^* \left[(U - c)^2 F'\right]' + \left[N^2 - k^2(U - c)^2\right] F^* F \right\} dz = 0. \qquad (4.238)$$

with the first term is integrated by parts

$$\int_0^H F^* \left[(U - c)^2 F'\right]' dz = \cancel{F^*(U - c)^2 F \big|_0^H} - \int_0^H F^{*'}(U - c)^2 F' dz$$

and with the term canceled by invoking boundary conditions (4.237.a, b) gives

$$\int_0^H |(U - c)^2 |S|^2 dz = \int_0^H N^2 |F|^2 dz. \qquad (4.239)$$

where

$$|S|^2 = |F'|^2 + k^2|F|^2 > 0. \tag{4.240}$$

This result involves positive real values except for the term $(U - c)^2$ involving wave speed $c = c_r + ic_i$ allowed to be a complex number, such that its real part c_r would give stable wave solutions

$$\hat{\psi} \sim e^{ik(x-ct)} \tag{4.241}$$

while the imaginary part would lead to growing and decaying solutions

$$\hat{\psi} \sim e^{ikx}e^{-kc_it}.$$

Real and imaginary parts of Eq. (4.240) are separated as

$$\int_0^H [(U - c_r)^2 - c_i^2]|S|^2 dz = \int_0^H N^2|F|^2 dz > 0,$$
$$2c_i \int_0^H (U - c_r)|S|^2 dz = 0. \tag{4.241, a.b}$$

The first one (4.241.a) has both sides completely composed of real and positive numbers. The second Eq. (4.241.b) would require the integral to vanish if $c_i \neq 0$. This result on the other hand would require that $(U - c_r)$ to change sign at least once in the integration domain to make the integral zero, so that the range of c_i is limited by the U_{min} and U_{max} extremes of the velocity profile $U(z)$

$$U_{min} < c_r < U_{max}. \tag{4.242}$$

Re-writing (4.241.a)

$$\int_0^H [U^2 + c_r^2 - 2Uc_r - c_i^2]|S|^2 dz > 0$$

by making use of (4.241.b) gives

$$\int_0^H [U^2 - c_r^2 - c_i^2]|S|^2 dz > 0. \tag{4.243}$$

Utilizing the limits imposed in (4.242) the following proposition is true:

$$\int_0^H [U(z) - U_{min}][U_{max} - U(z)]|S|^2 dz \geq 0, \tag{4.244}$$

which implies that

Fig. 4.16 Howard semi-circle limits, with unstable waves enclosed within the shaded region

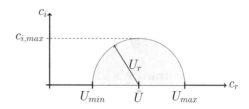

$$\int_0^H [U(z) - U_{min}][U_{max} - U(z)]|S|^2 dz$$

$$= \int_0^H \left\{ U(z)U_{max} + U(z)U_{min} - U_{min}U_{max} - U^2(z) \right\} |S|^2 dz$$

$$= \int_0^H \left\{ U(z)[U_{max} + U_{min}] - U_{min}U_{max} - (c_r^2 + c_i^2) \right\} |S|^2 dz \qquad (4.245)$$

$$= \left\{ c_r[U_{max} + U_{min}] - U_{min}U_{max} - (c_r^2 + c_i^2) \right\} \int_0^H |S|^2 dz \geq 0$$

where Eq. (4.241.b) is inserted in the fourth line, making part of the integrand independent of z, to bring it outside the integral in last line of (4.245). Since $\int_0^H |S|^2 dz > 0$, we have the algebraic inequality

$$\left\{ c_r[U_{max} + U_{min}] - U_{min}U_{max} - (c_r^2 + c_i^2) \right\} \geq 0$$

and by slight rearrangement

$$[c_r - \frac{1}{2}(U_{max} + U_{min})]^2 + c_i^2 \leq \frac{1}{2}[(U_{max} - U_{min})]^2 \qquad (4.246)$$

which is the *Howard Semi-Circle Theorem*. Defining

$$\bar{U} = \frac{1}{2}(U_{max} + U_{min}) \quad \text{and} \quad U_r = \frac{1}{2}[(U_{max} - U_{min})]$$

(4.246) can be expressed as

$$[c_r - \bar{U}]^2 + c_i^2 \leq U_r^2$$

describing the semi-circle in Fig. 4.16.

Taking up Eq. (4.232)

$$(U - c)^2 \hat{\psi}'' - (U - c)U'' \hat{\psi} + \left[N^2 - k^2(U - c)^2 \right] \hat{\psi} = 0.$$

a new variable transform is defined

$$\hat{\psi} = P^{\frac{1}{2}}Q \quad \text{where} \quad P = (U(z) - c) \quad \text{and} \quad Q = Q(z) \tag{4.247}$$

with various terms evaluated as

$$\hat{\psi} = P^{\frac{1}{2}}Q$$
$$\hat{\psi}' = \frac{1}{2}P^{-\frac{1}{2}}QU' + P^{\frac{1}{2}}Q'$$
$$\hat{\psi}'' = -\frac{1}{4}P^{-\frac{3}{2}}Q(U')^2 + \frac{1}{2}P^{-\frac{1}{2}}[Q'U' + QU''] + \frac{1}{2}P^{-\frac{1}{2}}Q'U' + P^{\frac{1}{2}}Q''.$$
$$\tag{4.248.a - c}$$

With these transformations substituted in Eq. (4.232),

$$(U - c)^2 \left\{ -\frac{1}{4}P^{-\frac{3}{2}}Q(U')^2 + \frac{1}{2}P^{-\frac{1}{2}}[Q'U' + QU''] + \frac{1}{2}P^{-\frac{1}{2}}Q'U' + P^{\frac{1}{2}}Q'' \right\}$$
$$- (U - c)U''P^{\frac{1}{2}}Q + \left[N^2 - k^2(U - c)^2 \right] P^{\frac{1}{2}}Q = 0$$

and dividing by $(U - c)P^{\frac{1}{2}}$

$$(U - c)Q'' - \frac{1}{2}QU'' - U''Q + \left[\frac{N^2 - (U')^2/4}{U - c} - k^2(U - c) \right] Q = 0.$$

further noting that

$$[(U - c)Q']' = (U - c)Q'' + Q'U'$$

one obtains

$$[(U - c)Q']' + \left[\frac{N^2 - (U')^2/4}{U - c} - \frac{1}{2}U'' - k^2(U - c) \right] Q. \tag{4.249}$$

Here, the phase speed $c = c_r + ic_i$ once again is a complex number. Multiplying by the conjugate Q^*, integrating (4.249) with the first term by parts, and using boundary conditions (4.237.a, b) transformed by (4.248) gives

$$\int_0^H \left\{ -(U - c)|Q'|^2 + \left[(U - c^*)\frac{N^2 - (U')^2/4}{|U - c|^2} - \frac{1}{2}U'' - k^2(U - c) \right] |Q|^2 \right\} dz = 0. \tag{4.250}$$

The real part of (4.250) divided by $U - c_r$ is

$$\int_0^H \left\{ |Q'|^2 + k^2|Q|^2 + \frac{1}{2}\frac{U''}{(U - c_r)}|Q|^2 \right\} dz = \int_0^H \frac{N^2 - (U')^2/4}{(U - c_r)^2 + c_i^2}|Q|^2 dz.$$
(4.251.a)

while the imaginary part is

$$c_i \int_0^H \left\{ |Q'|^2 + k^2|Q|^2 + \frac{N^2 - (U')^2/4}{(U - c_r)^2 + c_i^2}|Q|^2 \right\} dz = 0.$$
(4.251.b)

These equations can be used to evaluate stability of the flow. A trivial possibility is to accept $c_i = 0$ to satisfy Eq. (4.251.b). This would imply that the flow is unconditionally stable, invalidating any further search for unstable conditions.

On the other hand, if we accept that $c_i \neq 0$, then (4.251.b) is instructive to investigate stability. All terms in this equation are positive except the last one, and for the equation to be satisfied, this last term has to be negative in some region of the flow. Specifically, the factor $N^2 - (U')^2/4$ would have to pass through zero at least once in some part of the domain to make the integral zero. The condition $R \geq 1/4$ to be realized in some part of the flow domain ensures *sufficient condition* for stability.

As foreseen by Miles (1961) and Howard (1961) theory, this result imposes a criticality condition based on the *gradient Richardson Number* defined as

$$R = \frac{N^2}{U'^2}$$
(4.252.a)

with a critical value

$$R_c = \frac{1}{4}.$$
(4.252.b)

For Richardson number above critical value $R > R_c$ the flow would be stable, while for smaller values $R < R_c$, instability can be expected. In essence, $R < R_c = 1/4$ is therefore a necessary condition for instability, but not sufficient.

These results are also in agreement with earlier statements (4.218) and (4.224) regarding critical layer formation in a simple linear shear velocity profile with constant shear $U' = \alpha$ presented in Sect. 4.5.6.

The maximum growth rate for an unstable wave is expected to be

$$k^2 c_i^2 \leq \max\left(\frac{1}{4}\left(\frac{dU}{dz}\right) - N^2(z) \right).$$
(4.253)

In this text we only touched upon the sufficient and necessary conditions for instability. Boundaries have been found to have stabilizing effects. Actual regions of stability for specific shear and stratification profiles and geometrical boundaries have been worked out by Taylor (1931) and Goldstein (1931), Drazin (1958), Hazel(1972) and others finding slightly differing values of the critical Richardson Number R_c value for the specific application.

More reasonable criteria have been used to detect instabilities in data analyses of observations, based on *overall Richardson number* defined as

$$R_0 = \frac{gh(\Delta\rho/\rho)}{(\Delta U)^2} = \frac{g'h}{(\Delta U)^2}, \tag{4.254}$$

where Δ is the difference in properties measured over sampling intervals, h being the vertical scale of the problem, and $g' = g\Delta\rho/\rho$ defined as reduced gravity.

In the above, results obtained from Eq. (4.251.b) are evaluated. Considering both Eqs. (4.251.a, b), added together, except for multiplier c_i gives

$$\int_0^H \left\{ |Q'|^2 + k^2|Q|^2 + \frac{1}{4}\frac{U''}{(U - c_r)}|Q|^2 \right\} dz = 0 \tag{4.255.a}$$

and by subtracting

$$\int_0^H \frac{1}{4}\frac{U''}{(U - c_r)}|Q|^2 dz = \int_0^H \frac{N^2 - (U')^2/4}{(U - c_r)^2 + c_i^2}|Q|^2 dz. \tag{4.255.b}$$

The first Eq. (4.255.a) is a reduction of the original Eqs. (4.251.a, b), according to which $U'' = d^2U/dz^2$ has to change sign at least once in the domain for flow to be unstable, so as to make (4.255.a) satisfied. This amounts to requirement of having inflection points in shear velocity profile $U(z)$, according to well known theory of shear flows.

The second Eq. (4.255.b) establishes necessary but not sufficient conditions, alternatively based on reversibility in sign of either U'' or $N^2 - (U')^2/4$ in the domain.

Vorticity conservation statement in Chap. 1, Eq. (1.104) can be used to explain Eqs. (4.255.a, b):

$$\frac{D\omega}{Dt} = \omega \cdot \nabla\mathbf{u} + \omega\nabla \cdot \mathbf{u} + \frac{1}{\rho^2}(\nabla\rho \times \nabla p) + \nu\nabla^2\omega \tag{4.256}$$

with terms on the right hand side expressing vortex stretching, compressibility, overturning tendencies and viscous effects are ignored in the present case of 2-D, homogeneous, inviscid flows, with remaining material rates of change of basic state $\hat{\zeta} = -dU/dz = -U''$ and perturbation ζ vorticity components in the $x - z$ plane

$$\frac{D\hat{\zeta}}{Dt}^{\;\;0} + \frac{D\zeta}{Dt} = 0,$$

which in linearized form for a progressive wave reads as

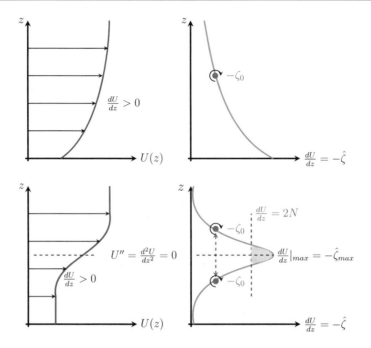

Fig. 4.17 Velocity profile $U(z)$ (left) and corresponding mean shear flow vorticity $\frac{dU}{dz} = -\hat{\zeta}$ respectively with (upper row) monotonous increase and (lower row) with an inflection point in the velocity profile. A perturbed vortex element moves around with the waves in monotonous shear flow (upper row), while a vortex element in a shear flow with inflection does not have to return its original position (lower row), leading to possible instability. In the case of an inflected profile (lower row), there is a chance for instability to grow if $\frac{dU}{dz}|_{max} > \frac{dU}{dz} > 2N$, or equivalently if $R < 1/4$, in the shaded region

$$\left(\frac{\partial}{\partial t} + U(z)\frac{\partial}{\partial x} \right) \zeta = ik[U(z) - c]\zeta = 0.$$

A vortex element displaced in the vertical direction in a monotonous $U(z)$ distribution of shear profile will return to its original position as shown in Fig. 4.17 (upper row), which suggests that the flow is stable. Twins of vortex elements exchanged between positions on two sides of a shear maximum in Fig. 4.17 (lower row) do not have to return their original positions, indicating unstable conditions.

It is a well known principle in basic fluid dynamics (e.g. potential flow theory) that inviscid fluids can not support local vorticity maxima or minima, and therefore $U'' = 0$ is an indicator for unstable motions.

Rayleigh's theorem states that, for a monotonous shear profile considered in Fig. 4.17 (upper row), the flow is always stable, regardless of the sign of basic shear flow vorticity dU/dz, so long as an inflection point does not occur in the flow domain, satisfying either of the following cases:

$$U'' > 0 \quad \text{in} \quad 0 < z < H$$
$$U'' < 0 \quad \text{in} \quad 0 < z < H.$$

$(4.259.a, b)$

The above cases excluding inflection points based on (4.251.a) can only happen if the integral is valued anything but zero, in effect equivalent to the case $c_i = 0$ assuring stability.

A valuable observation that could also be obtained from Fig. 4.17 is the different requirements for instability: (*i*) an inflection point $U'' = 0$ in the shear velocity profile for $U > c_r$ according to (4.255.a), (*ii*) equivalently having zero crossings of $N^2 - (U')^2/4 = 0$ leading to the criteria $R < 1/4$ near the inflection point according to (4.255.b) and Fig. 4.17. Since both these criteria are inconclusive as to being necessary but not sufficient to satisfy (4.255.a), one additional check could be made to see if the maximum shear vorticity exceeds a value twice the stratification parameter $dU/dz|_{max} > 2N$ for the instability to be realized in the region of interest.

In summary, we have strong evidence in favor of stability, if the shear velocity is either monotonous without any inflection, or is weaker than a critical magnitude $dU/dz|_{max} < 2N$, setting sufficient conditions for stability. On the contrary, these same criteria seem to set the necessary but not sufficient conditions for instabilities to occur, subject to certain limits indicated by the Howard semi-circle theorem.

4.5.8   Kelvin-Helmholtz Instability

In Sects. 4.5.7 and 4.5.8, instability of shear velocity in stratified flow has been examined. Perhaps at this stage, one could inquire what next happens as a consequence of instability? In particular, we have pointed out viscous processes, eventually taking over inviscid motion, such as near critical layers exemplified in Sect. 4.5.6.

Consider the relatively simpler case without any change in density across the interface. In this instance represented in Fig. 4.18, setting $\rho_1 = \rho_2$, the problem is equivalent to rolling up of a vortex sheet between two uniform flows of different magnitude $U_1 \neq U_2$. Given an initial disturbance, this sheet of vortices between the two fluids tends to form unstable waves propagating in the direction of currents. When the amplitude of waves grow by the contribution of nonlinear kinematic terms so as to be neglected in this account, wave steepening is followed by the formation of billows, and vortices exchanged between parallel flows creating mixing across the interface, which eventually thickens to become a layer rather than a single interface.

The more general case of discontinuity in velocity and density profiles $U_1 \neq U_2$, $\rho_1 \neq \rho_2$ is represented in Fig. 4.19. As we have seen in the previous section, when internal waves approach a critical layer and Richardson Number decreases below a critical value $R < R_c = 1/4$, interfacial rolls or "billows" are often formed, which then break up to release energy through mixing processes. In this case, both vorticity (or momentum) and mass is exhanged across the interface. In other words, we need to exchange mass particles as well vorticity elements between the layers of fluid,

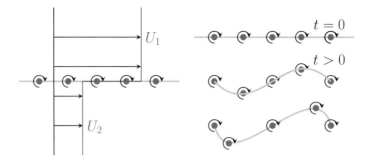

Fig. 4.18 a Two layers with uniform velocity (blue) at initial time $t = 0$ meeting at the interface (green line) creates a delta function of vorticity $-\zeta_\delta = (U_1 - U_2)\delta(z - z_0)$ distributed per unit length at the interface (red dots). **b** At later times $t > 0$, the sharp initial profiles are changed (violet) by Kelvin-Helmholtz instability processes that eventually leading to billows of vorticity distributed across the interface

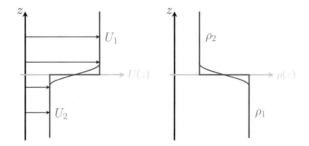

Fig. 4.19 An interface between two layers of **a** uniform velocity (left) and **b** density (right). Uniform velocity U_1, U_2 and density ρ_1, ρ_1 are assigned to the upper and lower layers. The green line indicates the original position of the interface at initial time $t = 0$ when properties are uniform in either layer (blue). At later times $t > 0$, the sharp initial profiles are changed (violet) by Kelvin-Helmholtz instability processes that produce mixing across the interface by collapsing of the growing waves

which then can be characterized as mixing processes. We can apply the concept of *overall Richardson Number* that we had seen in (4.254)

$$R_0 = \frac{g\frac{1}{\rho}\frac{\Delta\rho}{\Delta z}}{\left(\frac{\Delta U}{\Delta z}\right)^2} = g\frac{\Delta\rho}{\rho}\frac{\Delta z}{(\Delta U)^2} = \frac{g'h}{(\Delta U)^2} = \frac{1}{F_\Delta^2} \qquad (4.260)$$

where $\Delta U = U_1 - U_2$, $\Delta\rho = \rho_2 - \rho_1$ and replacing $h \sim \Delta z$ as the thickness of the mixed interface. It is noted that the overall Richardson Number is also related to the *interfacial Froude Number* F_Δ measuring the ratio of gravity forces to those of pressure, where

$$F_\Delta = \frac{\Delta U}{\sqrt{g'h}} \qquad (4.261)$$

and $g' = g\Delta\rho/\rho$ is the reduced gravity.

The following subsections are in support of theory and approximations used in describing Kelvin-Helmholtz type of instabilities.

4.5.8.1 Short Introduction to Potential Flow Theory

Two dimensional stratified shear flow at an interface is solved by making use of potential flow theory, not described earlier in the present GFD text. A short introduction on irrotational fluid dynamics is therefore necessary, as it has not been covered earlier in the present text.

Potential flow theory rests on a description of flow velocity field $\mathbf{u}$ by a the gradient of a scalar potential ϕ

$$\mathbf{u} = \nabla\hat{\phi}. \qquad (4.262)$$

By vector identities introduced in GFD-I, the curl of a gradient identically vanishes

$$\nabla \times \nabla\hat{\phi} = 0 \qquad (4.263)$$

and consequently, potential flow is referred to as being *irrotational*, meaning that it is free from vorticity. The *Potential Flow Theory* can thus be applied to flows where no vorticity is generated; or to cases where a greater part of the fluid is not influenced by vorticity balances expressed by terms on the right hand side of the vorticity equation (4.256).

If in addition to being irrotational, the flow is also incompressible,

$$\nabla \cdot \mathbf{u} = 0, \qquad (4.264)$$

then the set of equations are exceptionally simple. A consequence of the above simplifications is the versatility of the theory to obtain the 3D velocity field by solving the Laplace Equation, which is

$$\nabla^2\hat{\phi} = 0, \qquad (4.265)$$

a result of the preceding Eqs. (4.262)–(4.264).

4.5.8.2 Oscillations of an Interface

For the present analysis of an interface with uniform current U_i and density ρ_i in each layer ($i = 1, 2$), the flow can be assumed irrotational, except at the interface. Velocity potentials for the layers in the $x - z$ plane are

$$\begin{aligned}\hat{\phi}_1 &= U_1 x + \phi_1 \\ \hat{\phi}_2 &= U_2 x + \phi_2.\end{aligned} \qquad (4.266.a, b)$$

It is clear that the proposed solutions would satisfy (4.265) with the perturbations ϕ_1 and ϕ_2 being solutions to the Laplace equation

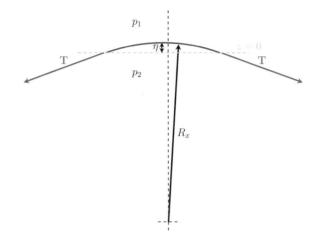

Fig. 4.20 Surface tension force T on a arc of an interface with radius of curvature R_x, separating two fluids of pressure fields p_1 and p_2

$$\nabla^2 \phi_1 = 0,$$
$$\nabla^2 \phi_2 = 0.$$
$$(4.267.a.b)$$

Kinematic boundary conditions are applied at the interface, with perturbed elevation η measured from the undisturbed position at $z = 0$:

$$\frac{\partial \eta}{\partial t} + U_1 \frac{\partial \eta}{\partial x} = w|_{z=0} = \frac{\partial \phi_1}{\partial x}|_{z=0}$$
$$\frac{\partial \eta}{\partial t} + U_2 \frac{\partial \eta}{\partial x} = w|_{z=0} = \frac{\partial \phi_2}{\partial x}|_{z=0}.$$
$$(4.268.a, b)$$

Basic elements of physics are built into these sets of equations for homogeneous fluids. At the interface, pressure matching conditions due to important surface tension effects are also included, accounting for curvature of the interface,

$$p_2 - p_1 = -T \left(\frac{1}{R_x} + \frac{1}{R_y} \right)$$
$$(4.269)$$

where T is the surface tension force applied in the horizontal directions x and y, with R_x and R_y measuring the surface curvature in respective directions, as shown in Fig. 4.20.

Approximating the curvature effects to first order (see shallow water equations developed in GFD-I), Eq. (4.269) becomes

$$p_2 - p_1 \simeq -T \left(\frac{\partial^2 \eta}{\partial x^2} + \frac{\partial^2 \eta}{\partial y^2} \right).$$
$$(4.270)$$

Although three dimensional equations have been used up to the present, it is now sufficient to proceed with 2D flows in the $x - z$ plane oriented with uniform flow. Dynamical boundary conditions have to be applied at the interface. Remembering to apply *Bernoulli equation* for each homogeneous layer of constant uniform velocity and density, a linearized version is developed as follows. Potential Flow Theory is applied to each irrotational fluid layer, by adding a time dependent term $\partial \hat{\phi}/\partial t$ as part of the Bernouilli constant required in steady versions of the same equation reviewed earlier in GFD-I,

$$\frac{\partial \hat{\phi}}{\partial t} + \frac{1}{2}\mathbf{u} \cdot \mathbf{u} + \frac{p}{\rho} + gz = \text{constant} \qquad (4.271)$$

where the quadratic term is defined as

$$\frac{1}{2}\mathbf{u} \cdot \mathbf{u} = \frac{1}{2}\left[\left(\frac{\partial \hat{\phi}}{\partial x}\right)^2 + \left(\frac{\partial \hat{\phi}}{\partial y}\right)^2 + \left(\frac{\partial \hat{\phi}}{\partial z}\right)^2\right]. \qquad (4.272)$$

The only term involving basic state flow component in (4.272) is first expanded as

$$\left(\frac{\partial \hat{\phi}}{\partial x}\right)^2 = \left(U + \frac{\partial \phi}{\partial x}\right)^2 = U^2 + 2U\frac{\partial \phi}{\partial x} + \left(\frac{\partial \phi}{\partial x}\right)^2.$$

The nonlinear perturbation term $u^2 = (\partial \phi/\partial x)^2$ is canceled out in (4.5.8.2), while the constant U^2 can be conveniently incorporated into the Bernouilli constant in (4.271) to produce linearized form of Bernouilli equations

$$\frac{\partial \phi_1}{\partial t} + U\frac{\partial \phi_1}{\partial x} + \frac{p_1}{\rho_1} + gz = 0$$
$$\frac{\partial \phi_2}{\partial t} + U\frac{\partial \phi}{\partial x} + \frac{p_2}{\rho_2} + gz = 0. \qquad (4.274.a, b)$$

With pressure referenced to interface level η relative to the mean position $z = 0$, and combining Eqs. (4.268.a, b), (4.270) and (4.274.a, b) one obtains

$$p_2 - p_1 = \rho_1\left(\frac{\partial \phi_1}{\partial t} + U_1\frac{\partial \phi_1}{\partial x} + g\eta\right) - \rho_2\left(\frac{\partial \phi_2}{\partial t} + U_2\frac{\partial \phi_2}{\partial x} + g\eta\right) = -T\frac{\partial^2 \eta}{\partial x^2}.$$
$$(4.275)$$

Solutions in perturbation variables are selected to be periodic in x and exponentially decaying in z, to satisfy Eqs. (4.275) and (4.267.a, b),

$$\eta = Ae^{i(kx-\omega t)}$$
$$\phi_1 = B_1 e^{-kz} e^{i(kx-\omega t)} \qquad (4.276.a - c)$$
$$\phi_2 = B_2 e^{kz} e^{i(kx-\omega t)}$$

noting that ϕ_1 and ϕ_2 readily satisfy Laplace equations (4.267.a, b) as long as the vertical decay rate and the wave-number is represented by the same value k. In addition the solutions have to be bounded

$$\phi_1 \to 0 \text{ as } z \to +\infty$$
$$\phi_2 \to 0 \text{ as } z \to -\infty$$

(4.277.a.b)

so that substitution in (4.268.a, b) and (4.275) yields

$$i(\omega - kU_2)A = -kB_1$$
$$i(\omega - kU_1)A = -kB_2$$
$$\rho_2[i(kU - \omega)B_2 - gA] - \rho_1[i(kU_1 - \sigma)B_1 - gA] = Tk^2A$$

(4.278.a − c)

from which B_1, B_2 and A are eliminated to give the dispersion relation

$$\rho_2(\omega - kU_2)^2 + \rho_1(\omega - kU_1)^2 = gk(\rho_2 - \rho_1) + Tk^3,$$

(4.279)

and solving for the wave frequency ω and speed $c = \omega/k$ yields

$$c = \frac{\omega}{k} = \frac{\rho_1 U_1 + \rho_2 U_2}{\rho_1 + \rho_2} \pm \left\{ \frac{g}{k} \frac{\rho_2 - \rho_1}{\rho_1 + \rho_2} + \frac{Tk}{\rho_1 + \rho_2} - \frac{\rho_1 \rho_2}{(\rho_1 + \rho_2)^2} (U_2 - U_1)^2 \right\}^{1/2}.$$

(4.280)

By defining

$$c = \frac{\omega}{k}$$
$$\bar{U}_\rho = \frac{(\rho_1 U_1 + \rho_2 U_2)}{(\rho_1 + \rho_2)}$$
$$\bar{\rho} = \frac{\rho_1 + \rho_2}{2}$$
$$\Delta U = U_2 - U_1$$
$$\Delta \rho = \rho_2 - \rho_1$$

(4.281.a − e)

where c is the wave phase speed, $\bar{U}_\rho$ is the density weighted mean velocity, $\bar{\rho}$ is the mean density and ΔU and $\Delta \rho$ are the velocity and density differences. With these definitions inserted, Eq. (4.280) takes the following form

$$c = \frac{\omega}{k} = \bar{U}_\rho \pm \left\{ \frac{1}{2} \left(\frac{g}{k} \frac{\Delta \rho}{\bar{\rho}} + \frac{Tk}{\bar{\rho}} \right) - \frac{\rho_1 \rho_2}{4\bar{\rho}^2} (\Delta U)^2 \right\}^{1/2}.$$

(4.282)

Upon using the shorthand for c_0^2 in (4.282), we see that the wave speed has a Doppler shift component of density weighted mean current velocity U_ρ added as

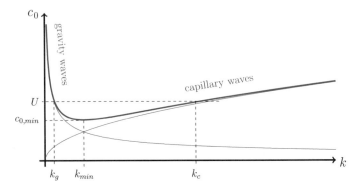

Fig. 4.21 Dispersion relation for gravity-capillary waves on a fluid interface, with terms in (4.283) plotted separately and combined together to give speed c_0 of interfacial waves excluding effects of shear currents, as a function of wave number k

directional phase speed $c_\star$ with corrections to the capillary-gravity wave speed c_0 in the following:

$$c = \bar{U}_\rho \pm \left\{ c_0^2 - \frac{\rho_1 \rho_2}{4\bar{\rho}^2} (\Delta U)^2 \right\}^{1/2}$$

$$= \bar{U}_\rho \pm c_\star$$

(4.283)

where the constant $c_\star$ is appropriately defined.

4.5.8.3 Stability of an Interface Without Current

Equations (4.282) or (4.283) can be applied to different settings. In the absence of currents, the problem would be reduced to the free motion of the interface by gravity and surface tension effects:

$$c_0^2 = \frac{1}{2} \left\{ \frac{g}{k} \frac{\Delta \rho}{\bar{\rho}} + \frac{Tk}{\bar{\rho}} \right\}.$$

(4.284)

The two terms in Eq. (4.284) represent gravity and capillary waves, with each term plotted separately and combined together in Fig. 4.21 to give celerity c_0 of interfacial waves, excluding effects of currents.

Returning to (4.284), which leaves out surface tension, the relationship is reduced to $c_{01} = \sqrt{g\Delta\rho/2k\bar{\rho}}$ characterizing deep water gravity waves in the ocean. On the other hand, if stratification $\Delta\rho/\bar{\rho} \to 0$ is left out, Eq. (4.284) is reduced to capillary wave dispersion relation $c_{02} = \sqrt{Tk/2\bar{\rho}}$ influenced solely by surface tension. The transition of wave speeds from gravity to capillary waves occurs at minimum of the curve as a function of k in Fig. 4.21.

$$k_{min} = \sqrt{\frac{g\Delta\rho}{T}},$$

$$c_{0,min} = \frac{2\sqrt{g\Delta\rho T}}{\bar{\rho}}. \tag{4.285.a, b}$$

With typical values of sea water surface tension $T = 0.25 \times 10^{-3} N/m$, gravity $g = 9.81 \, m/s^2$, $\bar{\rho} = 10^3 \, kg/m^3$, $\Delta\rho = 1 \, kg/m^3$, calculations give following values of minimum wavelength $\lambda_{min} = 2\pi/k_{min} = 0.3 \, cm$ and celerity $c_{0,min} = 1 \, cm/s$, indicating very short waves in the capillary range.

A single layer with a free surface, no current

Without having an upper fluid, setting $\rho_1 = 0$ and $\rho_2 = \rho$ for the fluid density, replacing $\Delta\rho = \rho$, the following dispersion relation for free surface waves is obtained:

$$c_1^2 = \frac{g}{k} + \frac{Tk}{\rho}, \tag{4.286}$$

similarly displayed in Fig. 4.21. With this case, one obtains

$$k_{min} = \sqrt{\frac{g\rho}{T}},$$

$$c_{1,min} = 2\sqrt{\frac{gT}{\rho}}. \tag{4.287.a, b}$$

With typical values of sea water surface tension $T = 0.25 \times 10^{-3} N/m$, gravity $g = 9.81 \, m/s^2$, $\rho = 10^3 \, kg/m^3$, the values $\lambda_{min} = 2\pi/k_{min} = 0.1 \, m$ and $c_{1,min} = 0.5 \, m/s$.

The above cases (4.284) and (4.285.a, b) apply to interface motions in two-layer fluid with surface tension and without currents. The case (4.286) applies to free surface ocean waves where gravity and surface tension play parts. Without surface tension component ($T = 0$), the first term approaches $c_1 \to g/k$, which is the same as the deep water dispersion relation of surface waves.

4.5.8.4 Stationary Waves Relative to Uniform Current

Based on Eq. (4.283) we can study response of the interface under steady conditions with $\omega = 0$, also letting the flow be uniform $U = U_1 = U_2$, and the second term in brackets vanishes as $\Delta U = 0$ so that $c = \omega/k = 0 = U \pm c_0$, allowing only stationary waves $\omega = 0$, $c_0 = U$.

Examining velocity magnitudes below threshold values of $U < c_{0,min}$, it is obvious that no standing gravity-capillary waves can be generated; instead decaying disturbances of steady form can be allowed, such as near a small body placed at the interface.

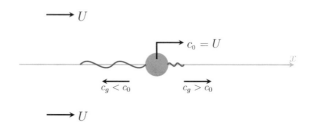

Fig. 4.22 Interfacial waves excluding effects of shear currents

On the other hand, for uniform velocity exceeding the threshold, $U > c_{0,min}$, there appears two roots k_g and k_c for gravity and capillary branches on the two sides of the wave number k_{min} in Fig. 4.21. These two branches of steady (standing) wave solutions can exist near a small object moving with the same speed U at the interface. Since the solutions are steady $\omega = 0$, the requirement of zero phase velocity $c = U \pm c_0 = 0$ implies $c_0 = U > 0$ for positive currents. Although the phase speed is zero in fixed coordinates, c_0 thus appears as the wave speed in the frame relative to the current. Since absolute phase speed is zero $c = 0$, stationary transfers of energy, for instance due to the disturbance created by an object can be investigated further by calculating group velocity c_g calculated relative to the current

$$c_g = \frac{d(kc_0)}{dk} = c_0 + k\frac{dc_0}{dk}. \tag{4.288}$$

Comparing Eq. (4.288) with Fig. 4.21, it becomes clear that the steady solution $c_0 = U$, evaluated at $k = k_g$ gives stationary gravity waves $c_g < U$ propagating energy upstream, and at $k = k_c$ gives stationary capillary waves $c_g > c_0$ propagating energy downstream, relative to the current $U = c_0$, as shown in Fig. 4.22.

4.5.8.5 Layers with Density and Current Differences

Consider the more general case with $U_1 \neq U_2$ and $\rho_2 \neq \rho_1$ where $\rho_2 > \rho_1$, according to (4.283). The flow would be expected to become unstable when

$$\frac{\rho_1\rho_2}{4\bar{\rho}^2}(\Delta U)^2 > c_{0,min}^2 = \frac{4g\Delta\rho T}{\bar{\rho}} \tag{4.289}$$

and replacing $c_{0,min}$ from (4.284) and (4.285.b), the criterion becomes

$$\Delta U > 4\sqrt{\frac{g\bar{\rho}\Delta\rho T}{\rho_1\rho_2}}. \tag{4.290}$$

for two layers of density. Alternatively for a single layer (4.286) and (4.287.b) gives

$$\Delta U >= 4\sqrt{\frac{gT}{\rho}}. \tag{4.291}$$

For either (4.290) and (4.291), typical orders of magnitude of the velocity differences at the interface are estimated as $\Delta U > 10 cm/s$ for incipient instability under the assumed conditions.

4.5.8.6 Case without Surface Tension

Letting surface tension $T \to 0$ in Eq. (4.282), the criterion for instability, i.e. the square root term, $c_\star^2 < 0$ becoming imaginary becomes

$$\frac{1}{2}\frac{\rho_1\rho_2}{(\bar{\rho})^2}(\Delta U)^2 > \frac{g}{k}\frac{\Delta\rho}{\bar{\rho}} \tag{4.292}$$

and letting $\rho_\star = \rho_1\rho_2/\bar{\rho}$ and selecting h as the vertical scale of the problem, criterion of instability is expressed in terms of overall Richardson number as defined in (4.260).

$$R_o = \frac{g\frac{\Delta\rho}{\rho_\star}h}{(\Delta U)^2} < \frac{1}{2}kh. \tag{4.293}$$

This typical form of Richardson Number dependence on the non-dimensional measure kh appears akin to the critical Richardson Number criteria $R_c < 1/4$ that we have reviewed earlier.

4.5.8.7 Gravitational Instability with Negative Buoyancy and Without Current

In earlier sections, denser fluid was assumed to be below the interface, $\rho_2 > \rho_1$ and the case of two-layer fluid without any currents has been found to be unconditionally stable, with phase speed $c_0^2 > 0$, according to (4.284) or (4.286).

On the other hand, with the choice of heavier fluid sitting on top, $\rho_1 > \rho_2$, gravitational instability is almost certain to set in. Re-defining of density difference as $\Delta\rho_\star = \rho_1 - \rho_2 > 0$,

$$c_0^2 = \frac{1}{2}\left\{-\frac{g}{k}\frac{\Delta\rho_\star}{\bar{\rho}} + \frac{Tk}{\bar{\rho}}\right\}. \tag{4.294}$$

Equation (4.292) gives unstable conditions when

$$c_0^2 < 0, \quad k^2 < \frac{g}{T}\Delta\rho_\star \tag{4.295}$$

i.e. for longer waves either when the surface tension T is too small, or negative density difference is too large to create gravitational overturning (Fig. 4.23).

These tests for instability could suggest a teacup experiment, with the cup first filled with hot water and warm tea carefully added on top, surface tension marginally preventing instability, in the case of immiscible fluids.

Fig. 4.23 Teacup
experiment with two layers,
with the denser fluid on top
is restrained by surface
tension at the interface

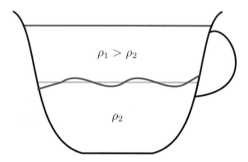

4.5.8.8 Gravitational Instability with Negative Buoyancy and Current

Returning to (4.284) for the case of negative buoyancy, i.e. heavier fluid on top,
$\rho_1 > \rho_2$, the criterion for instability becomes

$$Tk < \frac{\rho_1\rho_2}{2\bar{\rho}}(\Delta U)^2 + \frac{g}{k}\Delta\rho_\star. \tag{4.296}$$

By letting $\rho_\star = \frac{\rho_1\rho_2}{\bar{\rho}}$

$$Tk^2 - \frac{1}{2}\rho_\star(\Delta U)^2 k - \frac{g}{k}\Delta\rho_\star < 0. \tag{4.297}$$

Solving for k and taking the positive root for instability, it can be shown that both
surface tension and shear have joint influence on the interface stability.

$$k < \frac{1}{T}\left(\frac{1}{4}\rho_\star(\Delta U)^2 + \sqrt{\left[\frac{1}{4}\rho_\star(\Delta U)^2\right]^2 + g\rho_\star T}\right). \tag{4.298}$$

Exercises

Exercise 1

Internal waves in a vertically bounded channel
A fluid with constant stratification N is limited in the vertical by boundaries at $z = 0$
and $z = H$, with amplitude

$$w = \hat{w}(z)e^{i(kx-\omega t)}, \tag{4.46}$$

of the vertical motion given by (4.44)–(4.46)

$$\frac{d^2\hat{w}}{dz^2} + \left(\frac{N^2}{\omega^2} - 1\right)k^2\hat{w} = 0,$$

subject to boundary conditions

$$\hat{w}(0) = \hat{w}(H) = 0.$$

Describe the flow by solving the above equations.

Exercise 2

Internal Gravity Waves at a Density Interface
We have seen internal gravity waves in a uniformly stratified medium with constant N. We can also seek solutions of (4.44) or (4.46) with $N = N(z)$ representing variable stratification.

In this simple example consider a fluid with a density interface near a "thermocline" or "pycnocline", typical of the summer stratification in the upper ocean and try to develop internal wave solutions near the interface.

Boundary Layers in Stratified Fluids

5

5.1   A Review of Boundary Layers in Homogeneous and Stratified Fluids

Introductory theory of fluid dynamics applied to homogeneous density fluids reviewed in GFD-I have been extended to stratified ones in the present volume.

At this point, a word of caution once again seems to be in order. The analyses presented so far have utilized various levels of abstraction and approximations at relevant scales of observed motions in atmosphere and ocean. The scope does not necessarily cover entire spectrum of possible flow regimes and configurations subject to multiple scales in temporal and spatial evolution. As reviewed in preface to GFD-I, fluid dynamics, described by few 'simple' rules of mechanics is always prone to yield surprises that supersedes our understanding of earth's climate.

The same overall view expressed in the previous paragraph is also true in the case of non-homogeneous fluids. In earlier chapters alternative uses of quasi-geostrophic and Boussinesq approximations and simplified equilibrium thermodynamics have been used to describe motions extending from planetary to local scales appropriate for particular applications. Often the basic theory have been tuned to scales of interest, applying certain approximations in the governing equations.

Boundary layer regimes can exist everywhere in the fluid environment, also in other areas of physics where similar approximations are utilized. We owe most initial development to engineering sciences, studying thermo-hydrodynamics since the age of early industrial development.

In the present chapter, boundary layer theory is briefly presented, first applied to homogeneous fluids, followed by basic concepts of boundary layers extended to exemplary cases of stratified fluids in the environment. Introduction to thermal/buoyancy boundary layers is made, possibly excluding many lively applications, by making a small foray into a much greater and complicated domain.

© Springer Nature Switzerland AG 2021
E. Özsoy, *Geophysical Fluid Dynamics II*, Springer Textbooks in Earth Sciences, Geography and Environment,
https://doi.org/10.1007/978-3-030-74934-7_5

5.1.1   Boundary Layer Scaling

The flow of a viscous, incompressible fluid near solid boundaries is represented by
3D momentum and continuity equations

$$\frac{D\mathbf{u}}{Dt} = -\frac{1}{\rho}\nabla p + \nu\nabla^2\mathbf{u} + \mathbf{F} \qquad\qquad (5.1.a)$$

$$\nabla \cdot \mathbf{u} = 0 \qquad\qquad (5.1.b)$$

where ν is the kinematic viscosity and $\mathbf{F}$ a body force, e.g. gravity or any conservative force, and defining a modified pressure p, as shown in the GFD-I textbook.
Alternatively, (5.1.a) can be put in vorticity form

$$\frac{D\omega}{Dt} = \omega \cdot \nabla\mathbf{u} + \nu\nabla^2\omega \qquad\qquad (5.2)$$

where

$$\omega = \nabla \times \mathbf{u} \qquad\qquad (5.3)$$

is defined as the vorticity. In (5.1.a) $\nu\nabla^2\mathbf{u}$ is a diffusion term for momentum. In (5.2)
the first term represents vortex stretching and the second term is vorticity diffusion
that derives from the diffusive momentum term in (5.1.a).

For motion near a solid, *no-slip boundary conditions* at the solid boundary are
applied, requiring material points of fluid (particles) initially located on the boundary
stay at their initial positions. If motion is started from rest, shear stresses will start
to build up and vorticity will be generated at the boundary. Vorticity generated at
the boundary will then diffuse into the fluid as indicated by Eq. (5.2), which is very
similar to the thermal diffusion part of Eq. (1.15). The development of motion near
a solid boundary of an object is shown in Fig. 5.1, depending on the viscous effects
diffusing vorticity from the boundaries.

Vorticity will diffuse to a distance L from the body, within a diffusion time scale
of L^2/ν, while fluid particles moving relative to the body with speed U will traverse
a length L along the body in travel time L/U. Considering motion at distance L,
comparable to size of the body, the ratio between time scales of diffusion and motion
is obtained as $\mathrm{Re} = UL/\nu$, the *Reynolds number*.

If vorticity diffusion dominates the system, i.e. slow motion with respect to the
diffusion time scale $\mathrm{Re} \sim 1$, then the region influenced by vorticity generated at
boundaries will be comparable to the size of solid object. In the other extreme,
$\mathrm{Re} \gg 1$, the diffusion region will be confined near boundaries, such that the main
flow will not feel the viscous effects of the solid boundaries.

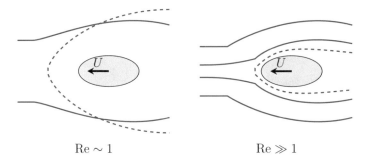

Fig. 5.1 Viscous flow regimes near a solid boundary: (left) viscous and inertial effects are of comparable magnitude (Re ~ 1), (right) boundary layer development near the solid boundary (Re ≫ 1)

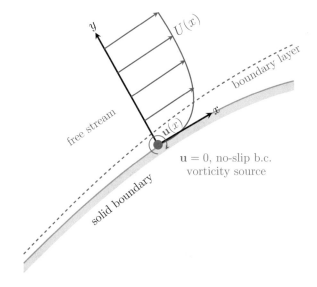

Fig. 5.2 Boundary layer development near a solid boundary, consisting of a free stream velocity $U(x)$ along the solid boundary with a velocity profile $u(x)$ modified within a boundary layer by viscous effects

As the flow velocity U or the Reynolds Number is increased, thickness of the region of influence for vorticity diffusion becomes smaller and less significant for the interior flow. However, this does not mean that viscous effects are unimportant. In this regime, viscous forces are increased and only confined within a *boundary layer* near the solid boundary as shown in Fig. 5.2. This separation in regions of influence has the advantage that boundary layer and interior flow regions can be separately analyzed.

5.1.2 Boundary Layer Equations

For simplicity, 2-D versions of (5.1.a, b) are considered in non-dimensional form

$$\frac{\partial u}{\partial t} + u\frac{\partial u}{\partial x} + v\frac{\partial u}{\partial y} = -\frac{\partial p}{\partial x} + \frac{1}{Re}\left(\frac{\partial^2 u}{\partial x^2} + \frac{\partial^2 u}{\partial y^2}\right)$$

$$1 \quad 1(1) \quad \delta\left(\frac{1}{\delta}\right) \quad 1 \quad \delta^2\left(1 \quad \frac{1}{\delta^2}\right)$$

$$\frac{\partial v}{\partial t} + u\frac{\partial v}{\partial x} + v\frac{\partial v}{\partial y} = -\frac{\partial p}{\partial y} + \frac{1}{Re}\left(\frac{\partial^2 v}{\partial x^2} + \frac{\partial^2 v}{\partial y^2}\right) \qquad (5.4.a-c)$$

$$\delta \quad 1(\delta) \quad \delta\left(\frac{\delta}{\delta}\right) \quad \frac{1}{\delta} \quad \delta^2\left(\delta \quad \frac{\delta}{\delta^2}\right)$$

$$\frac{\partial u}{\partial x} + \frac{\partial v}{\partial y} = 0$$

$$1 \quad \frac{\delta}{\delta}$$

where the orders of magnitude are indicated below each term, using the following scales:

$$u \sim U, \quad v \sim \delta U, \quad p \sim \rho U^2$$
$$x \sim L, \quad y \sim \delta L, \quad t \sim L/U \qquad (5.5)$$

with velocity component u along the boundary and v perpendicular to it, with a small ratio between them $v/u = O(\delta)$. The parallel and perpendicular distances are accordingly scaled as $x \sim L$, $y \sim \delta L$, such that $\delta << 1$, and the y-axis is oriented perpendicular to the boundary.

On the other hand, since the two terms of the continuity equation must be balanced, if we take $u \sim U$, then it must be that $v \sim \delta U$, to equate both terms up to $O(1)$. Having established the scales for these terms, we than look at the momentum equation (5.4.a). Here, all the left hand side terms are $O(1)$. Since the pressure gradient term on the right hand side is one of the main driving forces, it must be $O(1)$, yielding the appropriate pressure scaling. Considering the last two diffusion terms, it is evident that the first of these terms is $O(1)$ while the second one is $O(1/\delta^2) \gg 1$. It is clear that this term should be most significant near the boundary. Based on earlier discussion, it is expected that the viscous forces are reduced far away from the boundary, and in fact they should just balance the inertial forces within the boundary layer. As a result, we set

$$\frac{1}{Re} \sim \delta^2. \qquad (5.6)$$

With these approximations, it becomes evident that all terms of the momentum equation (5.4.b) are $O(\delta)$, balanced by the perpendicular pressure gradient

$$\frac{\partial p}{\partial y} \sim \delta, \tag{5.7}$$

so that by integration of (5.7) the pressure difference Δp across the boundary layer, one obtains

$$\Delta p \sim O(\delta^2). \tag{5.8}$$

This last condition states that the pressure does not change by an appreciable amount across the boundary layer, so that the pressure in interior flow is "impressed" on the boundary layer; i.e. pressure is expected to be the same within and outside the boundary layer. The boundary layer equations can thus be written as

$$\frac{\partial u}{\partial t} + u\frac{\partial u}{\partial x} + v\frac{\partial u}{\partial y} = -\frac{1}{\rho}\frac{\partial p}{\partial x} + \nu\frac{\partial^2 u}{\partial y^2}$$
$$\frac{\partial u}{\partial x} + \frac{\partial v}{\partial y} = 0 \tag{5.9.a, b}$$

For the interior flow, all viscous terms are unimportant, so that excluding these terms from (5.4.a, b), the same equations describe the interior flow. A velocity component $u(x)$ is defined following the boundary, assuming its curvature to be small. The interior momentum equation along the boundary can be written as

$$\frac{\partial U}{\partial t} + U\frac{\partial U}{\partial x} = -\frac{1}{\rho}\frac{\partial p}{\partial x} \tag{5.10}$$

where the pressure p in Eq. (5.9.a) and (5.10) are virtually the same (within $O(\delta^2)$).

In case of steady flow, Eq. (5.10) is further simplified by substituting pressure gradient from this equation into (5.9.a), so that the boundary layer equations take the following form:

$$u\frac{\partial u}{\partial x} + v\frac{\partial u}{\partial y} = U\frac{dU}{dx} + \nu\frac{\partial^2 u}{\partial y^2}$$
$$\frac{\partial}{\partial x} + \frac{\partial v}{\partial y} = 0. \tag{5.11.a, b}$$

Defining a stream function ψ

$$u = \frac{\partial \psi}{\partial y}, \quad v = -\frac{\partial \psi}{\partial x} \tag{5.12}$$

automatically satisfies (5.11.b). By substituting (5.12) into (5.11.a) a single equation in the unknown stream-function ψ is obtained

$$\psi_y\psi_{xy} - \psi_x\psi_{yy} = UU_x + \nu\psi_{yyy} \tag{5.13}$$

which is the vorticity equation (5.2) written in a different way. Vorticity can be obtained from this stream-function, by making use of

$$\omega = \nabla \times \mathbf{u} = \nabla \times (\mathbf{k} \times \nabla \psi) = \mathbf{k} \nabla^2 \psi. \tag{5.14}$$

5.1.3　Similarity Solutions

The driving force in (5.11.a) is due to the external velocity field $U(x)$, and therefore the solution at any point within the boundary layer is expected to depend on the local value of $U(x)$ as well as the coordinates x, y. It is also expected that the boundary layer thickness varies according to initial conditions along x as a function of the external velocity $U(x)$. Earlier considerations have indicated that the non-dimensional layer thickness varies as $\delta \sim 1/\sqrt{\mathrm{Re}}$. With this foresight, one can transform the equations to non-dimensional boundary layer coordinates:

$$\xi = \frac{x}{L} \quad \eta = \frac{y}{L} \frac{\sqrt{\mathrm{Re}}}{g(x)} \tag{15.15.a, b}$$

where L is a length scale and $g(x)$ is yet unknown function linking the x and y coordinates. With the above substitutions and noting that

$$\xi_x = \frac{1}{L}, \qquad \xi_y = 0$$

$$\eta_x = -\frac{y\sqrt{\mathrm{Re}}}{Lg^2} g_x = -\eta \frac{g_x}{g} = -\frac{\eta}{L} \frac{g_\xi}{g}, \qquad \eta_y = \frac{1}{L} \frac{\sqrt{\mathrm{Re}}}{g(\xi)} \tag{5.16.a − d}$$

the terms in (5.14) are calculated as

$$\psi_x = \psi_\xi \xi_x + \psi_\eta \eta_x = \frac{1}{L} \left(\psi_\xi - \eta \frac{g_\xi}{g} \psi_\eta \right)$$

$$\psi_y = \psi_\xi \xi_y + \psi_\eta \eta_y = \frac{\sqrt{\mathrm{Re}}}{Lg} \psi_\eta$$

$$\psi_{xy} = \psi_{x\eta} \eta_y = \frac{\sqrt{\mathrm{Re}}}{Lg} \frac{1}{L} \left(\psi_{\xi\eta} - \eta \frac{g_\xi}{g} \psi_{\eta\eta} - \frac{g_\xi}{g} \psi_\eta \right) \tag{5.17.a − c}$$

$$\psi_{yy} = \left(\frac{\sqrt{\mathrm{Re}}}{Lg} \right)^2 \psi_{\eta\eta}$$

$$\psi_{yyy} = \left(\frac{\sqrt{\mathrm{Re}}}{Lg} \right)^3 \psi_{\eta\eta\eta}$$

Substituting these in (5.14) and noting that $g_x = g_\xi/L$, and $u_x = u_\xi/L$ yields

$$\psi_{\xi\eta} \psi_\eta - \frac{g_\xi}{g} (\psi_\eta)^2 - \psi_\xi \psi_{\eta\eta} = \frac{L^2}{\mathrm{Re}} g^2 U U_\xi + \nu \sqrt{\frac{\mathrm{Re}}{g}} \psi_{\eta\eta\eta}. \tag{5.18}$$

Similarity solutions of the following form can be sought for the above equation, setting

$$\psi = \frac{L}{\sqrt{Re}} f(\eta) U(\xi) g(\xi). \tag{5.19}$$

Note that this is equivalent to requiring

$$u = \frac{\partial \psi}{\partial y} = \frac{\sqrt{Re}}{Lg} \psi_\eta = \frac{L}{\sqrt{Re}} f_\eta U \tag{5.20}$$

i.e. velocity in the boundary layer is proportional to the interior velocity U by a factor f_η, a function of the transformed coordinate η perpendicular to the boundary. Noting also that

$$\psi_\xi = \frac{L}{Re} f(Ug)_\xi$$

$$\psi_{\xi\eta} = \frac{L}{Re} f_\eta(Ug)_\xi$$

$$\psi_\eta = \frac{L}{Re} f_\eta(Ug)$$

$$\psi_{\eta\eta} = \frac{L}{Re} f_{\eta\eta}(Ug)$$

$$\psi_{\eta\eta\eta} = \frac{L}{Re} f_{\eta\eta\eta}(Ug)$$

substitution of (5.19) into (5.18), after cancellation of some terms, yields

$$(f_\eta)^2 U_\xi g - (Ug)_\xi f f_{\eta\eta} = gU_\xi + \nu \frac{Re}{Lg} f_{\eta\eta\eta}. \tag{5.21}$$

Here the Reynolds number $Re = U_* L/\nu$ will be defined with respect to an arbitrary velocity scale U_*, so that (5.21) can be written as

$$f_{\eta\eta\eta} + \underbrace{\left[\frac{L}{U_*} g(Ug)_\xi \right]}_{\alpha} f f_{\eta\eta} + \underbrace{\left[\frac{L}{U_*} g^2 U_\xi \right]}_{\beta} (1 - (f_\eta)^2) = 0 \tag{5.22}$$

For the solution to be *self-similar*, i.e. f to be a function of η only, terms in brackets have to be constant. Assuming a *separable* form of the solution to (5.19), it is time to find out the conditions under which such a solution is valid. Letting α and β respectively to represent the constants corresponding to the brackets in (5.22), the problem is reduced to

$$f_{\eta\eta\eta} + \alpha f f_{\eta\eta} + \beta(1 - f_\eta^2) = 0, \tag{5.23}$$

which is known as the *Falkner–Skan equation*.

Note that by making variable transformations assuming a similarity form of the solution, the two-dimensional problem governed by partial differential equation is reduced into a one-dimensional one governed by an ordinary differential equation (ODE).

Equation (5.23) is a nonlinear ODE, which in general can be solved through numerical techniques to determine $f(\eta)$. Then the remaining function $g(\xi)$ can be solved from either of the following equations:

$$\frac{L}{U_*}g(Ug)_\xi = \alpha$$
$$\frac{L}{U_*}g^2_\xi{}^U = \beta$$

(5.24.a, b)

for given $U(\xi)$, α and β. Note that (5.24.a, b) can be combined into

$$Ugg_\xi = \frac{1}{L}Ugg_x = \frac{U_*}{L}(\alpha - \beta)$$

(5.25)

where $g_\xi = g_x/L$ has been substituted, and by integrating we obtain

$$g^2 = 2(\alpha - \beta)\int_{x_0}^{x}\frac{dx}{(U/U_*)} + g^2(x_0).$$

(5.26)

The solution for $g(x)$ depends on the constant $(\alpha - \beta)$ and the interior velocity U/U_*. On the other hand, Eq. (5.25) requires that $U(x)$ be a simple form, so that the right hand side of (5.25) is always constant. Solutions for simple flow geometries are given in the legendary textbooks of Schlichting [1,2].

5.1.3.1 Example: Boundary Layer on a Flat Plate

Consider a uniform flow $U = U_*$ approaching a flat plate as shown in Fig. 5.3.

From Eq. (5.25) we see that

$$(g^2)_\xi = \text{constant},$$

(5.27)

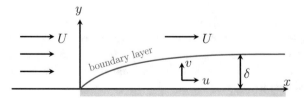

Fig. 5.3 Boundary layer development on a plate by free stream current U, for large Reynolds Number. The external flow is one dimensional, while 2D currents are generated within a boundary layer thickness of δ

and for an initial condition given as $\psi(0) = g(0) = 0$, we have

$$g^2 = \xi = x/L$$

or   (5.28.a, b)

$$g = \sqrt{x/L}$$

which gives the two constants as

$$\alpha = 1/2, \beta = 0 \tag{5.29.a, b}$$

by virtue of (5.24.a, b).

The Falkner–Skan equation (5.23) thus reduces to

$$f_{\eta\eta\eta} + \frac{1}{2} f f_{\eta\eta} = 0 \tag{5.30}$$

where

$$\eta = \frac{y}{L}\sqrt{\mathrm{Re}}\,\frac{1}{g(x)} = \frac{y}{L}\sqrt{\frac{U_* L}{\nu}}\,\frac{1}{\sqrt{x/L}}$$

$$= y\sqrt{\frac{U_*}{\nu x}}. \tag{5.31}$$

From (5.31) it can be seen that the boundary layer thickness (y/L) is scaled as $1/\sqrt{\mathrm{Re}} \sim \delta$. Consequent to (5.19), (5.20), (5.28) and (5.31), we find

$$\psi = \frac{L}{\sqrt{\mathrm{Re}}} f(\eta) U(\xi) g(\xi) = \sqrt{\nu U_* x}\, f(\eta)$$

$$u = \frac{\partial \psi}{\partial y} = \frac{\partial \psi}{\partial \eta}\frac{\partial \eta}{\partial y} = U_* f_\eta(\eta)$$

$$v = -\frac{\partial \psi}{\partial x} = -\left(\frac{\partial \psi}{\partial \xi} + \frac{\partial \psi}{\partial \eta}\frac{\partial \eta}{\partial \xi}\right) = \frac{1}{L}\left(\eta\frac{g_\xi}{g}\psi_\eta - \psi_\xi\right) = \frac{1}{2}\sqrt{\frac{\nu U_*}{x}}(\eta f_\eta(\eta) - f(\eta))$$

$$\tag{5.32.a - c}$$

The solutions $f(\eta)$ and $\eta f_\eta(\eta) - f(\eta)$ are found by solving (5.30) numerically, subject to the boundary conditions

$$\begin{aligned} f(0) &= 0 \\ f_\eta(0) &= 0 \\ f_\eta(\infty) &= 1 \end{aligned} \tag{5.33.a - c}$$

corresponding to the dimensional counterparts

$$\begin{aligned} u = v &= 0 \ \text{on} \ y = 0 \\ u &= U_* \ \text{as} \ y \to \infty. \end{aligned} \tag{5.34.a - b}$$

The solutions $U/U_* = f_\eta(\eta)$ and v as a function of η are illustrated in Fig. 5.4.

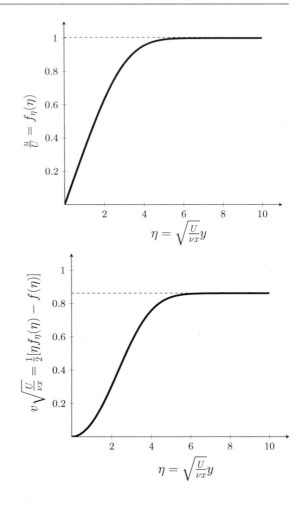

Fig. 5.4 Blasius solution for Falkner–Skan equations, non-dimensional variations of (top) along stream velocity u and (bottom) vertical velocity v across the boundary layer

5.2 Thermal Boundary Layers

5.2.1 Equations for Inhomogeneous Boundary Layers

To illustrate the buoyancy boundary layers, it is necessary to consider buoyancy forces resulting from thermal conduction between the fluid and a solid boundary.

It is assumed that the fluid is incompressible but inhomogeneous, so that the density is allowed to vary as a function of time and 3D space ($\rho = \rho(x, y, z, t)$). In addition to the momentum and continuity equations, thermodynamics is needed, represented by Eq. (1.15) developed in Chap. 1,

$$\rho c_p \frac{DT}{Dt} = \nabla \cdot K \nabla T + \rho (Q + \Phi) \tag{5.35}$$

and complement the system of equations with an *equation of state*

$$\rho = \rho(p, T). \tag{5.36}$$

In case of sea-water, salinity also enters (5.36), modified by salt. Introducing salinity as an additional unknown necessitates a conservation equation similar to (5.35) for salt.

In the following, however, the influence of salinity on density is ignored for the moment, working with the closed system of Eqs. (5.1.a, b), (5.35) and (5.36) to solve for unknowns $\mathbf{u}$, p, ρ, and T.

A linearized version of the equation of state (5.37), working around a mean temperature and density of T_0 and ρ_0 respectively is used

$$\rho = \rho_0 \left[1 - \alpha(T - T_0)\right] \tag{5.37}$$

where α is the thermal expansion coefficient to represent density change as a function of temperature.

In GFD-I the body force $\mathbf{F} = \rho_0 g$ has been introduced as a conservative force. In (5.1.a) however, the density is variable and therefore the total gravity force is no longer conservative, and as a result, the body force is written as

$$\rho \mathbf{F} = \rho \mathbf{g} = \rho_0 \mathbf{g} + (\rho - \rho_0)\mathbf{g} \tag{5.38}$$

and incorporate the constant first term $\rho_0 \mathbf{g}$ in the modified pressure, so that Eq. (5.1.a) becomes

$$\frac{D\mathbf{u}}{Dt} = -\frac{1}{\rho_0}\nabla p + \nu\nabla^2\mathbf{u} + \left(\frac{\rho - \rho_0}{\rho_0}\right)\mathbf{g} \tag{5.39}$$

where p is the modified pressure. On the other hand last term of (5.39) by virtue of (5.37) is expressed as

$$\frac{\rho - \rho_0}{\rho_0}\mathbf{g} = -\alpha(T - T_0)\mathbf{g} = -\alpha\theta\mathbf{g} \tag{5.40}$$

and where $\theta = T - T_0$ and T_0 is a constant value of temperature. Inserting $T = \theta + T_0$ into (5.35), also neglecting internal and frictional heat sources, letting $Q = \Phi = 0$, the governing hydro and thermodynamic equations become

$$\frac{D\mathbf{u}}{Dt} = -\frac{1}{\rho_0}\nabla p + \nu\nabla^2\mathbf{u} - \alpha\mathbf{g}\theta$$

$$\frac{D\theta}{Dt} = \nabla \cdot k\nabla\theta \tag{5.14.a - c}$$

$$\nabla \cdot \mathbf{u} = 0,$$

where $k = K/\rho c_p$ is the thermal diffusivity, K is the thermal conductivity. Note that Eqs. (5.41.a, b) are coupled because of buoyancy forces.

Fig. 5.5 Thermal boundary
layer development near a
slanted wall, influenced by
the difference of temperature
between the wall T_w and that
of the fluid T_0

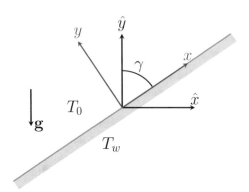

Next, consider the flow adjacent to a wall, through which there is some thermal conduction of heat, with the wall slanted with respect to the vertical plane with an angle γ, as shown in Fig. 5.5.

The coordinates $\hat{x}$, $\hat{y}$ are such that $\hat{y}$ is the vertical coordinate and therefore the gravity vector $\mathbf{g} = -g\hat{\mathbf{j}}$ pointing in the negative $\hat{y}$ direction. We denote x, y as the coordinates fixed in the boundary with x pointing along the boundary. The interior temperature T_0 is assumed to be different from the wall temperature T_w, so that there will be a diffusion of heat between the two media. As the heat diffuses into the water from the wall (or vice versa), the buoyancy forces $(-\mathbf{g}\alpha\theta)$ will generate motions in the fluid. Closer to the wall, the fluid will be more influenced by buoyancy, and therefore, the current will increase near the wall. However, if friction forces are to be important, then we must apply the no-slip boundary condition at the wall which means that the flow has to vanish at the wall, i.e. the velocity has a maximum at some intermediate distance from the wall. On the other hand, the diffused heat will be swept away by this flow and therefore will limit the influence region of thermal diffusion from the wall. Interaction between thermal and momentum diffusion fluxes will lead to boundary layer structure.

Transformation between (x, y) and $(\hat{x}, \hat{y})$ coordinates is given by

$$x = \hat{x}\sin\gamma + \hat{y}\cos\gamma$$
$$y = -\hat{x}\cos\gamma + \hat{y}\sin\gamma. \tag{5.42.a, b}$$

Components of $\mathbf{g}$ in (x, y) coordinates are

$$\mathbf{g} = -g\cos\gamma\,\mathbf{i} - g\sin\gamma\,\mathbf{j}. \tag{5.43}$$

Next, assuming the flow to be steady $(\partial/\partial t = 0)$, the following scaling is applied

$$x \sim L\,,\ y \sim \delta L\,,\ u \sim U\,,\ v \sim \delta U$$
$$p \sim \rho_0 U^2\,,\ \theta \sim \theta_0. \tag{5.44}$$

The scaled equations (5.41.a–c) are obtained as

$$u\frac{\partial u}{\partial x} + v\frac{\partial u}{\partial y} = -\frac{\partial p}{\partial x} + \left(\frac{\nu}{UL}\right)\left(\frac{\partial^2 u}{\partial x^2} + \frac{1}{\delta^2}\frac{\partial^2 u}{\partial y^2}\right) + \left(\frac{\alpha\theta_0 gL}{U^2}\right)\theta\cos\gamma$$

$$\delta^2\left(u\frac{\partial v}{\partial x} + v\frac{\partial v}{\partial y}\right) = -\frac{\partial p}{\partial y} + \delta^2\left(\frac{\nu}{UL}\right)\left(\frac{\partial^2 v}{\partial x^2} + \frac{1}{\delta^2}\frac{\partial^2 v}{\partial y^2}\right)\delta\left(\frac{\alpha\theta_0 gL}{U^2}\right)\theta\sin\gamma$$

$$u\frac{\partial\theta}{\partial x} + v\frac{\partial\theta}{\partial y} = \left(\frac{k}{UL}\right)\left(\frac{\partial^2}{\partial x^2} + \frac{1}{\delta^2}\frac{\partial^2\theta}{\partial y^2}\right)$$

$$\frac{\partial u}{\partial x} + \frac{\partial v}{\partial y} = 0.$$

$$(5.45.a-d)$$

Definition of dimensionless parameters

$$\text{Grasshoff \# : Gr} = \frac{\alpha\theta_0 gL^3}{\nu^2} = \text{Ratio}\left(\frac{\text{buoyancyforces}}{\text{viscousforces}}\right)$$

$$\text{Reynolds \# : Re} = \frac{UL}{\nu} = \text{Ratio}\left(\frac{\text{buoyancyforces}}{\text{viscousforces}}\right) \qquad (5.46.a-c)$$

$$\text{Prandtl \# : Pr} = \frac{\nu}{K} = \text{Ratio}\left(\frac{\text{momentum diffusion}}{\text{thermal diffusion}}\right)$$

allows to re-write (5.45.a–d) as

$$u\frac{\partial u}{\partial x} + v\frac{\partial u}{\partial y} = -\frac{\partial p}{\partial x} + \frac{1}{\text{Re}}\left(\frac{\partial^2 u}{\partial x^2} + \frac{1}{\delta^2}\frac{\partial^2 u}{\partial y^2}\right) + \frac{\text{Gr}}{\text{Re}^2}\theta\cos\gamma$$

$$\delta^2\left(u\frac{\partial v}{\partial x} + v\frac{\partial v}{\partial y}\right) = -\frac{\partial p}{\partial y} + \frac{\delta^2}{\text{Re}}\left(\frac{\partial^2 v}{\partial x^2} + \frac{1}{\delta^2}\frac{\partial^2 v}{\partial y^2}\right) + \frac{\delta\text{Gr}}{\text{Re}^2}\theta\sin\gamma$$

$$u\frac{\partial\theta}{\partial x} + v\frac{\partial\theta}{\partial y} = \frac{1}{\text{Pr Re}}\left(\frac{\partial^2\theta}{\partial x^2} + \frac{1}{\delta^2}\frac{\partial^2\theta}{\partial y^2}\right)$$

$$\frac{\partial u}{\partial x} + \frac{\partial v}{\partial y} = 0$$

$$(5.47.a-d)$$

The solution depends on the magnitudes of dimensionless parameters δ, Pr, Gr, Re determined by relations between them.

Note that we have used θ_0 in the definition of *Grasshoff Number*. An appropriate definition of θ_0 on the other hand is

$$\theta_0 = T_w - T_0 \qquad (5.48)$$

i.e. the temperature difference between the wall and the interior fluid.

The thin boundary layer structure of the flow with ($\delta \ll 1$) is determined by the various dimensionless numbers.

Since viscous terms are important in (5.47.a), we let

$$\frac{1}{\text{Re}} \sim O(\delta^2). \tag{5.49}$$

In this way, the $O(1)$ third term is incorporated on the right hand side of (5.47.a), in order to account for friction. As buoyancy is the driving force for the boundary layer flow, the last term of (5.47.a) has to be $O(1)$, leading to the estimate

$$\text{Gr} \sim O(\text{Re}^2), \tag{5.50.a}$$

based on the fact that

$$\text{Gr}/\text{Re}^2 = \alpha(T_w - T_0)gL/U^2 = O(1). \tag{5.50.b}$$

With the above assumptions made in (5.47.b), the balance of forces in the boundary layer dictates that

$$\frac{\partial p}{\partial y} \sim O(\delta) \tag{5.51}$$

at most, with a note that, in the absence of buoyancy this would only be up to $O(\delta^2)$ in homogeneous boundary layers considered earlier. Once again the pressure difference across the boundary layer Δp is found to be small, $(\Delta p \sim O(\delta))$, leading to the conclusion that the external pressure field is impressed on the boundary layer.

A first look at Eq. (5.47.c) shows that the last term is much larger than the first term on the right hand side, in order to balance terms on the left hand side, if diffusion and convection were both to be important. In fact, letting $1/\text{Re}^2 \sim O(\delta^2)$, this term balances the left hand side, suggesting that $\text{Pr} \sim O(1)$.

The Prandtl number $\text{Pr} = \nu/k$ is the ratio of the diffusivity constants for momentum and heat. For water and air in the laminar and turbulent regimes typically $\text{Pr} = O(1)$. In the laminar regime it has been found that $\text{Pr} = 0.73$ for air and water at $15\,^\circ\text{C}$. In the turbulent regime, corresponding values of Prandtl Number have often been given as $\text{Pr} \simeq 0.5 - 1$, based on selected values of turbulent diffusivity for water and air. Cases of $\text{Pr} < 1$ and $\text{Pr} > 1$ could also occur depending on applications, as illustrated in Fig. 5.6.

In fact, the Prandtl number can be related the ratio of momentum and thermal boundary thickness

$$\text{Pr} = \frac{\nu}{k} = \frac{\frac{UL}{K}}{\frac{UL}{\nu}} \sim \frac{\frac{1}{\delta_T^2}}{\frac{1}{\delta_M^2}} = \left(\frac{\delta_M}{\delta_T}\right)^2 \tag{5.52}$$

where δ_M and δ_T are the momentum and thermal boundary layer thicknesses respectively. In the limits $\delta_M > \delta_T$ or $\delta_T > \delta_M$ (either boundary layer is much thicker than the other, profiles of temperature and velocity will look like those illustrated

Fig. 5.6 The ratio $(\delta_M/\delta_T)^2$ between thickness of the momentum and boundary layers varies as a function of Prandtl Number Pr, demonstrated by cases of (upper) Pr < 1 and (lower) Pr > 1

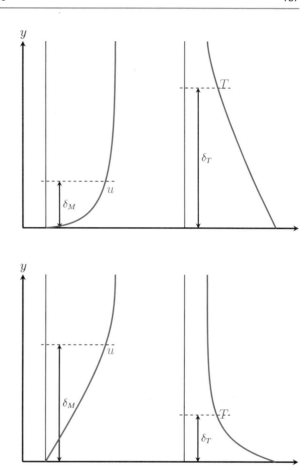

in Fig. 5.6. For Pr $\sim O(1)$, however, both boundary layers would have comparable thickness.

With above simplifications applied to boundary layer, we reduce Eqs. (5.47.a–d) in dimensional form as follows:

$$u\frac{\partial u}{\partial x} + v\frac{\partial u}{\partial y} = -\frac{1}{\rho_0}\frac{\partial p}{\partial x} + \nu\frac{\partial^2 u}{\partial y^2} + \alpha g\theta\cos\gamma$$

$$u\frac{\partial \theta}{\partial x} + v\frac{\partial \theta}{\partial y} = k\frac{\partial^2 \theta}{\partial y^2} \qquad (5.53.a-c)$$

$$\frac{\partial u}{\partial x} + \frac{\partial v}{\partial y} = 0.$$

By virtue of the assumption that pressure is applied uniformly across the boundary layer and the interior flow, followed by the assumption that buoyancy and friction are negligible in the interior flow, it is observed that

$$U \frac{\partial U}{\partial x} = -\frac{1}{\rho_0} \frac{\partial p}{\partial x} \qquad (5.54)$$

for the interior flow, with a velocity component U applied along the boundary layer coordinate x.

5.2.2   Free Convection—Homogeneous Interior Fluid

We consider the case of *free convection*, i.e. when there is no flow in the interior region, ($U = 0$). By virtue of (5.54), the pressure gradient in (5.53.a) can be dropped, letting $\partial p / \partial x = 0$. In this case, the flow is driven by heat conduction from the wall. Furthermore, it can be assumed that the interior flow (exterior of the boundary layer) is homogeneous, letting $\rho = \rho_0 = $ constant, $T = T_0 = $ constant. The configurations of heating or cooling boundaries are shown in Fig. 5.7.

Appropriate normalization of temperature is made by defining $\hat{\theta}$

$$\hat{\theta} = \frac{\theta}{\theta_0} = \frac{T - T_0}{T_w - T_0} \qquad (5.55)$$

Fig. 5.7 Thermal boundary layers on slanted boundaries in the cases of (upper figure) heating $T_w > T_0$ and (lower figure) cooling $T_w < T_0$

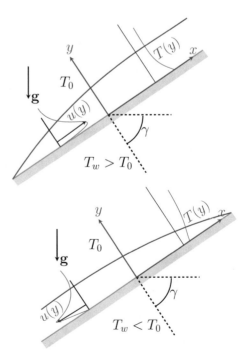

defined to stay within the range are

$$\hat{\theta} = 1 \text{ on } y = 0$$
$$\hat{\theta} = 0 \text{ as } y \to \infty.$$

$$(5.56.a, b)$$

Equation (5.53.a–c) is re-written as

$$u\frac{\partial u}{\partial x} + v\frac{\partial u}{\partial y} = -\frac{1}{\rho_0}\frac{\partial p}{\partial x} + \nu\frac{\partial^2 u}{\partial y^2} + [\alpha g(T_w - T_0)\cos\gamma]\hat{\theta}$$
$$u\frac{\partial\hat{\theta}}{\partial x} + v\frac{\partial\hat{\theta}}{\partial y} = k\frac{\partial^2\hat{\theta}}{\partial y^2}$$
$$\frac{\partial u}{\partial x} + \frac{\partial v}{\partial y} = 0$$

$$(5.57.a - c)$$

Similarity solutions can be obtained for the above problem if the coordinates are transformed as

$$\xi = x, \quad \eta = c\frac{y}{x^{1/4}}$$

$$(5.58)$$

and assume a stream function satisfying (5.57.c) i.e.

$$u = \frac{\partial\Psi}{\partial y}, \quad v = -\frac{\partial\Psi}{\partial x}.$$

$$(5.59.a, b)$$

The solution is given as

$$\Psi = 4\nu c x^{3/4} f(\eta)$$

$$(5.60.a)$$

where

$$c = \left(\frac{\alpha g(T_w - T_0)\cos\gamma}{4\nu^2}\right)^{1/4}.$$

$$(5.60.b)$$

The velocity components are calculated from (5.59.a, b) and (5.60.a, b)

$$u = 4\nu c^2 x^{1/2} f_\eta(\eta)$$
$$v = \nu c x^{-1/4}\{\eta f_\eta(\eta) - 3f(\eta)\}.$$

$$(5.61.a, b)$$

Substituting these transformations into Eq. (5.57.a–c) yields

$$f_{\eta\eta\eta} + 3ff_{\eta\eta} - 2\left(f_\eta\right)^2 + \hat{\theta} = 0$$
$$\hat{\theta}_{\eta\eta} + 3\Pr\nu\hat{\theta}_\eta = 0$$

$$(5.62.a, b)$$

where $\Pr = \nu/k$ is the Prandtl number. The boundary conditions to be applied are

$$u = v = 0 \text{ at } y = 0$$
$$u = 0 \text{ as } y \to \infty$$

requiring no-slip conditions at the wall and no motion in the interior region. The two other boundary conditions for temperature were written as part of (5.56.a, b), and combining, the boundary conditions are obtained in transformed variables

$$f(0) = 0$$
$$f_\eta(0) = 0$$
$$\hat{\theta}(0) = 1 \qquad\qquad (5.63.a - e)$$
$$f_\eta(\infty) = 0$$
$$\hat{\theta}(\infty) = 0.$$

The coupled ordinary differential equations (5.62.a, b) together with boundary conditions (5.63.a–e) have been solved by numerical methods, as a function of the Prandtl number Pr by Pohlhausen–Ostrach [3].

Note that by virtue of (5.58) and (5.65.b)

$$y = \frac{x^{1/4}}{c}\eta = \left(\frac{4\nu^2 x}{\alpha g(T_w - T_0)\cos\gamma}\right)^{1/4}\eta$$

where $\eta = O(1)$, so that the boundary layer thickness is roughly proportional to

$$\delta \sim \left(\frac{4\nu^2 x}{\alpha g(T_w - T_0)\cos\gamma}\right)^{1/4}. \qquad (5.64)$$

The solutions for the velocities u and v are illustrated in Fig. 5.8. Solution depends on wall slope through the coefficient c in (5.60.b). For a vertical wall $\gamma = 0$, c reaches a maximum value, which leads to the maximum boundary layer flow u.

The above solutions are actually only valid when the fluid is statically stable. For example the flow is confined to the underside of a heated sloping boundary (or upper surface of a cooled boundary) the flow is more stable as compared to the case when it is on the upper side of a heated (or lower side of a cooled) boundary, as shown in Fig. 5.9.

For heated boundary with flow on the upper side the flow begins to be unstable for $\gamma > 14°$.

5.2.3   Free Convection with Stratified Interior (Buoyancy Layer)

We next consider the case in which the interior temperature (or density) is not constant. We assume the interior fluid in static equilibrium, i.e. $T_0 = T_0(\hat{y})$, and for simplicity we consider the case

$$T_0 = T_* + s\hat{y} \qquad (5.65)$$

Fig. 5.8 Pohlhausen–Ostrach solutions for thermal boundary layers on slanted surfaces

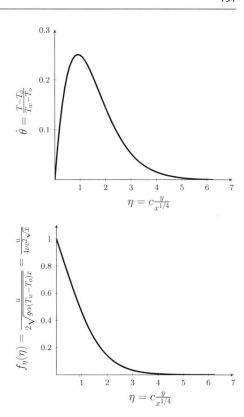

Fig. 5.9 Configurations for slanted boundary layers adjacent to up and down looking boundaries with heating or cooling

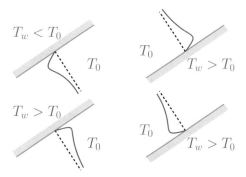

where T_* is a reference temperature and s is the temperature gradient in the vertical direction within the interior of the fluid adjacent to the slanted boundary, as demonstrated in Fig. 5.10.

In this case, Eqs. (5.53.a–c) describe the flow, except that we can not assume T_0=constant as we have done in writing Eqs. (5.53.a–c) by substituting $\theta = T - T_0$.

Fig. 5.10 Buoyancy
boundary layer configuration
with interior stratification
next to a slanted boundary

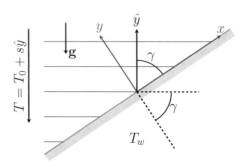

In fact, letting $T_0 = T_0(\hat{y})$ the following equations are obtained:

$$u\frac{\partial u}{\partial x} + v\frac{\partial u}{\partial y} = -\frac{1}{\rho_0}\frac{\partial p}{\partial x} + \nu\frac{\partial^2 u}{\partial y^2} + \alpha g \cos\gamma(T - T_0)$$

$$u\frac{\partial T}{\partial x} + v\frac{\partial T}{\partial y} = k\frac{\partial^2 T}{\partial y^2} \qquad (5.66.a-c)$$

$$\frac{\partial u}{\partial x} + \frac{\partial v}{\partial y} = 0$$

Extra simplifications are needed in order to solve (5.66.a–c). Firstly, free convection is assumed, i.e. no flow in the interior ($\partial p/\partial x = 0$). Furthermore, it is assumed that the flow is uniform in the x-direction i.e. $\partial u/\partial x = 0$, which requires that $\partial v/\partial y = 0$. But since no-slip conditions are applied $v(y = 0) = 0$, it is required that $v = 0$ everywhere. As a result, Eq. (5.66.a–c) is simplified into the following form:

$$0 = \nu\frac{\partial^2 u}{\partial y^2} + \alpha g \cos\gamma(T - T_0)$$

$$u\frac{\partial T}{\partial x} = k\frac{\partial^2 T}{\partial y^2} \qquad (5.67.a, b)$$

Defining the temperature anomaly in the boundary layer with respect to the fluid interior (exterior of the boundary layer)

$$\theta = T - T_0 \qquad (5.68)$$

where temperature T_0 as defined in (5.65) and (5.68) gives

$$T = \theta + T_0$$
$$= \theta + T_* + s\hat{y}, \qquad (5.69)$$

which allows to calculate the temperature gradient along the inclined boundary as

$$\frac{\partial T}{\partial x} = \overset{\sim\,0}{\cancel{\frac{\partial\theta}{\partial x}}} + \frac{\partial T_0}{\partial x} \simeq \frac{\partial T_0}{\partial x}. \qquad (5.70)$$

Dropping the smaller first term stipulates the approximation that interior temperature gradient is greater than the temperature anomaly gradient within the boundary layer i.e. ($\theta \ll T_0$). Therefore (5.69), (5.70), the following can be obtained

$$\frac{\partial T}{\partial x} \cong \frac{\partial T_0}{\partial x} = s\frac{\partial \hat{y}}{\partial x} = s\cos\gamma, \tag{5.71}$$

also noting by virtue of (5.69),

$$\frac{\partial^2 T}{\partial y^2} = \frac{\partial^2 \theta}{\partial y^2} + \cancel{\frac{\partial^2 T_0}{\partial y^2}}^{\;0}, \tag{5.72}$$

since $T_0 = T_* + s\hat{y}$ is a linear function in $\hat{y}$, and therefore in y. Finally, Eqs. (5.67) become

$$\nu\frac{\partial^2 u}{\partial y^2} = -(\alpha\cos\gamma)\theta$$
$$k\frac{\partial^2 \theta}{\partial y^2} = (s\cos\gamma)u. \tag{5.73.a, b}$$

In passing, it is noteworthy to recognize the similarity of Eqs. (5.73.a, b) to those developed in the context of Ekman boundary layer dynamics described in the GFD-I book, repeated here for comparison:

$$-fv = \nu\frac{\partial^2 u}{\partial y^2}$$
$$fu = \nu\frac{\partial^2 v}{\partial y^2}. \tag{5.74.a, b}$$

As regards the present problem, Brunt–Väisälä frequency N^2 for the interior fluid is calculated from the equation of state (5.37), i.e. $\rho_0 = \rho_*[1 - \alpha(T_0 - T_*)]$,

$$N^2 = -\frac{g}{\rho_*}\frac{\partial \rho_0}{\partial \hat{y}} = g\alpha\frac{\partial T_0}{\partial \hat{y}}. \tag{5.75}$$

By substituting (5.75) into (5.73.a, b) and re-defining *reduced gravity* or *buoyancy* by

$$\sigma = -\alpha g\theta = g\frac{\rho - \rho_0}{\rho_*} \tag{5.76}$$

one obtains

$$\nu\frac{\partial^2 u}{\partial y^2} = \cos\gamma\,\sigma$$
$$k\frac{\partial^2 \sigma}{\partial y^2} = -N^2\cos\gamma\,u. \tag{5.77.a, b}$$

The boundary conditions to be applied are

$$u(0) = 0$$
$$\sigma(0) = \sigma_w = -\alpha g(T_w - T_0)$$
$$u(\infty) = 0 \qquad\qquad\qquad (5.78.a - d)$$
$$\sigma(\infty) = 0.$$

Eliminating σ between (5.77.a, b) yields

$$\frac{\partial^4 u}{\partial y^4} + \frac{N^2 \cos^2 \gamma}{\nu k} u = 0 \qquad\qquad (5.79.a)$$

or

$$\frac{\partial^4 u}{\partial y^4} + \left(\frac{4}{l^4}\right) u = 0, \qquad\qquad (5.79.b)$$

where $l^4 = 4\nu k/(N^2 \cos^2 \gamma)$.

The roots of characteristic equation representing (5.79.b) are obtained as

$$m = \left(-\frac{4}{\rho^4}\right)^{1/4} = \frac{\sqrt{2}}{\rho} \left[e^{i(\pi+2n\pi)}\right]^{1/4}$$
$$= \frac{\sqrt{2}}{\rho} e^{i\left(\frac{\pi}{4} + n\frac{\pi}{2}\right)}$$
$$\qquad\qquad\qquad\qquad (5.81)$$
$$= \frac{\sqrt{2}}{\rho} \left\{e^{+i\frac{\pi}{4}}, e^{-i\frac{\pi}{4}}, e^{i\frac{5\pi}{4}}, e^{-i\frac{5\pi}{4}}\right\}$$
$$= \pm\frac{(1 \pm i)}{l}$$

so that the solution for u, $sigma$ are superpositions of

$$u, \sigma \sim e^{\pm(1\pm i)y/l}$$
$$\sim e^{\pm(1\pm i)\xi} \qquad\qquad (5.82)$$

where $\xi = y/l$ is the non-dimensional boundary layer vertical coordinate.

By virtue of boundary conditions (5.78.c, d) solutions $e^{(1+i)\xi}$ are to be rejected, since these terms grow as $\xi \to \infty$. The remaining terms result in the solutions

$$u = Re\left\{U_1 e^{-(1+i)\xi} + U_2 e^{(1-i)\xi}\right\}$$
$$\qquad\qquad\qquad\qquad (5.83.a, b)$$
$$\sigma = Re\left\{\sigma_1 e^{(1+i)\xi} + \sigma_2 e^{-(1-i)\xi}\right\}$$

By virtue of (5.78.a, b) the solutions become

$$u = Ae^{-\xi} \sin \xi$$
$$\qquad\qquad\qquad (5.84.a, b)$$
$$\sigma = \sigma_w e^{-\xi} \cos \xi$$

where the constant A is obtained through either of (5.77.a, b), yielding

$$A = \frac{\sigma_w}{N}\sqrt{\frac{k}{\nu}} = \frac{\sigma_w}{N\sqrt{\text{Pr}}} = \frac{N}{s\sqrt{\text{Pr}}}(T_0 - T_w) \tag{5.85}$$

through the use of (5.75) and (5.78.b). It is readily noted that

$$y = l\xi = \left(\frac{4\nu k}{N^2 \cos^2\gamma}\right)^{1/4}\xi \tag{5.86}$$

where $\xi = O(1)$, so that the boundary layer thickness is expected to be given as

$$\delta \sim \left(\frac{4\nu k}{N^2 \cos\gamma}\right)^{1/4} = \left(\frac{4\nu k}{\alpha g s \cos^2\gamma}\right)^{1/4}. \tag{5.87}$$

This result can be compared with Sect. 5.2.2, Eq. (5.64.9). It can be noted by virtue of (5.65) and (5.78.b) that

$$T_w = T_0 - \frac{\sigma w}{\alpha g} = T_* + s\hat{y} - \frac{\sigma w}{\alpha g}$$

$$= \left(T_* - \frac{\sigma w}{\alpha g}\right) + (s \cos\gamma)x. \tag{5.88}$$

Since the reduced gravity is already assumed to be $\sigma_w = $ constant in the solution, it is imperative to observe that there is a fixed temperature difference between the fluid and the boundary at every vertical level, i.e., the wall temperature being variable along the x-direction of the boundary layer. This is actually the requirement for this type of a solution to exist i.e. with uniform flow $\partial u/\partial x = 0$. By virtue of (5.76), the density is calculated from

$$\rho = \left(\frac{\rho_*}{g}\right)\sigma + \rho_0$$

$$= \frac{\rho_*}{g}\sigma_w e^{-y/l}\cos\left(\frac{y}{l}\right) + \rho_*\left[1 - \alpha(T_* + s\hat{y} - T_*)\right] \tag{5.89}$$

$$= \frac{\rho_*}{g}\sigma_w e^{-\xi}\cos\xi + \rho_*\left[1 - \alpha s\hat{y}\right]$$

Also noting that

$$y = \hat{x}\cos\gamma + \hat{y}\sin\gamma \tag{5.90}$$

we can plot lines of constant ρ, (i.e. $\hat{y}$ against $\hat{x}$), as shown in Fig. 5.11. The constant density lines essentially make a curvature near the boundary. It has also been shown that, even if there is no net flux of heat across the boundary i.e. $\partial T/\partial y = 0$, there

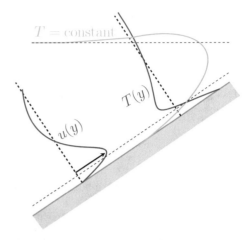

Fig. 5.11 Velocity and temperature profiles in a buoyancy boundary layer adjacent to a slanted boundary next to a basin with interior thermal stratification. The green line demonstrates deflection of isotherms near the boundary

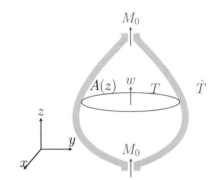

Fig. 5.12 Thermal diffusion and boundary layer convection in a stratified container

will still be a flow generated near the boundary. In this case, the density contours are normal to the slope and therefore the fluid against the wall is always lighter then the interior fluid at the same level, generating an up-slope current in the boundary layer (Turner, 1973, p. 245) [4].

5.2.4 Convection in Stratified Containers

Convection in a container can be created by external thermal forcing, supported by buoyancy boundary layers and through flow as shown in Fig. 5.12.

A total mass flux M_0 through the fluid is allowed, with corresponding interior flow component

$$M_0 = \int_A w \, dA \tag{5.91}$$

i.e., the vertical velocity w integrated over the variable cross sectional area $A(z)$ of the container.

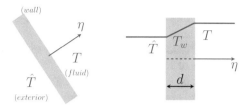

Fig. 5.13 Heat diffusion through the boundary of a container, showing (left) interior (T) and exterior ($\hat{T}$) temperatures and (right) the notation for thermal gradient of the wall temperature (T_w) through the boundary of wall thickness d

Heat diffusion through the thin solid wall obeys

$$k_w \frac{\partial^2 T_w}{\partial \eta^2} = 0 \tag{5.92}$$

assuming quasi-steady conditions exist at the side walls $\partial T / \partial t \simeq 0$, where k_w is the coefficient of thermal diffusion for the wall and η the coordinate normal to the wall, as shown in Fig. 5.13.

The solution to (5.92) with $T_w(\eta = -d) = \hat{T}$ and $T_w(\eta = 0) = T$ is

$$T_w = T + \frac{T - \hat{T}}{d} \eta \tag{5.93}$$

The heat flux through the wall is calculated by (*Fourier's Law*)

$$q_w = -k_w \frac{\partial T_w}{\partial \eta} = -\frac{k_w}{d}(T - \hat{T}) \tag{5.94}$$

This flux must be equal to the heat flux into the fluid, i.e.

$$q = -k \frac{\partial T}{\partial \eta} = q_w = -\frac{k_w}{d}(T - \hat{T}) \tag{5.95}$$

so that the fluid temperature gradient at the wall becomes

$$\frac{\partial T}{\partial \eta} = s_0(T - \hat{T}) \tag{5.96.a}$$

where

$$s_0 = \frac{k_w}{kd}. \tag{5.96.b}$$

The no-slip boundary condition is applied at the walls

$$\mathbf{u} = (u, v) = 0 \quad \text{on } \eta = 0. \tag{5.97}$$

The full equations governing the flow are again (5.41.a–c). Convection in the container is a complex problem. We will present a theory due to Rahm and Walin [5,6] considering the special case in which buoyancy boundary layers are formed at the periphery, i.e. in the case of strong stratification.

Strong stratification existing in the interior contrasts with the possibility of 3-D convection that could uniformly redistribute properties. In the present approximation, interior fluxes of volume and buoyancy instead are balanced by flows at buoyancy boundary layers developed near solid boundaries.

The solution is divided into boundary layer and interior parts. Since the interior is strongly stratified, it is assumed that

$$\frac{\partial T^I}{\partial x} \ll \frac{\partial T^I}{\partial z}, \quad \frac{\partial T^I}{\partial y} \ll \frac{\partial T^I}{\partial z}. \tag{5.98}$$

Furthermore, it is assumed that

$$\frac{k}{NL^2} \ll 1, \quad \frac{\nu}{NL^2} \ll 1, \quad s_0 \ll \left(\frac{\nu}{NL^2}\right)^{-1/2} \tag{5.99.a-c}$$

where the first two requirements represent favorable environment for the formation of buoyancy boundary layers i.e. $Re = \frac{UL}{\nu} \gg 1$ where the velocity scale $U \sim NL$, as demonstrated earlier by Eq. (5.85), so that $Re = NL^2/\nu \gg 1$, and since the Prandtl Number is typically $O(1)$, we also require $NL^2/k \gg 1$.

The third condition (5.99.a–c) is required only on the vertical side boundaries and is equivalent to

$$s_0 L \ll \left(\frac{\nu}{NL^2}\right)^{-1/2} = Re^{1/2} = \frac{1}{\delta}$$

from earlier boundary layer considerations in Sect. 5.1, i.e.,

$$\delta L \ll \frac{1}{s_0} = \frac{T - \hat{T}}{\frac{\partial T}{\partial \eta}} = L_T \tag{5.100}$$

i.e. the boundary layer thickness δL is smaller than the length scale L_T of the temperature variations in the interior. This is required for the linearization of the boundary layer equations as shown below.

The solution is divided into two parts as shown in Fig. 5.14, allowing to set

$$T = T^I + T^B$$
$$\mathbf{u} = \mathbf{u}^I + \mathbf{u}^B \tag{5.101.a-b}$$

where superscript I represents the interior and B is represents the boundary layer components. In the interior region, $T^B \ll T^I$ and $\mathbf{u}^B \ll \mathbf{u}^I$, so that the heat diffusion equation (5.98) becomes

$$\frac{\partial T^I}{\partial t} + w^I \frac{\partial T^I}{\partial z} = k \frac{\partial^2 T^I}{\partial z^2} \tag{5.102.a}$$

Fig. 5.14 Superposition of interior and boundary layer solutions at the slanted boundary of a container

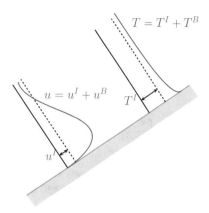

with interior functions of temperature T^I and vertical velocity w^I specified as

$$T^I = T^I(z, t)$$
$$w^I = w^I(z, t). \qquad\qquad (5.102.b, c)$$

The *buoyancy boundary layer* is designated by equations described in the last section. Various terms in the usual thermal boundary layer equations (5.66.a–c) are neglected, based on the strong stratification scaling, i.e. (5.99.a–c) in conformity with the buoyancy layer equations (5.67.a, b), namely

$$0 = \nu \frac{\partial^2 u}{\partial \eta^2} + \alpha g \cos\gamma (T - T^I)$$
$$u \frac{\partial T}{\partial \xi} = k \frac{\partial^2 T}{\partial \eta^2} \qquad\qquad (5.103.a, b)$$

Locally at least, the flow is assumed quasi-steady and the velocity is therefore independent of the ξ axis.

The solutions for the interior and the boundary layer equations (5.101.a, b) are superposed. However, the flow generated due to thermal convection is assumed to be stronger in the boundary layer as compared to the interior

$$|\mathbf{u}^B| \gg |\mathbf{u}^I| \qquad\qquad (5.103.c)$$

In fact, the only component of velocity $\mathbf{u}^I$ is w^I, which is the return flow in the interior to compensate for boundary layer flux, the only component u^B oriented along ξ axis aligned with the boundary, i.e. $u^B \gg w^I$. On the other hand, temperature disturbance caused by the boundary layer flow is much smaller than the temperature variations due to interior stratification $T^B \ll T^I$, as shown by Eq. (3.99.c) or (3.100), while the temperature gradients across the boundary-layer, perpendicular to the boundary, can not be neglected and are larger than that in the interior. Therefore, we can simplify (5.103) as

$$
\nu \left(\frac{\partial^2 u^I}{\partial \eta^2} + \frac{\partial^2 u^B}{\partial \eta^2} \right) + \alpha g \cos \gamma \underbrace{(T - T^I)}_{T^B} = 0
$$

$$
k \left(\frac{\partial^2 T^I}{\partial \eta^2} + \frac{\partial^2 T^B}{\partial \eta^2} \right) = (u^I + u^B) \left(\frac{\partial T^I}{\partial \xi} + \frac{\partial T^B}{\partial \xi} \right).
$$

$$(5.104.a - b)$$

Ignoring contributions of the cancelled terms, above equations are effectively linearized. Finally, referring to Eq. (5.71), it is noted that

$$
\frac{\partial T^I}{\partial \xi} = \frac{\partial T^I}{\partial z} \cos \gamma.
\tag{5.105}
$$

Combining these results yields the boundary layer equations, in the same way as (5.73.a, b), considered earlier,

$$
\nu \frac{\partial^2 u^B}{\partial \eta^2} + \alpha g \cos \gamma T^B = 0
$$

$$
k \frac{\partial^2 T^B}{\partial \eta^2} = u^B \frac{\partial T^I}{\partial z} \cos \gamma.
$$

$$(5.106.a - b)$$

Integrating (5.106.b) across the boundary layer determines the flux, denoted by m^B

$$
m_B = \int_0^\infty u^B d\eta
$$

$$
= \left(\frac{k}{\frac{\partial T^I}{\partial z} \cos \gamma} \right) \int_0^\infty \frac{\partial^2 T^B}{\partial \eta^2} d\eta = \left(\frac{k}{\frac{\partial T^I}{\partial z} \cos \gamma} \right) \left[\frac{\partial T^B}{\partial \eta} \right]_0^\infty
$$

$$\tag{5.107}$$

On the other hand, from (5.96.a) one can write at the boundary $\eta = 0$

$$
\left\{ \frac{\partial T}{\partial \eta} \right\}_{\eta=0} = \left\{ \frac{\partial T^I}{\partial \eta} + \frac{\partial T^B}{\partial \eta} \right\}_{\eta=0} = \left\{ s_0 (T^I + T^B - \hat{T}) \right\}_{\eta=0},
\tag{5.108}
$$

and also note that

$$
\left\{ \frac{\partial T^B}{\partial \eta} \right\}_{\eta \to \infty} \to 0,
\tag{5.109.a}
$$

By further note of the geometry,

$$
\frac{\partial T^I}{\partial \eta} = \frac{\partial T^I}{\partial z} \sin \gamma,
\tag{5.109.b}
$$

together with matching conditions (5.108) and (5.109.a, b), Eq. (5.107) becomes

$$m_B = -\left(\frac{k}{\frac{\partial T^I}{\partial z}\cos\gamma}\right)\left\{s_0(T^I - \hat{T}) - \frac{\partial T^I}{\partial \eta}\right\}_{\eta=0}$$
$$= -\frac{ks_0\left(T^I - \hat{T}\right)_{\eta=0}}{\frac{\partial T^I}{\partial z}\cos\gamma} + k\tan\gamma.$$

(5.110)

We next integrate this flux to determine the total boundary layer flux crossing any horizontal section of the container,

$$M_B = \oint_C m_B dl = k\oint_C \tan\gamma dl - \oint_C \frac{ks_0\left(T^I - \hat{T}\right)_{\eta=0}}{\frac{\partial T^I}{\partial z}\cos\gamma} dl$$

(5.111)

To simplify the first term of (5.111) we prefer to specify the cross-sectional area $A(z)$ around the closed perimeter $C(z)$ with geometrical details shown in Figs. 5.15 and 5.16

$$A = \mathbf{k}\cdot\frac{1}{2}\oint_C \mathbf{r}\times d\mathbf{r}$$

(5.112)

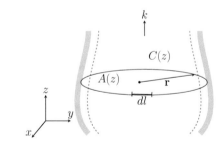

Fig. 5.15 The geometry of the container cross section with radial symmetry described in the text

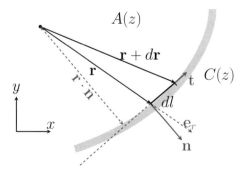

Fig. 5.16 Details of the container geometry in the horizontal, with coordinates specified in the $x-y$ plane

differentiating with respect to z, we obtain

$$
\frac{dA}{dz} = \frac{1}{2}\mathbf{k} \cdot \left[\oint_C \frac{d\mathbf{r}}{dz} \times d\mathbf{r} + \oint_C \mathbf{r} \times d\left(\frac{d\mathbf{r}}{dz}\right) \right]
$$

$$
= \mathbf{k} \cdot \oint_C \frac{d\mathbf{r}}{dz} \times d\mathbf{r}.
$$

(5.113)

Note that in the above, we have used the identity

$$
0 = \oint_C d(\mathbf{a} \times \mathbf{b}) = \oint_C \mathbf{a} \times d\mathbf{b} + \oint_C d\mathbf{a} \times \mathbf{b}
$$

$$
= \oint_C \mathbf{a} \times d\mathbf{b} - \oint_C \mathbf{b} \times d\mathbf{a}
$$

(5.114)

or equivalently

$$
\oint_C \mathbf{a} \times d\mathbf{b} = \oint_C \mathbf{b} \times d\mathbf{a}.
$$

(5.115)

Now (5.113) can be put into the following form:

$$
\frac{dA}{dz} = \mathbf{k} \cdot \oint_C \frac{d\mathbf{r}}{dz} \times d\mathbf{r} = \mathbf{k} \cdot \oint_C \left(\frac{dr}{dz}\mathbf{e}_r\right) \times \mathbf{t}\, dl
$$

$$
= \oint_C \frac{dr}{dz}\mathbf{k} \cdot (\mathbf{e}_r \times \mathbf{t})\, dl = -\oint_C \frac{dr}{dz}\mathbf{e}_r \cdot (\mathbf{k} \times \mathbf{t})\, dl
$$

$$
= \oint_C \left(\frac{dr}{dz}\mathbf{e}_r\right) \cdot \mathbf{n}\, dl = \oint_C \frac{d\mathbf{r}}{dz} \cdot \mathbf{n}\, dl = \oint_C \frac{d\mathbf{r}}{dz} \cdot \mathbf{n}\, dl = \oint_C \frac{d\,(\mathbf{r} \cdot \mathbf{n})}{dz}\, dl
$$

$$
= \oint_C \tan\gamma\, dl
$$

(5.116)

We can therefore replace the first term of (5.111) by (5.116) yielding

$$
M_B = \oint_C m_B dl = k\frac{dA}{dz} - \frac{1}{\frac{\partial T^I}{\partial x}} \oint_C k s_0 \left(T^I - \hat{T}\right)_{\eta=0} (\cos\gamma)^{-1} dl
$$

(5.117)

The integrated continuity equation for the whole container is

$$
M_B + w^I A(z) = M_0
$$

(5.118)

i.e. the total of the interior and boundary fluxes should equal to the mass flux M_0 passing through the system. Note that for the exceptional case $M_0 = 0$, the interior velocity would just be a return flow of the boundary layer transport $w^I = -M_B/A(z)$, otherwise given by (5.118). Since we have calculated M_B, we can evaluate w^I from (5.118)

$$w^I = \frac{M_0}{A} - \frac{k}{A}\frac{dA}{dz} + \frac{1}{A\frac{dT^I}{dz}} \oint_c ks_0(T^I - \hat{T})_{\eta=0}(\cos\gamma)^{-1}dl \qquad (5.119)$$

Substituting in (5.102), we obtain

$$\frac{\partial T^I}{\partial t} + \left(\frac{M_0}{A} - \frac{k}{A}\frac{dA}{dz}\right)\frac{\partial T^I}{\partial z} + \frac{ks_0}{A}\oint_c \frac{\left(T^I - \hat{T}\right)_{\eta=0}}{\cos\gamma}dl = k\frac{\partial^2 T^I}{\partial z^2} \qquad (5.120)$$

The above equation determines the interior temperature distribution for given mass flux M_0 diffusion coefficient k, ambient temperature $\hat{T}$ < wall thickness parameter s_0 and geometrical parameters $A(z)$ and $\cos\gamma$ of the container. Note that Eq. (5.120) automatically satisfies the boundary conditions an the inclined walls of the container. If the container is bounded by top and bottom boundaries, these boundary conditions are specified to solve for (5.120). Examples are provided in the following.

5.2.5 Thermal Stratification in a Cylinderical Container

5.2.5.1 Without Through-Flow, $M_0 = 0$

Consider a cylindrical container with height $2H$ and radius r_0 as described in Fig. 5.17. Let the temperature at its top surface be held constant with a value of T_2 and at its bottom surface of a temperature of T_1, and let the ambient temperature be $\hat{T}$.

We consider the steady case $\partial/\partial t = 0$, and no net flux through the container $M_0 = 0$. For simple geometry one has $\partial A/\partial z = 0$, $\cos\gamma = 1$, $A = \pi r_0^2$ so that Eq. (5.120) becomes

$$\frac{ks_0}{\pi r_0^2} \cdot 2\pi r_0(T^I - \hat{T}) = k\frac{\partial^2 T^I}{\partial z^2}$$

Fig. 5.17 Geometry of a cylindrical container with temperature boundary conditions

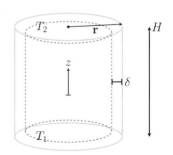

yielding

$$\frac{\partial^2 T^I}{\partial z^2} - \alpha^2 T^I = -\alpha^2 \hat{T} \qquad (5.121.a)$$

where

$$\alpha^2 = \frac{2s_0}{r_0}. \qquad (5.121.b)$$

The boundary conditions are

$$T^I = T_1 \quad \text{on } z = -H$$
$$T^I = T_2 \quad \text{on } z = +H \qquad (5.122.a, b)$$

The solution to (5.121.a) is

$$T^I = A_1 e^{-\alpha(z+H)} + A_2 e^{+\alpha(z-H)} + \hat{T} \qquad (5.123.a)$$

and the constants A_1, A_2 determined from (5.122.a, b) are

$$A_1 = \frac{\left(T_1 - \hat{T}\right) - \left(T_2 - \hat{T}\right)e^{-2\alpha H}}{1 - e^{-4\alpha H}} \qquad (5.123.b)$$

$$A_2 = \frac{\left(T_2 - \hat{T}\right) - \left(T_1 - \hat{T}\right)e^{-2\alpha H}}{1 - e^{-4\alpha H}} \qquad (5.123.c)$$

We can combine (5.123.a, b, c) to write the solution in an alternative way:

$$T^I = (T_2 - \hat{T})\frac{\sinh \alpha (z + H)}{\sinh \alpha H} - (T_1 - \hat{T})\frac{\sinh \alpha (z - H)}{\sinh 2\alpha H} + \hat{T} \qquad (5.124)$$

where the sinus hyperbolic is defined as

$$\sinh x \equiv \frac{e^x - e^{-x}}{2}. \qquad (5.125)$$

Note the two limits of the function $\sinh x$.

$$\text{For } x \gg 1, \quad \sinh x \simeq \frac{1}{2}e^x$$
$$\text{For } x \ll 1, \quad \sinh x \simeq x \qquad (5.126.a, b)$$

The solution (5.124) depends on the relative magnitude of the dimensionless parameter as illustrated in Fig. 5.18

$$\alpha H = \left(\frac{2s_0 H^2}{r_0}\right)^{1/2} = \left(\frac{2k_w H^2}{k r_0 d}\right)^{1/2} \qquad (5.127)$$

Fig. 5.18 Typical solutions of temperature distribution in a cylindrical container depending on the limits of the parameter αH

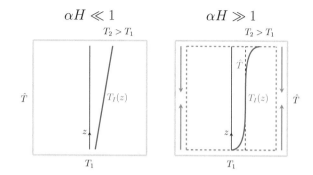

For $\alpha H \ll 1$, the solution (5.124) becomes

$$
\begin{aligned}
T^I &\cong \frac{T_2 - \hat{T}}{2H}(z + H) - \frac{T_1 - \hat{T}}{2H}(z - H) \\
&= \frac{T_2 - T_1}{2}\left(\frac{z}{H}\right) + \frac{T_2 - T_1}{2}
\end{aligned}
\tag{5.128}
$$

i.e. a linear profile of interior temperature. In this limit, diffusion dominates the interior according (5.127), compared to the boundary layer transport driven by heat diffusion from sides. In the other limit, $\alpha H \gg 1$ we can simplify (5.124) as

$$
T^I \cong
\begin{cases}
(T_2 - \hat{T})e^{+\alpha(z-H)} + \hat{T} & \text{near} \quad z \simeq H \\
(T_1 - \hat{T})e^{-\alpha(z+H)} + \hat{T} & \text{near} \quad z \simeq -H
\end{cases}
\tag{5.129}
$$

i.e. the temperature variable of the interior are confined near the top and bottom boundaries in e-folding distances of

$$
H_s = \frac{1}{\alpha} = \left(\frac{r_0}{2s_0}\right)^{1/2}
\tag{5.130}
$$

In this case the boundary layer transport dominates the diffusion in the interior.

When $\alpha H \ll 1$, boundary layer development is not accentuated and the ambient temperature $\hat{T}$ becomes relatively unimportant, with stratification maintained by diffusion between the top and bottom boundaries. On the other hand, when $\alpha H \gg 1$, the boundary layers transport fluid downward in the regions where $\hat{T} > T^I$. This vertical boundary layer transport is balanced by interior flow w^I and heating or cooling in turn by interior heat diffusion. The boundary layer and internal transports can be calculated from (5.117) and (5.118).

For a comparable unsteady problem, with initial conditions $T_1, T_2, \hat{T}$ imposed at $t = 0$, the time scale of evolution τ to establish the steady state could be estimated as

$$
\begin{aligned}
\tau_1 &\simeq H^2/k & \text{for} \quad \alpha H \ll 1 \\
\tau_2 &\simeq H_s^2/k = \frac{r_0}{2s_0 k} & \text{for} \quad \alpha H \gg 1
\end{aligned}
\tag{5.131}
$$

i.e. much smaller time is required to establish the stratification when $\alpha H \gg 1$ since $H_s \ll H$. In fact, in the case $\alpha H \gg 1$, by virtue of Eqs. (5.130) and (5.127) we have

$$\tau_2 \simeq \frac{H_s^2}{k} = \frac{r_0}{2s_0 k} = \frac{H^2}{(\alpha H)^2 k} \ll \frac{H^2}{k}\tau_1 \quad \text{for} \quad \alpha H \gg 1. \tag{5.132}$$

5.2.5.2 With Through-Flow $M_0 \neq 0$

If flow is allowed to pass through the cylinder ($M_0 \neq 0$) the governing equation (5.120) determines the steady-state interior temperature distribution

$$\frac{M_0}{\pi r_0^2}\frac{\partial T^I}{\partial z} + \frac{k s_0}{\pi r_0^2} 2\pi r_0 (T^I - \hat{T}) = k \frac{\partial^2 T^I}{\partial z^2}$$

or

$$\frac{\partial^2 T^I}{\partial z^2} - \left(\frac{M_0}{k\pi r_0^2}\right)\frac{\partial T^I}{\partial z} - \left(\frac{2s_0}{r_0}\right)T^I = -\left(\frac{2s_0}{r^\circ}\right)\hat{T} \tag{5.133}$$

with constant coefficients and the following definitions

$$\frac{\partial^2 \Delta}{\partial z^2} - a\frac{\partial \Delta}{\partial z} - b\Delta = 0 \tag{5.134}$$

$$\Delta = T^I - \hat{T}$$
$$a = \frac{M_0}{k\pi r_0^2} = \frac{M_0}{k A_0} \tag{5.135.$a - c$}$$
$$b = \frac{2s_0}{r_0}.$$

Substituting the transformation

$$\Delta = T^I - \hat{T} = \Delta_0 e^{\frac{a}{2}z} \tag{5.136}$$

gives

$$\frac{\partial^2 \Delta_0}{\partial z^2} - \left(b + \frac{a^2}{4}\right)\Delta_0 = 0 \tag{5.137}$$

and letting

$$\alpha^2 = b + \frac{a^2}{4} = \frac{2s_0}{r_0} + \frac{M_0^2}{4k^2 A_0^2} \tag{5.138}$$

the solution follows from (5.137) and (5.136) as

$$
\begin{aligned}
T^I &= A_1 e^{\frac{a}{2}z} e^{\alpha z} + A_2 e^{\frac{a}{2}z} e^{-\alpha z} + \hat{T} \\
&= A_1 e^{\left(\frac{a}{2}+\alpha\right)z} + A_2 e^{\left(\frac{a}{2}-\alpha\right)z} + \hat{T} \\
&= A_1 e^{\frac{a}{2}\left\{1+\sqrt{1+\frac{4b}{a^2}}\right\}z} + A_2 e^{\frac{a}{2}\left\{1-\sqrt{1+\frac{4b}{a^2}}\right\}z} + \hat{T} \\
&= A_1 e^{\gamma_1 z} + A_2 e^{\gamma_2 z} + \hat{T}
\end{aligned}
\tag{5.139}
$$

where

$$
\begin{aligned}
\gamma_1 &= \frac{M_0}{2kA_0}\left\{1+\left(1+\frac{8s_0 k^2 A_0^2}{r_0 M_0^2}\right)^{1/2}\right\} > 0 \\
\gamma_2 &= \frac{M_0}{2kA_0}\left\{1-\left(1+\frac{8s_0 k^2 A_0^2}{r_0 M_0^2}\right)^{1/2}\right\} < 0
\end{aligned}
\tag{5.140.a, b}
$$

The effect of a positive vertical flux $M_0 > 0$ on the interior distribution is that the diffusive region on top becomes thinner than the one near the bottom, as can be identified from equations (5.140.a, b) and as shown on the left side of Fig. 5.19. In the limit $M_0 \gg \frac{kA_0}{H}$ only the second part of the exponential is representative of the temperature distribution, indicated by the limits of Eqs. (5.140.a, b)

$$
\begin{aligned}
\gamma_1 &\simeq \frac{M_0}{2kA_0}\left\{1+\left(1+\frac{1}{2}\frac{8s_0 k^2 A_0^2}{r_0 M_0}+\cdots\right)\right\} \rightarrow \frac{M_0}{kA_0} \\
\gamma_2 &\simeq \frac{M_0}{2kA_0}\left\{1-\left(1+\frac{1}{2}\frac{8s_0 k^2 A_0^2}{r_0 M_0}+\cdots\right)\right\} \rightarrow \frac{2s_0 kA_0}{r_0 M_0}.
\end{aligned}
\tag{5.141.a, b}
$$

These results indicate that for $\gamma_1 H = \frac{M_0 H}{kA_0} \gg 1$, $\gamma_2 \sim O(1)$, so that the second exponential term dominates the solution as shown on the right hand side of Fig. 5.19.

$$
T^I \sim A_2 e^{\frac{A_0 s_0 k}{M_0} z} + \hat{T}.
\tag{5.142}
$$

Fig. 5.19 The development of internal stratification in a cylinder driven by a net flux M_0 and boundary fluxes

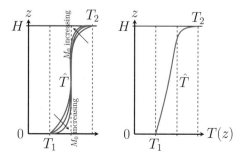

In this regime of $\alpha H \gg 1$, the advective contribution in (5.133) is shown to be larger than the diffusive one, so that

$$\frac{M_0}{k A_0} \frac{\partial T^I}{\partial z} + \frac{2 s_0}{r_0} T^I = \frac{2 s_0}{r_0} \hat{T}. \tag{5.143}$$

The solution in (5.142) indicates the diffusive term is only important near the top boundary. In this case inflow has been assumed to have temperature T_1, same as the bottom boundary.

5.2.6   Interior and Boundary Layer Flows

5.2.6.1 Steady-State Stratification in Radially Symmetric Containers
Equations (5.118)–(5.120) can be used to study thermal stratification in a container, established by interior and boundary layer flows. From (5.120),

$$\cancel{\frac{\partial T}{\partial t}} + \left(\frac{M_0}{A} - \frac{k}{A} \frac{dA}{dz} \right) \frac{\partial T^I}{\partial z} + \frac{k s_0}{A} \oint_C \frac{(T^I - \hat{T})}{\cos \gamma} dl = k \frac{\partial^2 T^I}{\partial z^2}, \tag{5.144}$$

steady state is approached when the unsteady term in (5.120) is made to vanish. For a container with radial symmetry, the simplified form is then rearranged as

$$A(z) \frac{d^2 T^I}{dz^2} + \frac{dA(z)}{dz} \frac{dT^I}{dz} - \frac{M_0}{k} \frac{dT^I}{dz} - s_0 \frac{T^I - \hat{T}}{\cos \gamma} C(z) = 0 \tag{5.145}$$

where $A(z)$ is the cross-sectional area and $C(z)$ is the perimeter of the container as a function of height z. By rearranging the first two terms,

$$\frac{d}{dz} \left(A(z) \frac{dT^I}{dz} \right) - \frac{M_0}{k} \frac{dT^I}{dz} - s_0 \frac{(T^I - \hat{T})}{\cos \gamma} C(z) = 0. \tag{5.146}$$

Defining an *integrating factor*

$$T^* \equiv (T^I - \hat{T}) \equiv \Delta(z) e^{\int \frac{M_0}{k A(z)} dz} \tag{5.147}$$

and re-arranging

$$A(z) \frac{dT^*}{dz} - \frac{M_0}{k} T^* = \left(\cancel{\frac{M_0}{k}} \Delta + A(z) \frac{d\Delta}{dz} \right) e^{\int \frac{M_0}{k A(z)} dz} - \cancel{\frac{M_0}{k}} \Delta e^{\int \frac{M_0}{k A(z)} dz} \tag{5.148}$$

serves to re-write (5.146) as

$$\frac{d}{dz}\left(A(z)\frac{d\Delta(z)}{dz}e^{\int \frac{M_0}{kA(z)}dz}\right) = \frac{s_0}{\cos \gamma}C(z)\Delta(z)e^{\int \frac{M_0}{kA(z)}dz}. \tag{5.149}$$

For a container with circular cross-section, the geometry is given as a function of the radius

$$A(z) = \pi r^2(z) \text{ and } C(z) = 2\pi r(z)$$

yielding

$$\frac{d}{dz}\left(\pi r^2(z)\frac{d\Delta(z)}{dz}e^{\int \frac{M_0}{k\pi r^2(z)}dz}\right) = \frac{s_0}{\cos \gamma}2\pi r(z)\Delta(z)e^{\int \frac{M_0}{k\pi r^2(z)}dz}. \tag{5.150}$$

5.2.6.2 Steady-State in Stratified Conical Container

Consider a conical container as shown in Fig. 5.20, with a mass flux M_0 and fluid with temperature entering through its apex at the bottom. The radius increases linearly as a function of height, reaching its maximum r_1 at H

$$r(z) = \tan \gamma \, z = \frac{r_1}{H}z \tag{5.151}$$

and (5.149) becomes

$$\frac{d}{dz}\left(\left(\frac{r_1}{H}\right)^2 z^2 \frac{d\Delta(z)}{dz}e^{\int \frac{M_0 H^2}{k\pi r_1^2 z^2}dz}\right) = \frac{s_0}{\cos \gamma}2\frac{r_1}{H}z\Delta(z)e^{\int \frac{M_0 H^2}{k\pi r_1^2 z^2}dz}. \tag{5.152}$$

Defining the constant

$$H^* = \frac{M_0 H^2}{k\pi r_1^2} = \frac{M_0 H^2}{kA_1} \tag{5.153}$$

Fig. 5.20 Geometry of a conical container

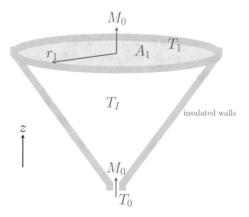

and evaluating the integral of the exponential term in (5.152) as

$$\int \frac{M_0 H^2}{k\pi r_1^2 z^2} dz = H^* \int \frac{1}{z^2} dz = -\frac{H^*}{z} \tag{5.154}$$

Equation (5.152) is reduced to

$$\frac{d}{dz}\left(z^2 \frac{d\Delta(z)}{dz} e^{-(H^*/z)}\right) = \frac{2s_0 H}{r_1 \cos \gamma} z\Delta(z) e^{-(H^*/z)}. \tag{5.155}$$

Using (5.151) and setting a constant

$$\frac{1}{H_0} = \frac{2s_0 H}{r_1 \cos \gamma} = \frac{2k_w H}{kdr_1 \cos \gamma} = \frac{2k_w}{kd \sin \gamma} \tag{5.156}$$

while also defining

$$\xi = z/H, \quad \mu = H^*/H = \frac{M_0 H}{kA_1}, \quad \sigma_0 = H/H_0 = \frac{2k_w H}{kd \sin \gamma} \tag{5.157}$$

Equation (5.155) can be written in dimensionless form as

$$\frac{d}{d\xi}\left(\xi^2 \frac{d\Delta(\xi)}{d\xi} e^{-\mu/\xi}\right) = \sigma_0 \, \xi\Delta(\xi) e^{-\mu/\xi} \tag{5.158}$$

which generates a second order ordinary differential equation (ODE) with variable coefficients

$$\xi^2 \frac{d^2\Delta(\xi)}{d\xi^2} + (2\xi + \mu)\frac{d\Delta(\xi)}{d\xi} - \sigma_0 \, \xi\Delta(\xi) = 0. \tag{5.159.a}$$

This simple looking equation in general does not avail itself to analytical solution, as it does not belong to well known forms of second degree ODE's with variable coefficients. In fact the form of (5.159) is an amalgamation of two different types, depending on selected settings of the constants.

Letting either of the constants to vanish, $\mu = 0$ and $\sigma_0 = 0$, different forms of reduced equations can be obtained. For instance selecting both constants to be zero, the system becomes comparable to the Euler equation (except for a missing term), of somewhat limited applicability:

$$\xi^2 \frac{d^2\Delta}{d\xi^2} + (2\xi)\frac{d\Delta}{d\xi} = 0. \tag{5.159.b}$$

As an option we can set $\mu = 0$ (or $M_0 = 0$), i.e. without through-flow, then

$$\xi^2 \frac{d^2 \Delta(\xi)}{d\xi^2} + (2\xi) \frac{d\Delta(\xi)}{d\xi} - \sigma_0\,\xi \Delta(\xi) = 0. \qquad (5.159.c)$$

This second order ODE however has to be solved either by Frobenius Series or by numerical methods. The form of the equation happens to be very close to an Euler Equation, were it not for the last term, involving extra dependence on ξ.

The other option setting $\sigma_0 = 0$ (or $s_0 = 0$) represents the case of isolated walls preventing heat diffusion from outside, resulting in

$$\xi^2 \frac{d^2 \Delta(\xi)}{d\xi^2} + (2\xi + \mu) \frac{d\Delta(\xi)}{d\xi} = 0, \qquad (5.159.d)$$

a case allowing direct integration, presented in the following.

5.2.6.3 Conical Container with Insulated Wall, $s_0 = 0$

The case with $\sigma_0 = 0$ ($s_0 = 0$) corresponds to steady-state thermal diffusion in a container with insulated side walls. Although a simplified version of the original Eq. (5.159.a), this example presents a better analytically tractable case. Returning to the dimensional equations for the present case ($s_0 = 0$),

$$k \frac{d^2 T^I}{dz^2} + \frac{k}{A} \frac{dA}{dz} \frac{dT^I}{dz} - \frac{M_0}{A} \frac{dT^I}{dz} = 0. \qquad (5.160.a)$$

is rearranged in the following form

$$\frac{d}{dz}\left(A \frac{dT^I}{dz}\right) = \frac{M_0}{k} \frac{dT^I}{dz} \qquad (5.160.b)$$

valid for any geometry of the container depending on the cross-sectional area $A(z)$.

Re-organizing in dimensionless variables (5.157) gives

$$\frac{d}{d\xi}\left(\xi^2 \frac{d\Delta}{d\xi}\right) + \mu \frac{d\Delta}{d\xi} = 0. \qquad (5.161)$$

which can be directly integrated to give

$$\xi^2 \frac{d\Delta}{d\xi} + \mu\Delta + c_0 = 0 \qquad (5.162)$$

where c_0 is a constant. Using an *integrating factor* a new variable $\hat{\Delta}(\xi)$ is defined as

$$\Delta(\xi) = \hat{\Delta}(\xi) e^{\mu/\xi} \qquad (5.163)$$

so that (5.162) is rewritten as

$$\xi^2 \frac{d(\hat{\Delta}e^{\mu/\xi})}{d\xi} + \mu\hat{\Delta}e^{\mu/\xi} + c_0 = 0 \tag{5.164}$$

and therefore

$$-\xi^2 \frac{\mu}{\xi^2}e^{\mu/\xi}\hat{\Delta} + \xi^2 e^{\mu/\xi}\frac{d\hat{\Delta}}{d\xi} + \mu\hat{\Delta}e^{\mu/\xi} + c_0 = 0 \tag{5.165}$$

the solution is obtained as

$$\begin{aligned}
\hat{\Delta} &= -c_0 \int \frac{1}{\xi^2}e^{-\mu/\xi}d\xi + c_1 \\
&= \frac{c_0}{\mu}e^{-\mu/\xi} + c_1
\end{aligned} \tag{5.166}$$

where differentials of $u = -\mu/\xi$ and $du = -d\xi/\xi^2$ are noted to convert and integrate $\int e^\mu du$. Remembering definitions made earlier in (5.147), (5.151), (5.153), (5.154), (5.157) and (5.165)

$$T^* = T^I - \hat{\mathcal{T}} = \Delta e^{\int \frac{M_0 H^2}{k\pi r_1^2 z^2}} = \Delta e^{-H^*/z} = \Delta e^{-\mu/\xi} = \hat{\Delta} \tag{5.167}$$

gives the solution, where the ambient temperature component $\hat{\mathcal{T}}$ is canceled in the present case of insulated walls. This looks very simple, after various steps of integration, with two constants in (5.166) determined by boundary conditions. However, we have to take care of the singularity at the apex of the cone $\xi = 0$, by appropriately selecting equivalent constants ae^μ and b so as to have

$$\begin{aligned}
T^* = T^I = \hat{\Delta} &= ae^\mu e^{-\mu/\xi} + b \\
&= ae^{\mu(1-\frac{1}{\xi})} + b.
\end{aligned} \tag{5.168}$$

The following boundary conditions specifying top and bottom temperatures are

$$\begin{aligned}
T^I &= T_0 \quad \text{at } z = 0, \\
T^I &= T_1 \quad \text{at } z = H.
\end{aligned} \tag{5.169.a, b}$$

Returning back to the solution (5.168), the constants a, b are determined by boundary conditions (5.169.a, b)

$$\begin{aligned}
T^I = \hat{\Delta} &= (T_1 - T_0)e^{\mu(1-\frac{1}{\xi})} + T_0 \\
&= (T_1 - T_0)e^{\frac{H^*}{H}\left(1-\frac{H}{z}\right)} + T_0
\end{aligned} \tag{5.170}$$

Fig. 5.21 The solution for interior temperature T^I representing thermal stratification in a conical container described by $e^{-\mu(1-1/\xi)}$ for different values of μ

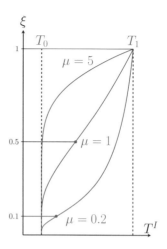

where

$$\mu = \frac{H^*}{H} = \frac{M_0 H}{k A_1}. \tag{5.171}$$

The solution as described in Fig. 5.21 involves the term $e^{-\frac{\mu}{\xi}}$ which might not be possible to visualize as $\xi \to 0$. In particular, note that $\lim_{\xi \to 0} e^{-\frac{\mu}{\xi}} \to 0$ while $\lim_{\xi \to 0} e^{+\frac{\mu}{\xi}} \to \infty$. With the first one of these limits, the correct evaluations of boundary conditions are obtained. Calculating first and second derivatives of the solution,

$$\frac{dT^I}{d\xi} = T_1 \frac{\mu}{\xi^2} e^{\mu(1-\frac{1}{\xi})}$$
$$\frac{d^2 T^I}{d\xi^2} = T_1 \left(\frac{\mu^2}{\xi^4} - 2\frac{\mu}{\xi^3} \right) e^{\mu(1-\frac{1}{\xi})} \tag{5.172.a, b}$$

It can be observed that $\frac{dT^I}{dz} > 0$ everywhere, but the second derivative vanishes at the inflection point of temperature at $z = H^*/2$ (or at $\xi = \mu/2$), provided that $H^*/2 < H$, $(\mu/2 > 1)$. To understand this feature physically we recall that even though the side walls are insulated, there will be a boundary layer flow In this buoyancy layer, the transport is calculated from (5.117), excluding the wall heat transfer with $s_0 = 0$

$$M_B = -k \frac{dA}{dz} \tag{5.173}$$

and from (5.118) or (5.119) we calculate the interior velocity based on (5.120), and inserting from (5.172.a, b) in terms of dimensional coordinates ξ:

$$w^I = \frac{M_0 - M_B}{A} = \left(\frac{M_0}{A} - \frac{k}{A}\frac{dA}{dz}\right) = \frac{k\frac{d^2 T^I}{dz^2}}{\frac{dT^I}{dz}}$$

$$= \frac{k}{H}\left(\frac{\mu - 2\xi}{\xi^2}\right). \tag{5.174}$$

Since $\frac{dT^I}{dz} > 0$ for all z, the vertical velocity w^I will vanish slightly at a lower layer than the height of the inflection point of the temperature, at the zero crossings as observed for solutions displayed in Fig. 5.22. Alternatively, we can interpret this to happen when boundary layer flux M_B exceeds M_0.

Further, by re-writing (5.118)

$$\frac{M_B}{M_0} = 1 - \frac{A(z)}{M_0}w^I \tag{5.175}$$

and extracting w^I from (5.174) and (5.172)

$$w^I = \frac{M_0}{A}\left(1 - \frac{k}{M_0}\frac{dA}{dz}\right) = \frac{M_0}{A(z)}\left(1 - 2\frac{kA_1}{M_0 H}\frac{z}{H}\right) \tag{5.176}$$

and combining gives

$$\frac{M_B}{M_0} = -2\frac{kA_1}{M_0 H}\frac{z}{H} = \frac{2}{\mu}\xi \tag{5.177}$$

which shows that the boundary layer flux increases linearly with height. The overall convection pattern in the conical container is described in Fig. 5.23.

Fig. 5.22 The solution for vertical velocity w^I in a conical container for different values of μ

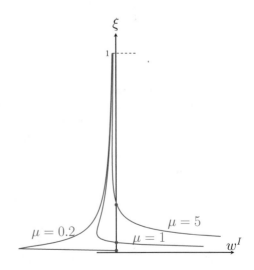

Fig. 5.23 Convection pattern in a conical container represented by the distribution interior temperature T^I and vertical velocity w^I, showing interior and boundary layer fluxes

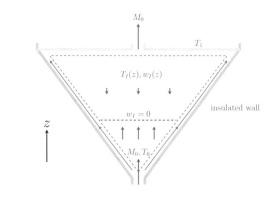

Fig. 5.24 The decay of thermal stratification in a cylindrical container

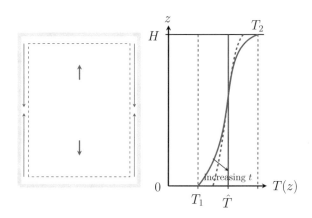

5.2.6.4 Decay of Stratification in a Container

For simplicity, consider a cylindrical container (with $A = A_0 = \pi r_0^2$) initially stratified with temperature profile $T_i^I(z)$ as shown in Fig. 5.24. Assume the horizontal top and bottom boundaries are insulated and there is no mean flux $M_0 = 0$. The governing equation for the temperature distribution at any time is obtained as

$$\frac{\partial T^I}{\partial t} + \frac{ks_0}{\pi r_0} 2\pi r_0 (T^I - \hat{T}) = K \frac{\partial^2 T^I}{\partial z^2} \tag{5.178.a}$$

or

$$\frac{\partial T^I}{\partial t} + k\alpha^2 (T^I - \hat{T}) = k \frac{\partial^2 T^I}{\partial z^2} \tag{5.179.a, b}$$

where $\alpha^2 = 2s_0/r_0$. The initial condition is

$$T^I(z, 0) = T_i^I(z). \tag{5.179.c}$$

We can consider two limits for the solution. For $\alpha H \ll 1$, we can neglect the heat input from the side walls and the solution is governed by time dependent diffusion in the interior alone, i.e.

$$\frac{\partial T'}{\partial t} = k\frac{\partial^2 T'}{\partial z^2} \tag{5.180}$$

and the time scale of decay is the diffusion time

$$\tau \sim H^2/k \tag{5.181}$$

On the other hand when $\alpha H \gg 1$, we can neglect diffusion in the interior and the decay process is governed by heat transfer from the side walls, i.e. the buoyancy layer transports alone

$$\frac{\partial T'}{\partial t} + k\alpha^2(T' - \hat{T}) = 0 \tag{5.182}$$

we can construct a solution by letting $\Delta = T' - haT$, $\hat{T} = constant$,

$$\frac{\partial \Delta}{\partial t} + k\alpha^2 \Delta = 0 \tag{5.183.a}$$

with the initial condition

$$\Delta = \Delta_i = T_i' - \hat{T} \quad \text{at } t = 0. \tag{5.183.b}$$

The solution is obtained as

$$\Delta = \Delta_i e^{-k\alpha^2 t} \tag{5.184.a}$$

or

$$T' - \hat{T} = (T_i' - \hat{T})e^{-(2ks_0/r_0)t}$$
$$= (T_i' - \hat{T})e^{-\frac{k}{H_s^2}t} \tag{5.184.b}$$

where $H_S = (r_0/2s_0)^{1/2}$. The initial profile shape remains the same, but decreases in amplitude with increasing time. The typical time scale for the decay is

$$\tau \sim \frac{H^2}{k} = \frac{r_0}{2ks_0} = \frac{H^2}{(\alpha H^2)\,k} \ll \frac{H^2}{k} \tag{5.185}$$

by virtue of Eq. (5.132). The decay is much faster than that would occur by pure diffusion.

However, note that this solution is not valid in the top and bottom boundary layers of thickness H_s, since in these regions we can not neglect the diffusive terms in spite of the fact that $\alpha H \gg 1$.

Fig. 5.25 Evolution of thermal stratification in a cylindrical container exposed to external temperature at the top

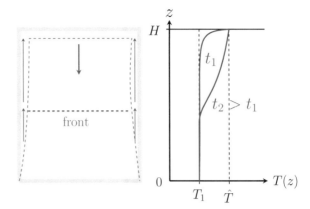

5.2.6.5 Stratification of an Initially Homogeneous Fluid in Container

Consider the evolution of thermal stratification in an insulated cylindrical container shown in Fig. 5.25, initially filled with homogeneous fluid, with initial temperature $T^I = T_i^I$ = constant and exposed to external temperature at top. At time $t = 0$, we apply an ambient temperature $\hat{T} > T^I$. Initially there will be thermal boundary layers developed on the sidewalls (see Sect. 5.2).

The boundary layer flux will be returned to the interior at the top, advecting warmer fluid with it and therefore starting stratification at the top. Once the interior gets stratified near the top, buoyancy layers will be formed in the region of stratification. At a later time, interior temperature profile deepens.

The front formed will move downward and reach the bottom at a later time, when the sidewall buoyancy layers develop while the interior is stratified. At a still later time, if the top and bottom are not insulated the fluid will have a uniform temperature of T. If however, the top and bottom are not insulated and have different temperatures T_1 and T_2 the fluid will evolve into a stratified condition as studied in Sect. 5.2.5.

5.3   Rayleigh–Bénard Thermal Convection

Convective instability:

Static instability of stratified fluids in the absence of viscous effects have been considered in earlier chapters. Gravitational instability of a two-layer system with lighter fluid in the bottom was found to depend on the presence of restoring forces, such as surface tension in the case of Kelvin-Helmholtz theory.

Alternatively, the effects of viscosity can damp out unstable motions before some threshold of instability. Consider a fluid between two horizontal plates, with differential heating from below, as shown in Fig. 5.26. Upward heat flux opposing gravity will decrease density near the bottom plate, potentially leading to gravitational instability.

In the absence of buoyancy forces, with heating too small to break electrochemical bonds between water molecules, the fluid could be stable, behaving almost

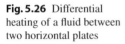

Fig. 5.26 Differential
heating of a fluid between
two horizontal plates

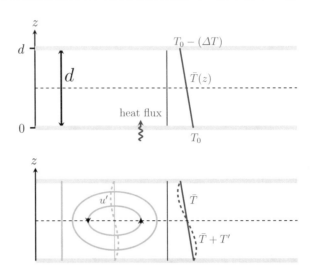

as a heat conducting solid. Under these stable conditions at initial time, a linear
temperature distribution as described in Fig. 5.26 is created as

$$\bar{T} = T_0 - \Delta T \left(\frac{z}{d}\right) = T_0 - \gamma z \tag{5.186}$$

satisfying the one-dimensional form of the thermal energy equation (1.59) or its
approximation (1.93) with only steady conduction terms

$$K \nabla^2 \bar{T} = 0, \quad \rightarrow \quad K \partial^2 \bar{T}/\partial z^2 = 0, \tag{5.187}$$

where $-dT/dz = \gamma = \Delta T/d$ the temperature gradient and K the thermal conductivity, $K\gamma$ the heat flux.

   If the heat flux is slightly increased so as to modify density according to equation
of state (1.94) in its linearized form (1.95), the bottom fluid will become lighter and
must rise to the top. Viscous effects tend to damp out this motion up to a certain critical
value of $\Delta T = \Delta T_c$. Then as water particles rise, the static condition (5.167) will
be perturbed to create small deviations of velocity $\mathbf{u}'$ and temperature T', disturbing
the basic state as shown in Fig. 5.26. In accordance, small circulation cells by rising
and sinking motions of fluid will be created.

   As noted in Chap. 1, in general, the equation of state is non-linear. Only its linearized form (1.95) is used here, keeping first order terms in temperature and without
taking account of salinity:

$$\rho = \rho_0 \left[1 - \alpha(T - T_0)\right] \tag{5.188}$$

where α is the linear coefficient of thermal expansion close to the ambient value T_0
of temperature.

In the above discussion, what matters is the density (buoyancy), linked to temperature through the equation of state (1.94), based on its linear form (5.169). Pure water has some peculiar properties explained by molecular bonds near $T_m = 4\,°C$ at temperature of maximum density, where the above linear coefficients become inadequate to approximate the nonlinear equation of state,

$$\begin{aligned} \alpha > 0 \ \text{ for } \ T > 4\,°C \\ \alpha < 0 \ \text{ for } \ T < 4\,°C \end{aligned} \qquad (5.189.a, b)$$

around temperature T_m of maximum density. Note, also that for values $T < T_m = 4\,°C$, heating from *above* could cause instability. This is the reason why temperate lakes in freezing environments often have stable layers of uniform temperatures $T_m \simeq 4\,°C$ established by gravitational adjustment processes.

If we now increase heating and let the fluid be slightly perturbed

$$\begin{aligned} T &= \bar{T} + T' \\ p &= \bar{p} + p' \\ \rho &= \bar{\rho} + \rho' \\ \mathbf{u} &= 0 + \mathbf{u}' \end{aligned} \qquad (5.190.a - d)$$

relative to the basic state,

$$\bar{\rho} = \rho_0 [1 - \alpha(\bar{T} - T_0)], \qquad (5.191)$$

by subtracting (5.172) from (5.169), we have:

$$\rho' = -\rho_0 \alpha T'. \qquad (5.192)$$

Next, considering perturbed state of the linearized momentum equation,

$$\frac{\partial \mathbf{u}'}{\partial t} = -g \frac{\rho'}{\rho_0} \mathbf{k} - \frac{1}{\rho_0} \nabla p' + \nu \nabla^2 \mathbf{u}' \qquad (5.193)$$

and substituting from (5.192),

$$-g \frac{\rho'}{\rho_0} \mathbf{k} = g \alpha T' \mathbf{k} \qquad (5.194)$$

gives

$$\frac{\partial \mathbf{u}'}{\partial t} = g \alpha T' \mathbf{k} - \frac{1}{\rho_0} \nabla p' + \nu \nabla^2 \mathbf{u}'. \qquad (5.195)$$

Dropping primes and separating vector components of the perturbed momentum Eq. (5.195) gives

$$\frac{\partial u}{\partial t} = -\frac{1}{\rho_0}\frac{\partial p}{\partial x} + \nu\nabla^2 u$$

$$\frac{\partial u}{\partial t} = -\frac{1}{\rho_0}\frac{\partial p}{\partial y} + \nu\nabla^2 v \qquad\qquad (5.196.a - c)$$

$$\frac{\partial w}{\partial t} = \alpha g T - \frac{1}{\rho_0}\frac{\partial p}{\partial z} + \nu\nabla^2 w.$$

Taking Laplacian gradients ∇^2 of (5.196.c) gives

$$\frac{\partial}{\partial t}\nabla^2 w = \alpha g\nabla^2 T - \frac{1}{\rho_0}\nabla^2\frac{\partial p}{\partial z} + \nu\nabla^4 w. \qquad\qquad (5.197)$$

Finally by combining Eqs. (5.196.a–c) by the following operations and thus Eliminating u, v by making use of continuity gives

$$\frac{\partial}{\partial z}\left(\frac{\partial}{\partial x}(5.196.a) + \frac{\partial}{\partial y}(5.177.b) + \frac{\partial}{\partial z}(5.196.c)\right)$$

yields

$$\frac{\partial}{\partial z}\frac{\partial}{\partial t}\underbrace{(\nabla\cdot\mathbf{u})}_{0} = \alpha g\frac{\partial^2 T}{\partial z^2} - \frac{1}{\rho_0}\frac{\partial}{\partial z}(\nabla^2 p) + \nu\nabla^2\underbrace{(\nabla\cdot\mathbf{u})}_{0} \qquad\qquad (5.198)$$

which then results in

$$-\frac{1}{\rho_0}\frac{\partial}{\partial z}\nabla^2 p = -\alpha g\frac{\partial^2 T}{\partial z^2}, \qquad\qquad (5.199)$$

Eliminating pressure p between (5.178) and (5.180) results in

$$\left(\frac{\partial}{\partial t} - \nu\nabla^2\right)\nabla^2 w = g\alpha\nabla_h^2 T \qquad\qquad (5.200)$$

where the second order horizontal Laplacian is denoted as

$$\nabla_h^2 = \frac{\partial^2}{\partial x^2} + \frac{\partial^2}{\partial y^2}. \qquad\qquad (5.201)$$

On the other hand, thermal energy equation describing the heat transfer in the fluid takes the form

$$\frac{\partial T}{\partial t} + \mathbf{u}\cdot\nabla T - K\nabla^2 T = 0 \qquad\qquad (5.202)$$

and upon subtracting the mean,

$$\nabla^2 \bar{T} = 0 \tag{5.203}$$

and neglecting the nonlinear terms of perturbed variables gives

$$\frac{\partial T'}{\partial t} + \mathbf{u} \cdot \nabla \bar{T} - \mathbf{u} \cdot \nabla T' - K \nabla^2 T' = 0, \tag{5.204}$$

and inserting the term representing mean vertical advection from (5.186)

$$\mathbf{u} \cdot \nabla \bar{T} = w \frac{\partial \bar{T}}{\partial z} = -\gamma w \tag{5.205}$$

where $\gamma = \Delta T / d$ representing temperature gradient, the linear perturbed heat equation

$$\frac{\partial T'}{\partial t} - K \nabla^2 T' = \gamma w. \tag{5.206}$$

Combining (5.200) and (5.206), finally dropping primes, one obtains the self-contained governing equation for temperature perturbations:

$$\left(\frac{\partial}{\partial t} - \nu \nabla^2 \right) \left(\frac{\partial}{\partial t} - K \nabla^2 \right) \nabla^2 T = g \alpha \gamma \nabla_h^2 T. \tag{5.207}$$

The sixth-order governing equation (5.207) for the perturbed state represents complex dynamics of the convection process. The boundary conditions are the *no-slip conditions*, requiring fluid particles to be motionless at the solid boundaries

$$u = v = w = 0 \quad \text{at} \quad z = 0 \quad \text{and} \quad z = d \tag{5.208}$$

which by virtue of continuity equation $\nabla \cdot \mathbf{u} = 0$ require that

$$w = 0 \quad \text{and} \quad \frac{\partial w}{\partial z} = 0 \quad \text{at} \quad z = 0 \quad \text{and} \quad z = d, \tag{5.209}$$
$$T = 0 \quad \text{at} \quad z = 0 \quad \text{and} \quad z = d.$$

Transforming dimensional coordinates to dimensionless variables on the left hand side of (5.210),

$$\begin{pmatrix} x' \\ y' \\ z' \\ \tau \end{pmatrix} = \begin{pmatrix} x/d \\ y/d \\ z/d \\ t/(d^2/\nu) \end{pmatrix}, \tag{5.210.a}$$

non-dimensional form of (5.188) is obtained where θ is the non-dimensional temperature. Using the above scales, and assigning nod-dimensional variables

$$
T = \gamma d \, \Theta
$$

$$
w = \frac{K}{d} \, W
$$

(5.210.b, c)

Equations (5.181) and (5.187) are converted to non-dimensional forms

$$
\left(\frac{\partial}{\partial \tau} - \nabla^2\right) \nabla^2 W = \mathrm{Ra} \nabla_h^2 \Theta
$$

$$
\left(\mathrm{Pr}\frac{\partial}{\partial \tau} - \nabla^2\right) \Theta = W
$$

(5.211.a, b)

When combined these give

$$
\left(\mathrm{Pr}\frac{\partial}{\partial \tau} - \nabla^2\right)\left(\frac{\partial}{\partial \tau} - \nabla^2\right) \nabla^2 \theta = \mathrm{Ra} \nabla_h^2 \theta
$$

(5.212)

where we find

$$
\mathrm{Pr} = \nu/K \quad \text{(Prandtl number)}
$$

$$
\mathrm{Ra} = \frac{g\alpha\gamma d^4}{K\nu} = \frac{g\alpha\Delta T d^3}{K\nu} \quad \text{(Rayleigh number)}.
$$

(5.213.a, b)

It is to be noted that the instability of perturbed flow is determined by 6th order partial differential equations representing viscous and diffusive dynamics. However, based on the homogeneous nature of the governing equations and boundary conditions, the theory presents an eigenvalue problem that is dependent on characteristic dimensional numbers Pr and Ra.

Trial solutions of the form

$$
\theta = F(x, y)Z(z)e^{\sigma\tau}
$$

(5.214)

are inserted in (5.193) and by applying separation of variables technique, one obtains

$$
(\nabla^2 - \sigma)(\nabla^2 - \mathrm{Pr}\,\sigma)(Z\nabla_h^2 F + \frac{d^2 Z}{dz^2} F) = \mathrm{Ra} Z \nabla_h^2 F
$$

(5.215)

and by rearranging,

$$
(\nabla^2 - \sigma)(\nabla^2 - \mathrm{Pr}\,\sigma)\left(\underbrace{\frac{\nabla_h^2 F}{F}}_{-a^2} + \frac{Z''}{Z}\right) FZ = \mathrm{Ra}\,\underbrace{\frac{\nabla_h^2 F}{F}}_{-a^2} FZ
$$

(5.216)

and setting a^2 as separation constant and denote

$$\nabla^2 F = \nabla_h^2 F + \frac{d^2 F}{dz^2} \quad \text{and} \quad Z'' = \frac{d^2 Z}{dz^2}. \tag{5.217}$$

Helmholtz equation is obtained for the horizontal part of the solution

$$\nabla_h^2 F + a^2 F = 0 \tag{5.218}$$

while the remaining part of the solution is obtained from

$$\left[\left(\frac{d^2}{dz^2} - a^2\right) - \text{Pr}\,\sigma\right]\left[\left(\frac{d^2}{dz^2} - a^2\right) - \sigma\right]\left(\frac{d^2}{dz^2} - a^2\right) Z = -a^2\,\text{Ra}\,Z \tag{5.219}$$

where a^2 is the separation constant, and a is the expected wave number of oscillatory solutions.

Typical cellular patterns of motion can be obtained as solutions in the horizontal plane, depending on the Prandtl and Rayleigh Numbers, and the yet undetermined separation constant a^2, as demonstrated in Fig. 5.27. For instance, as the Rayleigh number is increased, different patterns emerge, with rectangular cells obtained as

$$F(x, y) \sim \cos mx \cos ny, \quad m^2 + n^2 = a^2 \text{ for integers } m, n. \tag{5.220.a}$$

while hexagonal cells can be formed as

$$F(x, y) \sim \left[\cos\frac{a}{2}(\sqrt{3}x + y) + \cos\frac{a}{2}(\sqrt{3}x - y) + \cos ay\right]. \tag{5.220.b}$$

as shown in Fig. 5.27. The cells are oriented such that upward/downward vertical motions are concentrated at vertices shown with green/red colors in Fig. 5.27.

Often the linear theory is not sufficient to describe the actual convection pattern since the solution grows as

$$\sim e^{\sigma_r \tau}$$

Fig. 5.27 Pattern formation due to convection cells forced by differential heating of a fluid between two horizontal plates

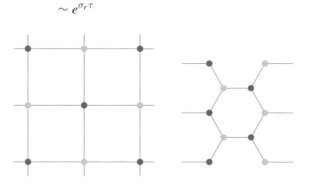

σ_r being the real part of σ, when the fluid is unstable, when nonlinear terms tend to become important. Extensive analyses including the conditions for neutral stability $\sigma = 0$ have been investigated by Chandrasekhar [7]. Steady solutions with neutral stability are investigated bys setting $\sigma = 0$,

$$\left(\frac{d^2}{dz^2} - a^2 \right)^3 Z = -a^2 \, \text{Ra} \, Z \tag{5.221}$$

with boundary conditions

$$Z = 0, \quad \frac{dZ}{dz} = 0, \quad \left(\frac{d^2}{dz^2} - a^2 \right)^2 Z = 0 \quad \text{at} \quad z = 0 \text{ and } z = 1.$$
$$\tag{5.222.a - c}$$

We notice that the solutions should depend on the numbers Pr, Ra and the separation constant a^2. The Prandtl number $\text{Pr} = \nu/K$ typically depends on fluid properties and for typical motion it has been rated as $\text{Pr} = 0.72$. It remains to search for the dependence on remaining two parameters, and specifically to seek the most unstable range. Under the theory of neutral instability, it has been shown that the Rayleigh number Ra reaches a minimum value as a function of a^2,

$$\text{Ra}_{min} = 1707.762 \text{ occurring at } a_{min} = 3.117 \tag{5.223.a, b}$$

which gives the criterion for instability

$$\text{Ra} < \text{Ra}_{min} \tag{5.224}$$

based on exact analytical and numerical solutions obtained by Chandrasekhar (1961).

Exercises

Exercise 1

Consider steady state stratification in a container with radial symmetry, to evaluate how geometrical properties influence thermal diffusion and boundary layer circulation. Derive a version of Eq. (5.144) that applies to thermal stratification for a particular shape of basin, with radius varying as a function of the vertical coordinate $r = r(z)$.

Exercise 2

Describe possible solutions for containers with monotonously increasing radius that satisfy equations described in Exercise 1.

Exercise 3

As demonstrated in Exercises 1 and 2, heat losses or gains by stratified containers can be redistributed effectively by boundary layer flows described in Sect. 5.2. The rapid transfer of heat by boundary layers competes with the slower diffusive processes maintaining stratification in the interior. External heating or cooling can be used advantageously to maintain or destroy stratification in a container.

In this guided exercise, the reader is motivated to investigate boundary layer control of stratification by loss of latent heat at the wet and free surfaces of the fluid and other components of heat flow balanced by diffusive fluxes in the fluid interior. Special applications in cooling water jugs, with apparent historical significance are brought to the attention of inquisitive readers, leaving exact solutions for further studies.

Exercise 4

In Exercise 3, equations for thermal stratification in containers of relatively simple geometry were described, leaving detailed solutions of specific problems for future investigations. In the present exercise, trial solutions will be attempted for time dependent development of thermal stratification in a conically shaped container, with linear increase in radius, corresponding to $q = 1$.

Exercise 5

This reading exercise is hoped to lead to investigations of stratified basins at larger scales of the earth geography. Can the various problems of thermal stratification development in containers analyzed in Exercises 1–4 help define and lead to investigation of real world problems at geophysical scales? As examples, interest of the reader is rekindled by some literature sources, to enable recent and upcoming research to bring new light to atmospheric and ocean dynamics of high plateaus, deep basins and abysses of bounded geometry.

References

1. Schlichting H (1968) Boundary layer theory. McGraw-Hill, New York
2. Schlichting H, Gersten K (2017) Boundary layer theory, 8th edn. Springer, Berlin (new edition)
3. Ostrach S (1952) An analysis of laminar free-convection flow and heat transfer about a flat plate parallel to the direction of the generating force. NACA Technical Note 2635, Lewis Flight Propulsion Laboratory, USA
4. Turner JS (1973) Buoyancy effects in fluids. Cambridge at the University Press
5. Rahm L, Walin G (1979a) Theory and experiments on the control of the stratification in almost-enclosed regions. J Fluid Mech 90:315–325
6. Rahm L, Walin G (1979b) On thermal convection in stratified Fluids. Geophys Astrophys Fluid Dyn 13:51–65
7. Chandrasekhar S (1961) Hydrodynamic and hydromagnetic stability. Oxford University Press, Oxford

Jets and Plumes

6

We present a much idealized theory of jets and plumes in this chapter. Jet flows are very similar to boundary layers considered in the last chapter. Free jets and wakes can be created away from solid boundaries by a momentum source, while buoyant plumes involve effects of density.

6.1   Momentum Jets

Pure momentum jets in a homogeneous fluid are created by injection of fluid from an initial source of momentum $\rho u_0^2 A_0$, where we define density ρ, kinematic viscosity ν, initial velocity u_0 and cross sectional area A_0 at the source. Ensuing flow depends on initial velocity and the environment where the discharge takes place.

Internal friction is an essential part of dynamics in a viscous fluid. For sufficiently small initial velocity, inviscid source/sink flow, as sketched in Fig. 6.1, is closely described by potential flow theory. At medium range of Reynolds number Re $= u_0 b_0 / \nu \leq O(1)$ (where b is the initial width), increased effects of internal friction become especially significant near the source, while the expected pattern remains close to that of potential flow shown in Fig. 6.2.

With source velocity increased to a range Re $= u_0 b_0 / \nu \gg 1$, the flow attains a boundary layer structure in a confined region aligned with the main axis, with smaller velocity outside the jet. As a result of viscous friction, some of the surrounding fluid is put into motion by the main stream of flow as shown in Fig. 6.2.

© Springer Nature Switzerland AG 2021   227
E. Özsoy, *Geophysical Fluid Dynamics II*, Springer Textbooks in Earth Sciences,
Geography and Environment,
https://doi.org/10.1007/978-3-030-74934-7_6

Fig. 6.1 Flow originating from an initial source for the medium range of Reynolds Number Re = $u_0 b_0 / \nu \leq O(1)$

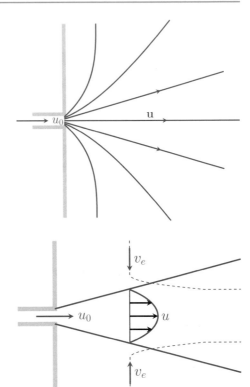

Fig. 6.2 Laminar jet flow originating from an initial source for the greater range of Reynolds Number Re = $u_0 b_0 / \nu \simeq O(1)$

6.1.1 Plane Laminar Jet

As an introduction, we first investigate the case of a plane laminar jet in two dimensions, making use of the relevant boundary layer equations (5.9.a,b) in Chap. 5. In this case we are also consider homogeneous fluids with density $\rho = $ constant.

$$\frac{\partial u}{\partial t} + u\frac{\partial u}{\partial x} + v\frac{\partial u}{\partial y} = -\frac{1}{\rho}\frac{\partial p}{\partial x} + \nu\frac{\partial^2 u}{\partial y^2}$$
$$\frac{\partial u}{\partial x} + \frac{\partial v}{\partial y} = 0 \qquad\qquad (6.1.a, b)$$

Considering steady-state ($\partial u/\partial t = 0$) *free-jets* without pressure forcing in the exterior ($\partial p/\partial x = 0$, equation (6.1.a) is integrated across the jet, to give

$$\int_{-\infty}^{\infty} u\frac{\partial u}{\partial x}dy + \int_{-\infty}^{\infty} v\frac{\partial u}{\partial y}dy = \nu\int_{-\infty}^{\infty}\frac{\partial^2 u}{\partial y^2}dy. \qquad (6.2)$$

The first term is

$$\int_{-\infty}^{\infty} u\frac{\partial u}{\partial x}dy = \int_{-\infty}^{\infty}\frac{1}{2}\frac{\partial u^2}{\partial x}dy = \frac{1}{2}\frac{d}{dx}\int_{-\infty}^{\infty} u^2 dy. \qquad (6.3.a)$$

The second term can be integrated with the help of continuity equation (6.1.b):

$$\int_\infty^\infty v \frac{\partial u}{\partial y} dy = [uv]_{-\infty}^\infty \rightarrow 0 - \int_{-\infty}^\infty u \frac{\partial v}{\partial y} dy$$

$$= \int_{-\infty}^\infty u \frac{\partial u}{\partial x} dy = \frac{1}{2} \frac{d}{dx} \int_{-\infty}^\infty u^2 dy \qquad (6.3.b)$$

where the first term vanishes because $u \to 0$ for $y \to \pm\infty$, whereas v approaches finite value away from the jet, as sketched in Fig. 6.2. The last term of (6.2) vanishes as

$$\nu \int_\infty^\infty \frac{\partial^2 u}{\partial y^2} dy = \nu \left[\frac{\partial u}{\partial y} \right]_{-\infty}^\infty = 0 \qquad (6.3.c)$$

because velocity u vanishes outside the jet, and similarly shear stresses $\nu \partial u / \partial y \to 0$ vanish as $y \to \pm\infty$. Therefore equation (6.2) becomes

$$\frac{d}{dx} \int_{-\infty}^\infty u^2 dy = 0 \qquad (6.4)$$

and the quantity

$$J = \rho \int_{-\infty}^\infty u^2 dy = \text{constant} \qquad (6.5)$$

is the integrated momentum J, which is found to be conserved along the jet.

As in boundary layer theory reviewed in Chap. 5, self-similar solutions can be expected, particularly because the free problem does not seem have a characteristic linear dimension. A self-similar velocity profile is specified with respect to dimensionless coordinate y/b, where $b(x)$ is the width of jet,

$$u = f \left(\frac{y}{b(x)} \right), \qquad (6.6.a)$$

a transform of the x-coordinate raised to power q is proposed, such that width $b \to 0$ as $x \to 0$,

$$b = x^q. \qquad (6.6.b)$$

In essence, a stream-function Ψ is specified according to coordinate transformations set as follows, where p and q are undetermined constants:

$$\Psi = \zeta^p f(\eta) \qquad (6.6.c)$$

$$\zeta = x \quad \text{and} \quad \eta = \frac{y}{b(\zeta)} = \frac{y}{\zeta^q}. \qquad (6.6.d)$$

Velocity components in transformed coordinates are then found as

$$u = \frac{\partial \Psi}{\partial y} = \frac{\partial \Psi}{\partial \zeta} \frac{\partial \zeta^{0}}{\partial y} + \frac{\partial \Psi}{\partial \eta} \frac{\partial \eta}{\partial y}$$

$$= \zeta^{p} f_{\eta}(\eta) \frac{1}{\zeta^{q}}$$

$$= \zeta^{p-q} f_{\eta}(\eta) \tag{6.7a}$$

$$v = \frac{\partial \Psi}{\partial x} = -\frac{\partial \Psi}{\partial \zeta} \frac{\partial \zeta}{\partial x} - \frac{\partial \Psi}{\partial \eta} \frac{\partial \eta}{\partial x}$$

$$= -p\zeta^{p-1} f(\eta) + \zeta^{p} f_{\eta}(\eta)(-q\eta\zeta^{-1}) \tag{6.7.b}$$

$$= -\zeta^{p-1} \left[pf(\eta) - q\eta f_{\eta}(\eta) \right].$$

Evaluating the momentum flux J to a constant gives

$$J = \rho \int_{-\infty}^{\infty} u^{2} dy = \rho \int_{-\infty}^{\infty} \zeta^{2(p-q)} [f_{\eta}(\eta)]^{2} \zeta^{q} dy$$

$$= \rho \zeta^{2p-q} \int_{-\infty}^{\infty} [f_{\eta}(\eta)]^{2} d\eta = \text{constant} \tag{6.8}$$

yields a requirement that p and q are related as

$$2p - q = 0 \tag{6.9.a}$$

and another relation between p and q is obtained by substituting (6.7.a,b) into (6.1.a),

$$2p - 2q - 1 = p - 3q. \tag{6.9.b}$$

Equations (6.9.a,b) establish the constants

$$p = \frac{1}{3} , \quad q = \frac{2}{3} \tag{6.9.c}$$

and the required transformations as

$$\zeta = x, \quad \eta = \frac{1}{3\nu^{1/2}} \frac{y}{x^{2/3}}, \quad \Psi = \nu^{1/2} x^{1/3} f(\eta). \tag{6.10}$$

Substituting these into (6.1.a), the set of governing equation and necessary boundary conditions take the form

$$f_{\eta\eta\eta} + ff_{\eta\eta} + (f_{\eta})^{2} = 0$$

$$f(0) = 0$$

$$f_{\eta\eta}(0) = 0 \tag{6.11.a - d}$$

$$f_{\eta}(\infty) = 0.$$

The first one of the boundary conditions set the center streamline has a value of zero, while the second one establishes symmetry of u velocity about the center-line and third one establishes that there is no flow external to the free jet. Equation (6.11.a) can be put into the form of

$$f_{\eta\eta} + (f f_\eta)_\eta = 0 \tag{6.12}$$

and integrating once, subject to boundary conditions (6.11.b–d) yields

$$f_{\eta\eta} + f f_\eta = 0. \tag{6.13}$$

To integrate (6.13) once more, we make use of a new set of transforms

$$\xi = \alpha \eta \quad \text{and} \quad f(\eta) = 2\alpha F(\xi) \tag{6.14.a, b}$$

where α is an arbitrary constant to be determined later. Equation (6.13) with boundary conditions

$$F_{\xi\xi} + 2F F_\xi = (F_\xi + F^2)_\xi = 0 \tag{6.15}$$

$$\begin{aligned} F(0) &= 0 \\ F'(\infty) &= 0. \end{aligned} \tag{6.16.a.b}$$

are integrated to give

$$F_\xi + F^2 = 1 \tag{6.17}$$

where the integration constant arbitrarily set to unity, considered as part of α.

We notice that (6.17) is the *Ricatti equation* which can not be directly integrated to give F in terms of ξ. Instead, ξ is integrated in terms of F,

$$\xi = \int \frac{dF}{1 - F^2} = \frac{1}{2} \ln\left(\frac{1+F}{1-F}\right) = \tanh^{-1} F. \tag{6.18}$$

or

$$F = \tanh \xi = \frac{e^\xi - e^{-\xi}}{e^\xi + e^{-\xi}} \tag{6.19}$$

Since

$$F_\xi = 1 - \tanh^2 \xi \tag{6.20}$$

u-velocity distribution of the jet is deduced with the help of (6.7.a)

$$u = \frac{2}{3}\alpha^2 x^{-1/3}(1 - \tanh^2 \xi). \tag{6.21}$$

The constant α can not be assigned any physical meaning unless its contribution to the momentum flux can be set as

$$J = \frac{8}{3}\rho\alpha^3\nu^{1/2}\int_{-\infty}^{\infty}(1 - \tanh^2\xi)^2 d\xi = \frac{16}{9}\rho\alpha^3\nu^{1/2} \tag{6.22.a}$$

with α evaluated as

$$\alpha = \left(\frac{9}{16}\frac{J}{\rho\nu^{1/2}}\right)^{1/3} = 0.83\left(\frac{J}{\rho\nu^{1/2}}\right)^{1/3}. \tag{6.22.b}$$

From (6.7.b), v-velocity can be obtained. Specifically, v is evaluated as $y \to \infty$,

$$v_e = -0.55\left(\frac{J\nu}{\rho x^2}\right)^{1/2} \tag{6.23}$$

i.e. $v \neq 0$ as $y \to \infty$, with negative velocity towards the jet center-line in the positive half-space of y, while the sign is reversed in the negative part. This is because fluid particles outside of the jet are carried away with the jet due to friction extending to the boundaries of the jet. *Entrainment velocity* is an essential element of jet dynamics, expected to occur also as part of *turbulent entrainment* process in the case of turbulent jets.

Jet horizontal velocity $u \to 0$ as $y \to \infty$ as the function $\tanh(\infty) \to 1$ and $F_\xi \to 0$ by virtue of (6.7.a). The net horizontal mass flux of the jet is calculated as

$$Q = \rho\int_{-\infty}^{\infty} u\, dy = 3.302\left(\frac{J\nu x}{\rho}\right)^{1/3} \tag{6.24}$$

which increases as $x \to \infty$ as a result of *entrainment* incorporating some exterior fluid into the jet. While initial momentum of the jet $J = \rho u_0^2 b_0$ is conserved ($J =$ constant), the volume flux Q is not conserved, increasing from its initial value of $Q = \rho u_0 b_0$ as a result of entrainment.

6.1.2   Plane Turbulent Jet

In the introduction to the present chapter it was noted that a free jet structure would develop from an outflow in the range of large Reynolds Number $Re = u_0 b_0/\nu \gg 1$. For some initial range the jet will remain laminar. With further increases in Re and the flow passing through a transition to turbulence, a turbulent jet flow will result. In the case of a planar jet, this transition has been found to occur near $Re \simeq 30$ (Schlichting 1968). The structure of the turbulent jet is sketched in Fig. 6.3.

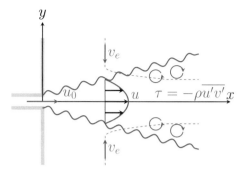

Fig. 6.3 Turbulent jet flow originating from an initial source for the large Reynolds Number Re = $u_0 b_0/\nu \gg O(1)$ case. The general pattern will look similar to the laminar case, but the momentum fluxes are greatly increased by turbulent eddies creating shear stress on the average. The boundary of the jet also takes an irregular form due to eddies, represented by the wavy surface in the above

Turbulent equations of motion are developed for the case of a plane turbulent jet (i.e. 2-D flow). Turbulent equations are actually obtained from either (5.3.a–c) or (5.9.a,b) by letting $u = \bar{u} + u'$, $v = \bar{v} + v'$ and averaging the equations over "appropriate" turbulence time scale T, with average properties indicated by overbars applied to terms within brackets in the following

$$\bar{()} = \frac{1}{T} \int_0^T ()dt$$

resulting in the averaged, steady-state boundary layer equations

$$\bar{u}\frac{\partial \bar{u}}{\partial x} + \bar{v}\frac{\partial \bar{u}}{\partial y} = -\frac{1}{\rho}\frac{\partial \bar{p}}{\partial x} + \nu\frac{\partial^2 \bar{u}}{\partial y^2} - \frac{\partial \overline{u'v'}}{\partial y} - \frac{\partial \overline{u'^2}}{\partial x}$$

$$0 = -\frac{1}{\rho}\frac{\partial \bar{p}}{\partial y} - \frac{\partial \overline{v'^2}}{\partial y} \qquad (6.25.a-c)$$

$$\frac{\partial \bar{u}}{\partial x} + \frac{\partial \bar{v}}{\partial y} = 0$$

where $\bar{u}$ and $\bar{v}$ are time averaged velocities and u' and v' represent fluctuating quantities due to turbulence. The last few terms in (6.25.a,b) result from averaging or correlations of fluctuating components $-\overline{u'v'}$ or $-\overline{u'u'}$ better referred to as *Reynold's stresses*. The same scaling rules for boundary layers, $v \ll u$ or $y \ll x$ are used here, neglecting some small terms in the momentum equations. Integrating (6.25.b) from y to a point located outside the jet gives

$$\bar{p} - p_\infty = -\rho\overline{v'^2}, \qquad (6.26.a)$$

since turbulence is expected to die outside of the jet (at $y \rightarrow \infty$), where the pressure has a constant value p_∞ outside the jet. Consequently, the pressure gradient term in (6.25.a) is replaced by

$$-\frac{1}{\rho}\frac{\partial \bar{p}}{\partial x} = -\frac{1}{\rho}\frac{\partial p_\infty}{\partial x}^{\,0} + \frac{\partial}{\partial x}(\overline{v'^2}) \qquad (6.26.b)$$

and substituting (6.26.b) into (6.25.a), from now on dropping overbars on the regular terms, while keeping them in correlation terms gives

$$u\frac{\partial u}{\partial x} + v\frac{\partial u}{\partial y} = \nu\frac{\partial^2 u}{\partial y^2} - \frac{\partial \overline{u'v'}}{\partial y} - \frac{\partial}{\partial x}(\overline{u'^2} - \overline{v'^2}). \qquad (6.27)$$

The second to last term on the right hand side of equation (6.27) represents gradients of turbulent *shear stress*, while the last term includes turbulent *normal stresses* responsible for transfer of momentum in the jet. The shear stress term (lateral turbulent transfer of momentum) is much more important than normal stresses (along jet transfer at momentum) jets, which justifies neglecting the last term. Remaining terms on the right hand side of (6.27) are combined as

$$\begin{aligned} \nu\frac{\partial^2 u}{\partial y^2} + \frac{\partial}{\partial y}(-\overline{u'v'}) &= \frac{1}{\rho}\frac{\partial}{\partial y}\left\{\mu\frac{\partial u}{\partial y} - \rho\overline{u'v'}\right\} \\ &= \frac{1}{\rho}\frac{\partial}{\partial y}(\tau_l + \tau_t) \end{aligned} \qquad (6.28)$$

in which the first term τ_l represents the laminar shear stress and the second term τ_t is the turbulent shear stress. For a free turbulent jet uninfluenced by solid boundaries turbulent stresses are much larger than their laminar counterparts, $\tau_t \gg \tau_l$ so that τ_l can be ignored completely, after which we identify shear stress as the turbulent one $\tau \equiv \tau_l$. Therefore turbulent free jet equations are simply written as

$$\begin{aligned} u\frac{\partial u}{\partial x} + v\frac{\partial u}{\partial y} &= \frac{1}{\rho}\frac{\partial \tau}{\partial y} \\ \frac{\partial u}{\partial x} + \frac{\partial v}{\partial y} &= 0 \end{aligned} \qquad (6.29.a, b)$$

The same way as it was done for laminar jets (Eqs. 6.2–6.4), momentum equation (6.29.a) for turbulent jet flow is integrated across the jet, to yield

$$\frac{d}{dx}\int_{-\infty}^{\infty}\rho u^2 dy + \rho[uv]_{-\infty}^{\infty} = [\tau]_{-\infty}^{\infty}. \qquad (6.30)$$

Boundary conditions similar to laminar jets are applied to limits of variables for turbulent jets. The u-velocity vanishes away from the jet center-line, and small but finite entrainment velocity v_e is created away from the jet, turbulent shear stress τ for obvious reasons vanishes in the exterior flow:

Fig. 6.4 Similarity solution fur turbulent jet flow originating from an initial source

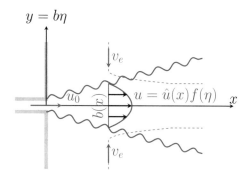

$$u \to 0, \quad v \to \pm v_e, \quad \tau \to 0 \quad \text{as} \quad y \to \pm\infty. \tag{6.31.a - c}$$

The entrainment velocity v_e is expected in this case by capture of ambient fluid parcels into the jet by turbulent eddies, rather than viscous friction in the laminar case.

Boundary conditions (6.31.a-c) applied to (6.30) leads to

$$\frac{d}{dx} \int_{-\infty}^{\infty} \rho u^2 dy = 0 \tag{6.32.a}$$

which then integrates to

$$J = \int_{-\infty}^{\infty} \rho u^2 dy = \text{constant.} \tag{6.32.b}$$

The velocity distribution is proposed to be self similar

$$\frac{u}{\hat{u}(y)} = f(\eta), \quad \eta = \frac{y}{b(x)} \tag{6.33.a, b}$$

such that the velocity profile $u(y)$ across the jet is normalized with respect the jet velocity at the center-line $\hat{u} = \hat{u}(x)$. Proposing a similarity solution in Fig. 6.4by making use of simple forms

$$\hat{u} \sim x^{-p}, \quad \text{and} \quad b \sim x^q \tag{6.34.a, b}$$

as done in the case of laminar jets, equation (6.32.a) then leads to

$$\frac{d}{dx} \int_{-\infty}^{\infty} \rho \hat{u}^2 f^2(\eta) \, b \, d\eta = \frac{d}{dx} \rho \hat{u}^2 b \int_{-\infty}^{\infty} f^2(\eta) d\eta = 0 \tag{6.31}$$

and substituting (6.33.a,b) into (6.31) gives

$$q - 2p = 0. \tag{6.32}$$

A second equation is needed to determine p, q. In the laminar case, a second equation was obtained from the momentum equation, but in the present case, the shear stress τ is not expressed in terms of other variables. The solution depends on how τ is parameterized, while only two equations (6.29.a,b) are available in the three unknowns u, v and τ. On dimensional grounds it is expected that

$$\frac{\tau}{\rho \hat{u}^2} = g(\eta). \tag{6.33}$$

We can then substitute (6.33.a,b) and (6.34.a,b) into the governing equations (6.29.a,b) to determine the functional dependence. Substituting (6.33), (6.33.a,b) and (6.34.a,b) into the governing equations (6.29.a,b), the individual terms are calculated as follows. Reminding that

$$u = \hat{u} f(\eta) \tag{6.38.a}$$

the u-velocity gradient is given by

$$\begin{aligned}
\frac{\partial u}{\partial x} &= \frac{\partial}{\partial x}[\hat{u}(x) f(\eta)] = \frac{\partial \hat{u}}{\partial x} f + \hat{u} \frac{\partial f}{\partial \eta} \frac{\partial \eta}{\partial b} \frac{\partial b}{\partial x} \\
&= \hat{u}_x f - \hat{u} f_\eta \frac{\eta}{b} b_x
\end{aligned} \tag{6.38.b}$$

v-velocity is given as

$$\begin{aligned}
v &= -\int_0^y \frac{\partial u}{\partial x} dy \\
&= \int_0^y (\hat{u} f_\eta \eta \frac{b_x}{b} - \hat{u}_x f) dy \\
&= \hat{u} b_x \int_0^\eta \eta f_\eta d\eta - \hat{u}_x b \int_0^\eta f d\eta \\
&= \hat{u} b_x (\eta f - \int_0^\eta f d\eta) - \hat{u}_x b \int_0^\eta f d\eta
\end{aligned} \tag{6.38.c}$$

and other terms are evaluated as

$$\frac{\partial u}{\partial y} = \frac{\partial}{\partial y}(\hat{u} f) = \hat{u} f_\eta \frac{\partial \eta}{\partial y} = \frac{\hat{u}}{b} f_\eta \tag{6.38.d}$$

$$\frac{1}{\rho} \frac{\partial \tau}{\partial y} = \frac{1}{\rho} \frac{\partial}{\partial y}(\rho \hat{u}^2 g) = \frac{\hat{u}^2}{b} g_\eta. \tag{6.38.e}$$

Substituting the above in (6.34.a) and rearranging,

$$g_\eta = \frac{b\hat{u}_x}{\hat{u}} \left(f^2 - f_\eta \int_0^\eta f d\eta \right) - b_x \left(\eta f_\eta - \eta f f_\eta + f_\eta \int_0^\eta f d\eta \right). \qquad (6.39)$$

In this equation, the left hand side is only a function of η, therefore the right hand side is also a function of η as well

$$\frac{b\hat{u}_x}{\hat{u}} = \text{constant}, \quad b_x = \text{constant} \qquad (6.40.a, b)$$

and substituting (6.34.a,b)

$$-p\frac{x^q x^{-p-1}}{x^{-p}} = -px^{q-1} = \text{constant}, \quad qx^{q-1} = \text{constant}. \qquad (6.41.a, b)$$

To satisfy these conditions together with (6.32), it is required that

$$q = 1 \quad p = \frac{1}{2}, \qquad (6.42.a, b)$$

which then leads to

$$\hat{u} \sim \frac{1}{x^{1/2}} \quad b \sim x \qquad (6.43.a, b)$$

The *entrainment velocity* is obtained at the limit $v \to v_e$ as $y \to \infty$, which then is parameterized as follows. The volume flux is calculated as

$$Q = \int_{-\infty}^{\infty} u \, dy \qquad (6.40)$$

differentiated to give twice the value of entrainment velocity v_e

$$\begin{aligned} \frac{dQ}{dx} &= \frac{d}{dx} \int_{-\infty}^{\infty} u dy \\ &= \int_{-\infty}^{\infty} \frac{\partial u}{\partial x} dy \\ &= -\int_{-\infty}^{\infty} \frac{\partial v}{\partial y} dy \\ &= -[v]_{-\infty}^{\infty} \\ &= 2v_e. \end{aligned} \qquad (6.41)$$

On the other hand, making use of (6.33.a,b) and (6.34.a,b),

$$v_e = \frac{1}{2}\frac{d}{dx}\int_{-\infty}^{\infty} \hat{u} f(\eta)\, b\, d\eta = \frac{1}{2}\frac{d}{dx}\hat{u}b \underbrace{\int_{-\infty}^{\infty} f(\eta)d\eta}_{\text{constant}}$$

$$\simeq \frac{1}{2}\underbrace{\int_{-\infty}^{\infty} f(\eta)d\eta}_{\text{constant}} \frac{d}{dx}(x^{-\frac{1}{2}}x) \tag{6.42}$$

$$\simeq \frac{d}{dx}x^{1/2}$$

$$\simeq x^{1/2}$$

to show that v_e has the same x-dependence as $\hat{u}$

$$v_e(x) = -\alpha_e \hat{u}(x) \tag{6.48}$$

where α_e is defined as entrainment coefficient, illustrating the fact that entrainment velocity is everywhere proportional to the center-line velocity. The dependence of the solutions on x has been demonstrated. By similarity hypothesis, we also assume that the dependence on y-coordinate is represented by $f(\eta) = f\left(\frac{y}{b}\right)$. The exact form of this function $f(\eta)$ depends on the lateral dependence to be derived from equation (6.29.a,b). However, we still have an unknown τ in these equations and the form of solution will depend on how we parameterize τ.

Tollmien's solution is one of the early attempts to obtain shear stress for a plane turbulent jet, making use of *Prandtl's mixing length hypothesis*

$$\tau = \rho\ell^2 \left(\frac{\partial u}{\partial y}\right)^2 \tag{6.49}$$

where ℓ is the so-called *mixing length*. Based on dimensional considerations it can be postulated that

$$\ell = \beta c_2 x \tag{6.50}$$

where β and c_2 are constants and

$$b = c_2 x \tag{6.51}$$

is the width of the jet as derived in (6.43.b) with the appropriate constant c_2, with an additional proportionality constant β for mixing length. The last term of (6.29.a) is calculated as

$$\frac{1}{\rho}\frac{\partial\tau}{\partial y} = \frac{1}{\rho}\frac{\partial}{\partial y}\left[\rho\beta^2 c_2^2 x^2 \left(\frac{\partial u}{\partial y}\right)^2\right] = a^3 x^2 \frac{\partial u}{\partial y}\frac{\partial^2 u}{\partial y^2} \tag{6.52.a}$$

where

$$a^3 = 2\beta^2 c_2^2. \tag{6.52.b}$$

Defining u-velocity as a function of transformed coordinates

$$\frac{u}{\hat{u}} = f(y/b) = f\left(\frac{y}{c_2 x}\right) = \bar{f}\left(\frac{y}{ax}\right) = \bar{f}(\hat{\eta}) \tag{6.53.a}$$

where

$$\hat{\eta} = y/ax = (c_2/a)\eta, \tag{6.53.b}$$

yielding the u-velocity by virtue of (6.43.a)

$$u = \hat{u}(x)\bar{f}(\hat{\eta}) = sx^{-1/2}\bar{f}(\hat{\eta}) \tag{6.54}$$

where s is another constant. A stream-function is then defined as

$$\begin{aligned}
\Psi &= \int u\, dy = sx^{-1/2} \int \bar{f}(\hat{\eta}) ax\, d\hat{\eta} \\
&= asx^{1/2} \int \bar{f}\, d\hat{\eta} = asx^{1/2} F(\hat{\eta}).
\end{aligned} \tag{6.55}$$

Inserting this expression and evaluating the various terms in (6.29.a) yields

$$2F_{\hat{\eta}\hat{\eta}} F_{\hat{\eta}\hat{\eta}\hat{\eta}} + FF_{\hat{\eta}\hat{\eta}} + F_{\hat{\eta}}^2 = 0 \tag{6.56}$$

with the corresponding boundary conditions

$$\begin{aligned}
u(0) &= \hat{u} &\rightarrow\quad F_{\hat{\eta}}(0) &= 1 \\
u(\infty) &= 0 &\rightarrow\quad F_{\hat{\eta}}(\infty) &= 0 \\
v(0) &= 0 &\rightarrow\quad F(0) &= 0 \\
\tau(0) &= 0 &\rightarrow\quad F_{\hat{\eta}\hat{\eta}}(0) &= 0 \\
\tau(\infty) &= 0 &\rightarrow\quad F_{\hat{\eta}\hat{\eta}}(\infty) &= 0.
\end{aligned}$$

Equation (6.56) is then integrated, first to yield

$$F_{\hat{\eta}\hat{\eta}}^2 + FF_{\hat{\eta}} = 0 \tag{6.57}$$

using some of the boundary conditions. Equation (6.57) remains to be integrated numerically to obtain solution. The constants a, s are left to be determined experimentally.

A second solution was obtained by Goertler by assuming a different version of *Prandtl's hypothesis*

$$\tau = \rho\varepsilon\frac{\partial u}{\partial y} \tag{6.58}$$

where ε is the *kinematic eddy viscosity*. Equation (6.58) formulates turbulent shear stress in a similar way as the laminar case with ν replaced by ε. However in the turbulent case ε is not kept as a constant but rather found to be proportional to velocity. By dimensional reasoning, Goertler proposed

$$\varepsilon = k\hat{u}b \tag{6.59}$$

where k is a constant. Letting $\hat{u}$ and b_s to be the centerline velocity and jet width respectively, they are expressed as

$$\hat{u} = \hat{u}\left(\frac{x}{s}\right)^{-1/2}, \quad b = b_s\frac{x}{s} \tag{6.60.a, b}$$

and therefore (6.59) becomes

$$\varepsilon = kb_s\hat{u}\left(\frac{x}{s}\right)^{1/2} = \varepsilon_s\left(\frac{x}{s}\right)^{1/2} \tag{6.60.c}$$

where

$$\varepsilon_s = kb_s\hat{u}. \tag{6.60.d}$$

Furthermore, defining a new constant σ

$$\eta = \sigma\frac{y}{x} \tag{6.60.e}$$

and in line with (6.59), stream-function is obtained as

$$\psi = \sigma^{-1}\hat{u}s^{1/2}x^{1/2}F(\eta). \tag{6.60.f}$$

Substituting these in equation (6.29.a), it is found to become

$$\frac{1}{2}F_\eta + \frac{1}{2}FF_{\eta\eta} + \frac{\varepsilon_s}{\hat{u}s}\sigma^2 F_{\eta\eta\eta} = 0 \tag{6.61.a}$$

with boundary conditions

$$\begin{aligned} F(0) &= 0 \\ F_\eta(0) &= 0 \\ F_\eta(\infty) &= 0. \end{aligned} \tag{6.61.b − d}$$

By letting

$$2\sigma^2 = \frac{\hat{u}_x s}{\varepsilon_s} \tag{6.62}$$

Equation (6.61.a) becomes

$$F_{\eta\eta\eta} + FF_{\eta\eta} + (F_\eta)^2 = 0. \tag{6.63}$$

By integrating equation (6.63) twice we have

$$F^2 + F_\eta = 1 \tag{6.64}$$

I.e. the same as the laminar jet (Eq. 6.17), with the solution

$$F = \tanh \eta. \tag{6.65}$$

Free constants are obtained by evaluating the momentum

$$J = \rho \int_{-\infty}^{\infty} u^2 dy = \frac{4}{3} \rho u_s^2 \frac{s}{\sigma} \tag{6.66}$$

so that the solution is written as

$$u = \frac{\sqrt{3}}{2} \sqrt{\frac{J\sigma}{\rho x}} (1 - \tanh^2 \eta)$$

$$v = \frac{\sqrt{3}}{4} \sqrt{\frac{J}{\rho \sigma x}} [2\eta(1 - \tanh^2) - \tanh \eta] \tag{6.67.a - c}$$

$$\eta = \sigma \frac{y}{x}.$$

Note that the Tollmien solution depends on a single empirical constant a and the Goertler solution also depends on a single empirical σ. Based on experimental results it is found that $\sigma = 7.67$ [1–3].

This free parameter in fact represents the turbulent entrainment into the jet and is therefore related to the entrainment coefficient α_e in Eq. (6.48). From Goertler's solution (6.67.b) we can find the entrainment velocity

$$v_e = \lim_{\eta \to \infty} v = -\frac{\sqrt{3}}{4} \sqrt{\frac{J}{\rho \sigma x}}. \tag{6.68}$$

On the other hand, from (6.63.a)

$$\hat{u} = u(\eta = 0) = \frac{\sqrt{3}}{2} \sqrt{\frac{J\sigma}{\rho x}}. \tag{6.69}$$

so that from (6.48)

$$\alpha_e = -\frac{v_e}{\hat{u}} = \frac{1}{2\sigma} \tag{6.70}$$

and using a value of $\sigma = 7.67$ we find $\alpha_e = 0.065$. The combination of various experiments done over the years have establish a typical value of

$$\alpha_e = 0.053 \tag{6.71}$$

commonly used for entrainment coefficient in the case of plane jets. This coefficient roughly establishes that the entrainment is about 5% of the center-line velocity at any point along the jet.

Through experiments, the lateral dependence of jet velocity was also found and various approximate functional fits were made. One of the commonly used profiles is the Gaussian distribution

$$\frac{u}{\hat{u}} = \exp\left(-\frac{y^2}{A^2 x^2}\right) \tag{6.72}$$

where $A = 0.12$ typically for plane jets.

6.1.3   Axisymmetric Turbulent Jet

Consider now an axisymmetric jet issuing from a round orifice. With the various approximations that are similar to the plane jet we write the jet equation in cylindrical coordinates

$$u\frac{\partial u}{\partial x} + v\frac{\partial u}{\partial r} = \frac{1}{\rho}\left(\frac{\partial \tau_{xx}}{\partial x} + \frac{1}{r}\frac{\partial}{\partial r}r\tau_{rx}\right)$$
$$\frac{\partial u}{\partial x} + \frac{1}{r}\frac{\partial}{\partial r}rv = 0 \tag{6.73.a, b}$$

where r is the radial distance measured from the jet centerline and x is distance along the axis. Integrating in polar coordinates the around the axis applying $\int_0^{2\pi}()d\theta$ yields a factor 2π due to radial symmetry.

We observe that the shear stress along the jet axis must be small compared to the one across the jet $\tau_{xx} << \tau_{rx}$, so that the contribution of first term on the r.h.s. is canceled out. Multiplying (6.73.a) by ρr and integrating in the radial direction gives

$$\int_0^\infty \rho u r \frac{\partial u}{\partial x}dr + \int_0^\infty \rho v r \frac{\partial u}{\partial r}dr = \int_0^\infty \frac{\partial r\tau_{rx}}{\partial r}dr \tag{6.74.a}$$

upon simplifying and making use of (6.73.b) in the first two terms of (6.73.a),

$$\int_0^\infty \rho u r \frac{\partial u}{\partial x}dr = \frac{d}{dx}\int_0^\infty \rho\frac{1}{2}u^2 r\,dr \tag{6.74.b}$$

and similarly

$$\int_0^\infty \rho r v \frac{\partial u}{\partial r}dr = \int_0^\infty \rho\frac{\partial ruv}{\partial r}dr - \int_0^\infty \rho u \frac{\partial rv}{\partial r}dr$$
$$= [\rho r u v]_0^{\infty}{}^{0} + \int_0^\infty \rho r u \frac{\partial u}{\partial r}dr \tag{6.74.c}$$
$$= \frac{d}{dx}\int_0^\infty \rho\frac{1}{2}u^2 r\,dr$$

and therefore, considering boundary conditions $u(\infty) = 0$, $v(0) = 0$ and $\tau(\infty) = \tau_{rx}(0) = 0$ momentum integral is obtained

$$\frac{d}{dx} \int_0^\infty \rho u^2 r\, dr = [r\tau_{rx}]_0^\infty = 0 \qquad (6.74.d)$$

with a multiplier 2π applied to obtain

$$\frac{dM}{dx} = \frac{d}{dx} \int_0^\infty 2\pi r \rho u^2 dr = 0 \qquad (6.74.e)$$

which is a statement of the conservation of axial momentum

$$M = \int_0^\infty 2\pi r \rho u^2 dr = \pi r_0^2 \rho u_0^2 = M_0 \qquad (6.74.e)$$

of the jet, where r_0 is the radius and u_0 is the mean axial velocity at the source (opening) where the jet is issued. As has been applied for planar jets a solution can be proposed as

$$\frac{u}{\hat{u}} = f\left(\frac{r}{b}\right) = f(\eta) \qquad (6.75.a)$$

$$\hat{u} \sim x^{-p}, \quad b \sim x^q. \qquad (6.75.b)$$

Substituting these in (6.75) yields a relation between p and q

$$q - p = 0. \qquad (6.76.a)$$

A second relation is obtained by substituting (6.75.a–c) into the governing equations (6.73.a,b) which yields a value based on self-similarity hypothesis as

$$q = 1 \qquad (6.76.b)$$

so that from (6.76.a)

$$p = 1 \qquad (6.76.c)$$

selecting

$$\hat{u} \sim x^{-1}, \quad b \sim x \qquad (6.77.a, b)$$

in the case of the circular jet. The volume flux is found from

$$Q = \int_0^\infty 2\pi r u\, dr \qquad (6.78)$$

consequently enabling to compute its rate of change, by making use of continuity equation (6.73.a)

$$\frac{dQ}{dx} = \frac{d}{dx} \int_0^\infty 2\pi r u \, dr = \int_0^\infty 2\pi r \frac{\partial u}{\partial x} dr = -\int_0^\infty 2\pi \frac{\partial r v}{\partial r} dr$$

$$= -2\pi [rv]_0^\infty = -2\pi r_\infty v_\infty = 2\pi r_\infty v_e = \text{constant} \qquad (6.79)$$

making use of axisymmetric nature of the jet $v(0) = 0$, and at distance from the jet center-line radial velocity approaches the entrainment velocity $v_\infty = -v_e$. Here, r_∞ and v_∞ are symbolic quantities. Since $u \to 0$ as $r \to 0$ the limit of the integral $\int_0^\infty 2\pi r u \, dr$ can be changed to $\int_0^{\bar{b}} 2\pi r u \, dr$ where $\bar{b}$ is the hypothetical width of the jet where $u \to 0$. Therefore we can write instead of (6.79)

$$\frac{dQ}{dx} = \frac{d}{dx} \int_0^\infty 2\pi r u \, dr = 2\pi \bar{b} v_e \qquad (6.80)$$

where v_e is the entrainment velocity. On the other hand, utilizing (6.75.a) and (6.77.a,b) gives

$$v_e = \frac{1}{2\pi \bar{b}} \frac{d}{dx} \int_0^\infty 2\pi r u \, dr = \frac{1}{2\pi \bar{b}} \frac{d}{dx} \int_0^\infty 2\pi b^2 \hat{u} f(\eta) \eta \, d\eta$$

$$= \frac{1}{\bar{b}} \underbrace{\int_0^\infty f(\eta) \eta \, d\eta}_{\text{constant}} \frac{d}{dx} b^2 \hat{u} \qquad (6.81)$$

which is simplified to

$$v_e \sim \frac{1}{x} \frac{d}{dx} (x^2 x^{-1}) \sim \frac{1}{x} \sim \hat{u}. \qquad (6.82)$$

Alternatively, we propose the entrainment velocity be proportional to the axial velocity

$$v_e = \alpha_e \hat{u}. \qquad (6.83)$$

As in the plane-jet problem, Tollmien and Goertler type solutions for the round-jet can be obtained, depending on the type of shear stress parameterization.

If Tollmien type of Prandtl mixing length theory is used

$$\tau = \rho l^2 \left(\frac{\partial u}{\partial r} \right)^2 \qquad (6.84)$$

we arrive at

$$\left(F_{\eta\eta} - \frac{F_\eta}{\eta} \right)^2 = F F' \qquad (6.85)$$

which can be solved by series or numerical methods, with appropriate boundary conditions.

Instead with Goertler type of linear parameterization where turbulent viscosity simlpy replaces molecular viscosity

$$\tau = \rho\varepsilon\frac{\partial u}{\partial r} \tag{6.86}$$

the following equation is obtained

$$FF_\eta = F_\eta - \eta F_{\eta\eta} \tag{6.87.a}$$

By considering symmetry properties and boundary conditions of the axisymmetric jet is noted that, if $F(\eta)$ is a solution to equation $F(\sigma\eta) = F(\xi)$ is also a solution [1]. Where σ is an integration constant and

$$\xi = \sigma r/x \tag{6.87.b}$$

replaces the independent variable in equation (6.86)

$$F\frac{dF}{d\xi} = \frac{dF}{d\xi} - \xi\frac{d^2 F}{d\xi^2} \tag{6.87.c}$$

with a solution obtained as

$$F(\xi) = \frac{\xi^2}{1 + \xi^2/4}. \tag{6.87.d}$$

Solutions in terms of flow variables u, v are

$$u = \frac{\varepsilon}{x}\sigma^2\frac{1}{\xi}\frac{dF}{d\xi} = \frac{\varepsilon}{x}\frac{2\sigma^2}{\left(1 + \xi^2/4\right)^2}$$

$$v = \frac{\varepsilon}{x}\sigma\left(\frac{dF}{d\xi} - \frac{F}{\xi}\right) = \frac{\varepsilon}{x}\sigma\frac{\xi - \xi^3/4}{\left(1 + \xi^2/4\right)^2} \tag{6.88.a, b}$$

The conservation of axial momentum is evaluated by integrating (6.74.e) by making use of (6.88.a)

$$M = \int_0^\infty 2\pi r\rho u^2 dr = \frac{16}{3}\pi\rho\sigma^2\varepsilon^2 = \pi r_0^2\rho u_0^2 = M_0 \tag{6.89.a}$$

From which the value of σ could be related to the initial momentum and turbulent diffusivity, whereas an experimental value $\sigma = 18.5$ havs been reported. It is difficult to relate σ and α_e for axisymmetric jets as $v \to 0$ for $\xi \to \infty$, which is a result of radial symmetry and as indicated by (6.88.b). One possibility is to define an entrainment velocity at a reasonable distance from the jet axis in proportion to $\eta =$

y/b, arbitrarily defined as the outer edge of the jet, which allows estimate α_e based on experiments. A typical value has been reported as

$$\alpha_e = 0.026. \tag{6.89.b}$$

6.2　Plumes and Buoyant Jets

6.2.1　Simple Jet with Passive Diffusion of Concentration or Temperature

In the previous section, we have studied diffusion of momentum which is initially supplied to the jet at the source of the jet. If the jet has a second quantity such as temperature or solute concentration in addition to its initial momentum, this property will also transported by the jet. In this section at least, this additional property assumed not to influence the density of fluid, so that buoyancy forces do not enter the momentum equation. In this section, turbulent processes are assumed. In case of a plane jet, the momentum and continuity equations (6.28 and 6.29.a,b) are employed

$$u\frac{\partial u}{\partial x} + v\frac{\partial u}{\partial y} = \frac{1}{\rho}\frac{\partial}{\partial y}(-\overline{\rho u'v'})$$
$$\frac{\partial u}{\partial x} + \frac{\partial v}{\partial y} = 0 \tag{6.90.a, b}$$

which were integrated across the jet to give the momentum flux (6.32.a and 6.31):

$$\frac{d}{dx}\int_{-\infty}^{\infty} \rho u^2 dy = \frac{d}{dx}I_1\rho\hat{u}^2(x)b(x) = 0 \tag{6.91.a}$$

where

$$I_1 = \int_{-\infty}^{\infty} f^2(\eta)d\eta = \text{constant} \tag{6.91.b}$$

$$f(\eta) = \frac{u(x,y)}{\hat{u}}, \quad \eta = \frac{y}{b} \tag{6.91.c}$$

demonstrating that jet momentum is conserved. Similarity arguments iare used

$$\hat{u} \sim x^{-1/2}, \quad b \sim x \tag{6.92.a, b}$$

to evaluate the constant I_1 in (6.91.a) either by particular parameterization of $\tau = -\rho\overline{u'v'}$ or the formulation of $f(\eta)$. For instance, a Gaussian profile derived from experiments can be used so that

$$f(\eta) = \frac{u}{\hat{u}} = e^{-\left(\frac{y}{b(x)}\right)^2} = e^{-\eta^2} \tag{6.93}$$

where $\hat{u}$ is the center-line velocity and b is defined as the width of the jet, such that velocity is decreased to $u/\hat{u} = e^{-1} = 0.37$ at $\eta = 1$. From (6.91.b) and (6.93),

$$I_1 = \int_{-\infty}^{\infty} e^{-2\eta^2} d\eta = \sqrt{\frac{\pi}{2}} \tag{6.94}$$

As noted, if the plume carries a passive quantity such as temperature T, a steady-state thermal diffusion equation has to accompany (6.90.a,b):

$$u\frac{\partial T}{\partial x} + v\frac{\partial T}{\partial y} = \frac{\partial}{\partial y}(-\overline{v'T'}) \tag{6.95}$$

where $-\overline{v'T'}$ represents the lateral flux of heat by turbulent fluctuations. The ambient fluid is assumed to have constant temperature T_a which differs from the temperature in the jet

$$\Delta T = T - T_a \tag{6.96.a}$$

and therefore (6.95) can be written as

$$u\frac{\partial \Delta T}{\partial x} + v\frac{\partial \Delta T}{\partial y} = \frac{\partial}{\partial y}(-\overline{v'T'}) \tag{6.96.b}$$

Integrating the left hand side of this equation across the jet yields

$$\int_{-\infty}^{\infty} \left(u\frac{\partial \Delta T}{\partial x} + v\frac{\partial \Delta T}{\partial y} \right) dy$$

$$= \int_{-\infty}^{\infty} \frac{\partial}{\partial x}(u\Delta T)\, dy - \int_{-\infty}^{\infty} \Delta T \frac{\partial u}{\partial x}\, dy + \int_{-\infty}^{\infty} v\frac{\partial \Delta T}{\partial y}\, dy$$

$$= \frac{d}{dx}\int_{-\infty}^{\infty} u\Delta T\, dy + \int_{-\infty}^{\infty} \Delta T\frac{\partial v}{\partial y}\, dy + \int_{-\infty}^{\infty} v\frac{\partial \Delta T}{\partial y}\, dy$$

$$= \frac{d}{dx}\int_{-\infty}^{\infty} u\Delta T\, dy + [v\Delta T]_{-\infty}^{\infty} \overset{0}{-} \int_{-\infty}^{\infty} v\frac{\partial \Delta T}{\partial y}\, dy + \int_{-\infty}^{\infty} v\frac{\partial \Delta T}{\partial y}\, dy$$

$$\tag{6.97.a}$$

by making use of the continuity equation (6.90.b), integrating by parts and observing that $\Delta T \to T_a - T_a = 0$ as $y \to \pm\infty$. The right hand side of (6.96) is

$$\int_{-\infty}^{\infty} \frac{\partial}{\partial y}(-\overline{v'T'})\, dy = [-\overline{v'T'}]_{-\infty}^{\infty} = 0 \tag{6.97.b}$$

since the turbulent flux $-\overline{v'T'} \to 0$ as $y \to \pm\infty$. Therefore,

$$\frac{d}{dx}B = \frac{d}{dx}\int_{-\infty}^{\infty} u\Delta T\, dy = 0 \tag{6.98}$$

expresses the fact that buoyancy flux B is conserved along the jet. Based on the similarity hypothesis, velocity and temperature profiles are given respectively by (6.93) and

$$\frac{\Delta T}{\Delta \hat{T}} = \frac{T - T_a}{\hat{T} - T_a} = e^{-\left(\frac{y}{\lambda b}\right)^2} = e^{-\frac{\eta^2}{\lambda^2}} = g(\eta) \tag{6.99}$$

where $\hat{T}(x)$ is the center-line temperature. The coefficient λ accounts for differences in diffusion of heat as compared to momentum, i.e. the transports of momentum $-\overline{u'v'}$ and $-\overline{u'T'}$ are assumed similar in from although their magnitudes may differ. Substituting (6.99) and (6.93) into (6.97) yields

$$M = \frac{d}{dx} \int_{-\infty}^{\infty} \hat{u}(x) f(\eta) \Delta \hat{T}(x) g(\eta) b(x) \, d\eta = \frac{d}{dx}(I_2 \hat{u} \Delta \hat{T} b) = 0 \tag{6.100}$$

where

$$I_2 = \int_{-\infty}^{\infty} f(\eta) g(\eta) d\eta = \int_{-\infty}^{\infty} e^{-\left(1+\frac{1}{\lambda^2}\right)} d\eta = \sqrt{\frac{\pi}{\left(1 + \frac{1}{\lambda^2}\right)}}. \tag{6.101}$$

Substituting

$$\Delta \hat{T}(x) \sim x^s \tag{6.102}$$

together with (6.92.a,b) into (6.100) gives

$$\frac{d}{dx}(I_2 x^{-1/2} x^s x) \sim \frac{d}{dx}(x^{s+\frac{1}{2}}) = 0 \tag{6.103}$$

and therefore

$$s = -\frac{1}{2}, \quad \Delta \hat{T} \sim x^{-1/2} \tag{6.104}$$

which shows that the decay rate of temperature difference is similar to that for velocity. The thickness of the thermal boundary layer is $b_T = \lambda b$ which is different than the width b of the momentum boundary layer. The ratio between thermal and momentum boundary layer thickness is referred to as *Schmidt number*:

$$\lambda = \frac{b_T}{b} = \frac{\lambda b}{b} \simeq 1.2 \tag{6.105}$$

with a typical value as given above.

The main results discussed above for passive 2D jets do not change much following similar development for round turbulent jets. In the above discussion, it has been assumed that temperature does not change buoyancy or density. These effects can be ignored so far as they do not influence momentum. In fact, solutions in the present

Fig. 6.5 A pure plume
originating from a buoyancy
source

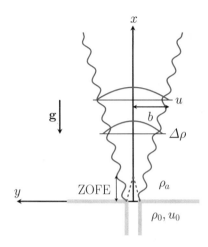

section are only valid if the Froude number at source is large enough to ignore such
influence:

$$F_0 = \frac{u_0}{\sqrt{g \frac{\Delta \rho_0}{\rho_a} b_0}} \rightarrow \infty, \tag{6.106}$$

quantifying the ratio of inertia to buoyancy forces. For smaller values of F_0 one must
take into account of buoyancy forces, considered in the next section.

6.2.2 Pure Plume

Consider a *buoyant plume* with density variations contrasted with a homogenous
ambient fluid with density ρ_a, allowing the motion to be driven by buoyancy alone,
without an initial source of momentum. The plume will rise vertically if the source
density is smaller than ρ_a, or sink vertically if it is larger than ρ_a), with buoyancy
forces in relation to gravity, as shown in Fig. 6.5.

Cosidering the 2-D case first (plane plume), the governing equations are

$$\frac{\partial u}{\partial x} + \frac{\partial v}{\partial y} = 0$$

$$u\frac{\partial u}{\partial x} + v\frac{\partial v}{\partial y} = -g\frac{\Delta \rho}{\rho_a} - \frac{\partial \overline{u'v'}}{\partial y} \tag{6.107.a - c}$$

$$u\frac{\partial \Delta \rho}{\partial x} + v\frac{\partial \Delta \rho}{\partial y} = -\frac{\partial \overline{v'\rho'}}{\partial y}$$

where

$$\Delta \rho = \rho - \rho_a \tag{6.108}$$

is the difference of plume density from the ambient. In comparing (6.90.a,b) and
(6.95) to (6.107.a–c), it can be noted that buoyancy force $-g\Delta\rho/\rho_a$ has been introduced on the right hand side of (6.107.a-c), which will be positive for a rising plume
$(\rho < \rho_a)$.

We note the similarity of (6.107.b,c) to equations (6.95) and (6.97.b), employing
the linearized equation of state (cf. 6.33)

$$\rho = \rho_a(1 - \alpha(T - T_a)) \qquad (6.109.a)$$

simply relatng density and temperature differences

$$\Delta\rho = \rho_a\alpha\Delta T. \qquad (6.109.b)$$

Integration of equations (6.107.a-c) across the plume proceeds as follows. Integration of the continuity equation (6.107.a) gives

$$\int_{-\infty}^{\infty} \frac{\partial u}{\partial x}dy + [v]_{-\infty}^{\infty} = \frac{d}{dx}\int_{-\infty}^{\infty} udy - 2v_e = 0 \qquad (6.110)$$

where v_e is the entrainment velocity. Next, by makimg use of (6.30) and integrating
(6.107.b), one obtains

$$\frac{d}{dx}\int_{-\infty}^{\infty} u^2dy + [uv]_{-\infty}^{\infty} = g\int_{-\infty}^{\infty} \frac{\Delta\rho}{\rho_a}dy + [-\overline{u'v'}]_{-\infty}^{\infty} \qquad (6.111.a)$$

and noting that $u \to \overline{u'v'} \to 0$ as $y \to \pm\infty$ this is simplified as

$$\frac{d}{dy}\int_{-\infty}^{\infty} u^2dy = g\int_{-\infty}^{\infty} \frac{\Delta\rho}{\rho_a}dy, \qquad (6.111.b)$$

which states that the jet momentum is not conserved, since the buoyancy forces
increase plume momentum. Integrating (6.107.c), with the help of (6.97.a,b) yields

$$\frac{d}{dx}\int_{-\infty}^{\infty} u\Delta\rho dy = 0 \qquad (6.112)$$

which states that the buoyancy flux is conserved along the plume. Proceeding with
experimentally verified self similar profiles as given in (6.93) and (6.99), the following is postulated

$$f(\eta) = \frac{u}{\hat{u}} = e^{-\eta^2} = e^{-\left(\frac{y}{b(x)}\right)^2}$$

$$g(\eta) = \frac{\Delta\rho}{\Delta\hat{\rho}} = \frac{\rho - \rho_a}{\hat{\rho} - \rho_a} = e^{-\frac{\eta^2}{\lambda^2}} = e^{-\left(\frac{y}{\lambda b(x)}\right)^2} \qquad (6.113.a, b)$$

where $\hat{u}$ and $\hat{\rho}$ are the jet velocity and density at the center-line. Substituting (6.113.a,b) into (6.110-6.112) and employing the entrainment assumption of $v_e = \alpha_e \hat{u}$ produces the following buoyant plume equations

$$\frac{d}{dx}(\sqrt{\pi}\hat{u}b) = 2\alpha_e \hat{u}$$

$$\frac{d}{dx}(\sqrt{\pi}\hat{u}^2 b) = \sqrt{\pi}\lambda g \frac{\Delta\hat{\rho}}{\rho_a} b \qquad\qquad (6.114.a - c)$$

$$\frac{d}{dx}\left(\sqrt{\frac{\pi}{1+\frac{1}{\lambda^2}}}\,\hat{u}\,\Delta\hat{\rho}\,b\right) = 0.$$

Seeking similarity solutions to equations (6.110.a-c)

$$\hat{u} \sim x^p \,,\ b \sim x^q \,,\ \Delta\rho \sim x^s \qquad\qquad (6.115.a - c)$$

yields

$$\frac{d}{dx}x^{p+q} = x^{p+q-1} \sim x^p$$

$$\frac{d}{dx}x^{2p+q} = x^{2p+q-1} \sim x^{s+p} \qquad\qquad (6.116.a - c)$$

$$\frac{d}{dx}x^{p+q+s} = 0$$

establishes the relationships

$$\require{cancel}\cancel{p} + q - 1 = \cancel{p}$$
$$2p + \cancel{q} - 1 = s + \cancel{q} \qquad\qquad (6.117.a - c)$$
$$p + q + s = 0$$

so that

$$p = 0$$
$$q = 1 \qquad\qquad (6.118.a - c)$$
$$s = -1$$

and therefore

$$\hat{u} = \text{constant}, \quad b \sim x, \quad \Delta\rho \sim \frac{1}{x} \qquad\qquad (6.119.a - c)$$

and with constant spreading rate of $b \sim x$,

$$\frac{db}{dx} = k_p \tag{6.120}$$

where k_p represents the constant expansion rate of the plume. Since $\hat{u} = $ constant, the entrainment coefficient α_{pe} of the plume is also constant and related to jet expansion as

$$\alpha_{pe} = \frac{\sqrt{\pi}}{2} k_p. \tag{6.121}$$

[*Comparison with 2-D jet:*
Differences in entrainment coefficients for plumes and jets can be investigated, based on the assumption of self-similar, Gaussian profiles. In the case of a jet, adoption of (6.114.a) yields

$$\hat{u}_j = c_j x^{-1/2}, \quad b_j = k_j x \tag{6.122.a, b}$$

$$\frac{d}{dx}(\sqrt{\pi} c_j x^{-1/2} k_j x) = \frac{\sqrt{\pi}}{2} c_j k_j x^{-1/2} \tag{6.123}$$

where c_j and k_j are appropriate constants, from which the corresponding jet entrainment coefficient is calculated as

$$\alpha_{ej} = \frac{\sqrt{\pi}}{4} k_j, \tag{6.124}$$

Hence proving that the entrainment into a plane is twice as large as of the jet, even if the spreading ratios were assumed to be same, wh,ile in reality, $k_p > k_j$ based on experiments. A further difference to be noted is plume center-line velocity $\hat{u}_p = $ constant versus that of the jet $\hat{u}_j = x^{-1/2}$, indicating a jet is decelerated faster, while $\Delta\hat{\rho}_p \sim x^{-1}$ for plume versus $\Delta\hat{\rho}_j \sim x^{-1/2}$ for jet, indicating a plume is diluted faster than a jet.]
Retuning to the pure plume and combining equations (6.119) with (6.121) for the plume one can write

$$\hat{u} = u_0 = \text{constant}, \quad b = \frac{2}{\sqrt{\pi}} \alpha_e x, \quad \frac{\Delta\hat{\rho}}{\rho_a} = \frac{c_p}{x} \tag{6.125.a - c}$$

where u_0 is the center-line velocity at source, c_p selected as an appropriate constant. Note, however, that the solutions become unrealistic near the origin $x \to 0$, $u = u_0 = $ constant, the width of the jet $b \to 0$ and $\Delta\hat{\rho} \to \infty$. In reality, concepts of *virtual origin* and *zone of flow establishment* are often adopted to overcome such difficulties as shown in Fig. 6.5.

Despite the singularity at origin, note however, that the fluxes of volume, momentum and buoyancy do not become singular as $x \to 0$

$$Q = \int_{-\infty}^{\infty} u \, dy = \int_{-\infty}^{\infty} \hat{u} b f(\eta) d\eta = \hat{u} b \int_{-\infty}^{\infty} f(\eta) d\eta$$

$$= u_0 \frac{2}{\sqrt{\pi}} \alpha_e x \int_{-\infty}^{\infty} f(\eta) d\eta \to 0, \quad x \to 0$$

$$M = \int_{-\infty}^{\infty} u^2 \, dy = \hat{u}^2 b \int_{-\infty}^{\infty} f^2(\eta)^2 d\eta = u_0^2 \frac{2}{\pi} \alpha_e x \int_{-\infty}^{\infty} f^2 d\eta \to 0, \quad x \to 0$$

$$B = \int_{-\infty}^{\infty} u \Delta \rho \, dy = \hat{u} b \Delta \hat{\rho} \int_{-\infty}^{\infty} f(\eta) g(\eta) d\eta$$

$$= u_0 \rho_a c_p \frac{2}{\sqrt{\pi}} \alpha_e \int_{-\infty}^{\infty} f g \, d\eta = B_0 = \text{constant}, \quad x \to 0$$

$$(6.126.a--c)$$

For the solution to be valid, the initial volume and momentum fluxes have to be specified as zero and initial buoyancy flux B_0 specified at the source defines the constant $\rho_a u_0 c_p$. Substituting (6.125.a–c) in the conservation equations (6.114.a–c) of volume, momentum and buoyancy fluxes, one gets

$$\frac{d}{dx}(\sqrt{\pi} \hat{u} b) = \sqrt{\pi} u_0 \frac{db}{dx} = 2\alpha_e u_0$$

$$\frac{d}{dx} \sqrt{\frac{\pi}{2}} \hat{u}^2 b = \sqrt{\frac{\pi}{2}} u_0^2 \frac{db}{dx} = \sqrt{2} \alpha_e u_0^2$$

$$= \sqrt{\pi} \lambda g \frac{\Delta \hat{\rho}}{\rho_a} b = \sqrt{\pi} \lambda g \frac{c_p}{\not{x}} \frac{2}{\sqrt{\pi}} \alpha_e \not{x} = 2 c_p \lambda g \alpha_e \qquad (6.127.a\text{--}c)$$

$$B_0 = \sqrt{\frac{\pi}{1 + \frac{1}{\lambda^2}}} \hat{u} b \Delta \hat{\rho} = \frac{2}{\sqrt{1 + \frac{1}{\lambda^2}}} u_0 \rho_a c_p \alpha_e$$

simplifying the above equations by making use of (6.127.b,c) gives

$$u_0^2 = \sqrt{2} c_p \lambda g$$

$$B_0 = \frac{2 u_0 \rho_a \alpha_e c_p}{\sqrt{1 + \frac{1}{\lambda^2}}} \qquad (6.128.a, b)$$

and eliminating u_0 between (6.128.a,b) the buoyancy flux can be evaluated from

$$B_0^2 = \frac{8\sqrt{2} \rho_a g \alpha_e \lambda^3 c_p^2}{1 + \lambda^2} \qquad (6.128.c)$$

for values of c_p and specified values of B_0, λ, ρ_a, and α_e, which are to be verified through experiment.

Note that the similarity solution presented above for a pure plume has a singularity at source $x = 0$, with volume and momentum fluxes $Q \to 0$ and $M \to 0$ and buoyancy flux with constant value $B = B_0$ according to (6.126.a–c), noting also that $\Delta\hat{\rho} \to \infty$ as $x \to 0$, according to (6.119.c). These results indicate that the plume is driven by an infinitesimal source of buoyancy B_0 at source, without initial sources of volume or momentum. We are only able to accept this solution by assigning a virtual origin at $x = 0$, which is physically unreasonable. Constants u_0 and ρ_0 had to be inserted as decoys in the above description. Actual initial conditions for velocity, width and buoyancy were not specified at the entrance, as falsely displayed in Fig. 6.5, in which case a similarity solution could not obtained for a pure plume. Solutions for buoyant plumes will address some of these questions in the next section.

Defining a local densimetric Froude number for the pure plume,

$$\mathrm{Fr}_L = \frac{u_0}{\left(\frac{\Delta\hat{\rho}}{\rho_a}gb\right)^{1/2}} \tag{6.128.d}$$

then making use of (6.125.a-c) and (6.128.c) and (6.128.a) it can be shown that the densimetric Froude number has a constant value

$$\mathrm{Fr}_L = \left(\frac{\pi}{2}\right)^{1/4}\left(\frac{\lambda}{\alpha_e}\right)^{1/2}. \tag{6.128.e}$$

There are two scaling coefficients in the Froude number, $(2/\pi)^{1/4}$ appearing due to the Gaussian velocity profiles and the factor $(\alpha_e/\lambda)^{1/2}$ accounting for entrainment difference between velocity and density profiles. Without these scaling factors the Froude number could be compared to the typical value of unity,that represents the balance for momentum jets.

6.2.3 Buoyant Jet

As discussed in the last section, pure plumes are created by infinitesimal source of buoyancy without a volume source of fluid motion. In case of realistic plumes, as discussed by Anwer (1969), [4] it is needed to specify relevant initial conditions at the source,

$$b = b_0, \quad \Delta\rho = \rho_a - \hat{\rho} = \rho_a - \rho_0, \quad \hat{u} = u_0 \text{ at } x = 0 \tag{6.129.$a - c$}$$

where ρ_0, u_0 and b_0 are respectively the density, velocity and width at the inlet. Starting with the above initial conditions, (6.114.a-c) govern the changes in volume, momentum and buoyancy fluxes applied to a *buoyant jet* rather than a pure plume

$$\frac{d}{dx}(b\hat{u}) = \frac{2}{\sqrt{\pi}}\alpha_e \hat{u}$$

$$\frac{d}{dx}(b\hat{u}^2) = \sqrt{2}\lambda g b\frac{(\rho_0 - \hat{\rho})}{\rho_a} \qquad (6.130)$$

$$\frac{d}{dx}[g(\rho_a - \hat{\rho})b\hat{u}] = 0$$

Dimensionless equations are generated by substituting the following dimensionless variables

$$X = \frac{2}{\sqrt{\pi}}\alpha_e \frac{x}{b_0}$$

$$B = \frac{b}{b_0}$$

$$P = \frac{\rho_a - \hat{\rho}}{\rho_a - \rho_0} \qquad (6.131.a - d)$$

$$U = \left(\sqrt{\frac{2}{\pi}}\frac{1}{g}\frac{\alpha_e}{\lambda}\frac{1}{b_0}\frac{\rho_a}{(\rho_a - \rho_0)}\right)^{1/2}\hat{u}$$

yielding

$$\frac{d}{dX}BU = U$$

$$\frac{d}{dX}BU^2 = BP \qquad (6.132.a - c)$$

$$\frac{d}{dX}PBU = 0.$$

The *source densimetric Froude number* is defined as

$$Fr = \left(\frac{2}{\pi}\right)^{1/4}\left(\frac{\alpha_e}{\lambda}\right)^{1/2}\frac{u_0}{\sqrt{g\frac{\rho_a - \rho_0}{\rho_a}b_0}}. \qquad (6.133)$$

As discussed in the last section, Froude number is scaled by two coefficients $(2/\pi)^{1/4}$ and $(\alpha_e/\lambda)^{1/2}$ representing adjustments of Gaussian velocity and density profiles. Then, by virtue of (6.133), (6.131.d) can be written as

$$U = Fr\frac{\hat{u}}{u_0} \qquad (6.134)$$

so that the initial conditions (6.129.a–c) become

$$B = 1, \quad P = 1, \quad U = Fr, \quad \text{at } X = 0 \qquad (6.135)$$

Equation (6.132.a,b) is readily integrated to give

$$PBU = \text{constant} = [PBU]_{x=0} = \text{Fr}. \tag{6.136}$$

Defining volume and momentum fluxes

$$Q = BU$$
$$M = BU^2 \tag{6.137.a, b}$$

Equations (6.132.a,b) take the following forms

$$\frac{dQ}{dX} = U = \frac{M}{Q}$$
$$\frac{dM}{dX} = BP = \frac{\text{Fr}}{U} = \text{Fr}\frac{Q}{M} \tag{6.138.a, b}$$

where use has been made of (6.137.a,b) and (6.136). The initial conditions for (6.138.a,b) are

$$Q = \text{Fr}, \quad M = \text{Fr}^2 \quad \text{at} \quad X = 0. \tag{6.138.c}$$

Equations (6.138.a,b) are combined into a single equation

$$\frac{dM}{dQ} = \text{Fr}\frac{Q^2}{M^2} \tag{6.140}$$

which then is integrated to get

$$\frac{M^3}{3} = \text{Fr}\frac{Q^3}{3} + C \tag{6.141}$$

where C is a constant, evaluated by making use of the initial conditions (6.138.c)

$$C = \frac{1}{3}(M^3 - \text{Fr } Q^3) = \frac{1}{3}(\text{Fr}^6 - \text{Fr Fr}^3) = \frac{\text{Fr}^4}{3}(\text{Fr}^2 - 1) \tag{6.142}$$

Substituting in (6.141) yields as the solution M as a function of Q.

$$M^3 = \text{Fr } Q^3 + \text{Fr}^4(\text{Fr}^2 - 1). \tag{6.143}$$

Equation (6.138.a) reconstructs the solution X in terms of the dependent variables

$$X = \int_0^X dX = \int_{\text{Fr}}^Q \frac{Q}{M}dQ = \int_{\text{Fr}}^Q \frac{Q\,dQ}{\left(\text{Fr } Q^3 + \text{Fr}^4\left(\text{Fr}^2 - 1\right)\right)^{1/3}}. \tag{6.144}$$

Letting

$$Q = \text{Fr} \, (\text{Fr}^2 - 1)^{1/3} t \tag{6.145}$$

yields

$$X = \int_{(\pi^2-1)^{-1/3}}^{t} \frac{\left(\text{Fr} \left(\text{Fr}^2 - 1\right)^{1/3}\right)^2 t \, dt}{\left[\text{Fr}^4 \left(\text{Fr}^2 - 1\right) t^3 + \text{Fr}^4 \left(\text{Fr}^2 - 1\right)\right]^{1/3}}$$

$$= \text{Fr}^{2/3} (\text{Fr}^2 - 1)^{1/3} \int_{(\pi^2-1)^{-1/3}}^{t} \frac{t \, dt}{\left(1 + t^3\right)^{1/3}} \tag{6.146}$$

which is suitably evaluated numerically. By virtue of (6.143) and (6.145)

$$M = [\text{Fr} \, Q^3 + \text{Fr}^4 (\text{Fr}^2 - 1)]^{1/3} = [\text{Fr}^4 (\text{Fr}^2 - 1) t^3 + \text{Fr}^4 (\text{Fr}^2 - 1)]^{1/3}$$
$$= \text{Fr}^{4/3} (\text{Fr}^2 - 1)^{1/3} (1 + t^3)^{1/3} \tag{6.147}$$

and by virtue of (6,137.a,b) and (6.136),

$$B = \frac{Q^2}{M} = \frac{\text{Fr}^2 \left(\text{Fr}^2 - 1\right)^{2/3} t^2}{\text{Fr}^{4/3} \left(\text{Fr}^2 - 1\right)^{1/3} \left(1 + t^3\right)^{1/3}} = \text{Fr}^{2/3} (\text{Fr}^2 - 1)^{1/3} \frac{t^2}{\left(1 + t^3\right)^{1/3}}$$

$$U = \frac{M}{Q} = \frac{\text{Fr}^{1/3} \left(1 + t^3\right)^{1/3}}{t}$$

$$P = \frac{\text{Fr}}{Q} = (\text{Fr}^2 - 1)^{-1/3} \frac{1}{t}. \tag{6.148.a - c}$$

The local densimetric Froude number along the buoyant jet is obtained with the use of (6.140), (6.131.a–d), as

$$\text{Fr}_L = \frac{\hat{u}}{\left(\frac{\Delta\hat{\rho}}{\rho_a} gb\right)^{1/2}} = \frac{\frac{1}{\text{Fr}} U u_0}{\left(\frac{\rho_a - \rho_0}{\rho_a} gb_0 P B\right)^{1/2}}$$

$$\frac{u_0}{\left(\frac{\rho_a - \rho_0}{\rho_a} gb_0\right)^{1/2}} \frac{1}{\text{Fr}} \frac{U}{(PB)^{1/2}} = \frac{u_0}{\left(\frac{\rho_a - \rho_0}{\rho_a} gb_0\right)^{1/2}} \frac{1}{\text{Fr}} \cdot \frac{1}{\text{Fr}^{1/2}} U^{3/2} \tag{6.149}$$

Denoting a constant part as

$$\text{Fr}_0 = \frac{u_0}{\left(\frac{\rho_a - \rho_0}{\rho_a} gb_0\right)^{1/2}} \tag{6.150}$$

the local densimetric Froude number is given as

$$
\begin{aligned}
\mathrm{Fr}_L &= \frac{\mathrm{Fr}_0}{\mathrm{Fr}^{3/2}} \cdot \mathrm{Fr}^{1/2} \frac{\left(1 + t^3\right)^{1/2}}{t^{3/2}} \\
&= \frac{\mathrm{Fr}_0}{\mathrm{Fr}} \frac{\left(1 + t^3\right)^{1/2}}{t^{3/2}}
\end{aligned}
\tag{6.151}
$$

which varies along the buoyant jet. By virtue of (6.150) and (6.133) the previously mentioned scaling ratio is

$$
\frac{\mathrm{Fr}_0}{\mathrm{Fr}} = \left(\frac{\pi}{2}\right)^{1/4} \left(\frac{\alpha_e}{\lambda}\right)^{1/2} .
\tag{6.152}
$$

By examining the solution it is noted that $Q \to \infty$ and $t \to \infty$ as $x \to \infty$, and

$$
\mathrm{Fr}_L \to \frac{\mathrm{Fr}_0}{\mathrm{Fr}} = \left(\frac{\pi}{2}\right)^{1/2} \left(\frac{\alpha_e}{\lambda}\right)^{1/2}
\tag{6.153}
$$

i.e. the buoyant jet becomes a *pure plume* far away from the source, as the initial momentum becomes insignificant. On the other hand as $x \to 0$, $Q \to \mathrm{Fr}$ and $t \to (\mathrm{Fr}^2 - 1)^{-1/3}$, the local Froude number at the source attains the value

$$
\mathrm{Fr}_L \to \frac{\mathrm{Fr}_0}{\mathrm{Fr}} \frac{\left(1 + \left(\mathrm{Fr}^2 - 1\right)^{-1}\right)^{1/2}}{\left(\mathrm{Fr}^2 - 1\right)^{-1/2}} = \frac{\mathrm{Fr}_0}{\mathrm{Fr}} (\mathrm{Fr}^2 - 1 + 1)^{1/2} = \mathrm{Fr}_0
\tag{6.154}
$$

The above solutions are valid only after the zone of flow establishment (ZOFE) which essentially replaces the virtual origin, in order to apply appropriate boundary conditions i.e. b_0, u_0, ρ_0 at some distance from the actual inlet. In fact the length of the ZOFE x_s is found experimentally as

$$
x_s = 3b_*
\tag{6.155}
$$

where b_* is the actual inlet width. The inlet width b_0 is related to b_* by the following analysis. The flux of buoyancy at any section of the plume is obtained by virtue of (6.113.a,b).

$$
\begin{aligned}
\bar{B} &= \int_{-\infty}^{\infty} g(\rho_a - \rho) u \, dy \\
&= g(\rho_a - \hat{\rho}) \hat{u} \int_{-\infty}^{\infty} e^{-\left(\frac{y}{\lambda b}\right)^2} e^{-\left(\frac{y}{b}\right)^2} dy \\
&= g(\rho_a - \hat{\rho}) \hat{u} \sqrt{\frac{\pi}{1 + \frac{1}{\lambda^2}}} b.
\end{aligned}
\tag{6.156}
$$

On the other hand, in the ZOFE, $\hat{\rho} = \rho_0$, $\hat{u} = u_0$ so that (6.156)

$$\bar{B} = g(\rho_a - \rho_0)u_0\sqrt{\frac{\pi}{1 + \lambda^2}}\lambda b \quad \text{for} \quad x < x_s \qquad (6.157)$$

while the initial buoyancy flux is

$$\bar{B}_0 = g(\rho_a - \rho_0)u_0 b_*, \qquad (6.158)$$

and equating 6.157)–(6.158) at $x = x_s$, relating b_* to $b = b_0$, we find

$$\left(\frac{b_*}{b_0}\right) = \lambda\sqrt{\frac{\pi}{1 + \lambda^2}}. \qquad (6.159)$$

6.2.4   Axisymmetric Pure Plumes and Buoyant Jets

The analysis of circular plumes and buoyant jets are similar to the planar case, based on the equations

$$r\frac{\partial u}{\partial x} + \frac{\partial rv}{\partial r} = 0$$

$$u\frac{\partial u}{\partial x} + v\frac{\partial u}{\partial r} = y\frac{\Delta\rho}{\rho_a} - \frac{1}{r}\frac{\partial r\overline{u'v'}}{\partial r} \qquad (6.160.a - c)$$

$$u\frac{\partial \Delta\rho}{\partial x} + v\frac{\partial \Delta\rho}{\partial r} = -\frac{1}{r}\frac{\partial r\overline{v'\rho'}}{\partial r}$$

Multiplying (6.160) by 2π and integrating across the jet,

$$\int_0^\infty 2\pi r\frac{\partial u}{\partial x}\,dr + \int_0^\infty 2\pi\frac{\partial rv}{\partial r}\,dr = \frac{d}{dx}\int_0^\infty 2\pi ru\,dr + [2\pi rv]_0^\infty = 0 \quad (6.161)$$

Assuming

$$[2\pi rv]_\infty = [2\pi bv]_b = -2\pi\alpha_e b\hat{u} \qquad (6.162)$$

and noting that $v(0) = 0$ by symmetry, (6.161) becomes

$$\frac{d}{dx}\int_0^\infty 2\pi ru\,dr = 2\pi\alpha_e b\hat{u} \qquad (6.163)$$

where α_e is the entrainment coefficient and b is the width of the jet. Integrating (6.160) multiplied by $2\pi r$ gives

$$\int_0^\infty 2\pi ru\frac{\partial u}{\partial x}\,dr + \int_0^\infty 2\pi rv\frac{\partial u}{\partial r}\,dr = \int_0^\infty 2\pi rg\frac{\Delta\rho}{\rho_a}\,dr \underset{}{-2\pi[r\overline{u'v'}]_0^\infty} \overset{0}{} \quad (6.164)$$

Then, integrating the second term by parts and utilizing (6.160.a), the left hand side is calculated as

$$
\begin{aligned}
&\int_0^\infty 2\pi r u \frac{\partial u}{\partial x}\,dr + \int_0^\infty 2\pi r v \frac{\partial u}{\partial r}\,dr \\
&= \frac{d}{dx}\int_0^\infty 2\pi r \frac{u^2}{2}\,dr + [2\pi r v u]_0^{\infty}{}^{\,0} - \int_0^\infty 2\pi u \frac{\partial r v}{\partial r}\,dr \\
&= \frac{d}{dx}\int_0^\infty 2\pi r \frac{u^2}{2}\,dr + \int_0^\infty 2\pi r u \frac{\partial u}{\partial x}\,dr \\
&= \frac{d}{dx}\int_0^\infty 2\pi r u^2\,dr
\end{aligned} \tag{6.165}
$$

so that (6.164) becomes

$$
\frac{d}{dx}\int_0^\infty 2\pi r u^2\,dr = \int_0^\infty 2\pi r g \frac{\Delta\rho}{\rho_a}\,dr. \tag{6.166}
$$

Next, the integration of (6.160.c) yields

$$
\int_0^\infty 2\pi r u \frac{\partial \Delta\rho}{\partial x}\,dr + \int_0^\infty 2\pi r v \frac{\partial \Delta\rho}{\partial r}\,dr = -[r\overline{v'\rho'}]_0^{\infty}{}^{\,0}. \tag{6.167}
$$

Transforming the left hand side by making use of (6.160) produces

$$
\begin{aligned}
&\int_0^\infty 2\pi r \frac{\partial \Delta\rho}{\partial x}\,dr + [2\pi r v \Delta\rho]_0^{\infty}{}^{\,0} - \int_0^\infty 2\pi \Delta\rho \frac{\partial r v}{\partial r}\,dr \\
&= \int_0^\infty 2\pi r u \frac{\partial \Delta\rho}{\partial x}\,dr + \int_0^\infty 2\pi \Delta\rho r \frac{\partial u}{\partial x}\,dr \\
&= \int_0^\infty 2\pi r \frac{\partial u \Delta\rho}{\partial x}\,dr.
\end{aligned} \tag{6.168}
$$

then (6.167) becomes

$$
\frac{d}{dx}\int_0^\infty 2\pi r u \Delta\rho\,dr = 0 \tag{6.169}
$$

Defining the volume, momentum and buoyancy fluxes as

$$
\begin{aligned}
q &= \int_0^\infty 2\pi r u\,dr = \int_A u\,dA \\
m &= \int_0^\infty 2\pi r u^2\,dr = \int_A u^2\,dA \\
p &= \int_0^\infty 2\pi r u \Delta\rho\,dr = \int_A u \Delta\rho\,dA
\end{aligned} \tag{6.170.a - c}
$$

and from (6.163), (6.166) and (6.169),

$$\frac{dq}{dx} = 2\pi\alpha_e b\hat{u}$$

$$\frac{dm}{dx} = \int_0^\infty 2\pi r g \frac{\Delta\rho}{\rho_a} \, dr \qquad (6.171.a-c)$$

$$\frac{dp}{dx} = 0.$$

Assuming Gaussian similarity profiles

$$\frac{u}{\hat{u}} = e^{-(\frac{r}{b})^2}, \qquad \frac{\Delta\rho}{\Delta\hat{\rho}} = \frac{\rho_a - \rho}{\rho_a - \hat{\rho}} = e^{-(\frac{r}{\lambda b})^2} \qquad (6.172.a.b)$$

and noting that

$$\int_0^\infty r e^{-(\frac{r}{b})^2} \, dr = \frac{1}{2} b^2 \qquad (6.173)$$

Equation (6.167.a–c) become:

$$\frac{d}{dx}\pi b^2 \hat{u} = 2\pi\alpha_e b\hat{u}$$

$$\frac{d}{dx}\frac{\pi}{2} b^2 \hat{u}^2 = \pi g \lambda^2 \left(\frac{\rho_a - \hat{\rho}}{\rho_a}\right) b^2 \qquad (6.174.a-c)$$

$$\frac{d}{dx}\pi b^2 \hat{u}(\rho_a - \hat{\rho})\frac{\lambda^2}{1 + \lambda^2} = 0.$$

Let a be the area of the plume

$$\pi b^2 = a \qquad (6.175)$$

allowing to write (6.174.a–c) as

$$\frac{d}{dx}(a\hat{u}) = 2\sqrt{\pi}\alpha_e a^{1/2}\hat{u}$$

$$\frac{d}{dx}(\frac{1}{2}a\hat{u}^2) = g\lambda^2 \left(\frac{\rho_a - \hat{\rho}}{\rho_a}\right) a \qquad (6.176.a-c)$$

$$\frac{d}{dx}\left(\frac{\lambda^2}{1 + \lambda^2}(\rho_a - \hat{\rho})\hat{u}a\right) = 0.$$

6.2.4.1 Axisymmetric Pure Plume

Seeking a similarity solution

$$\hat{u} \sim x^p \quad a \sim x^q, \quad \Delta\hat{\rho} = \rho_a - \hat{\rho} = x^s$$

Equations (6.176.a–c) are reduced to

$$\frac{d}{dx}x^{p+q} = x^{p+q-1} \sim x^{\frac{q}{2}+p}$$

$$\frac{d}{dx}x^{2p+q} = x^{2p+q-1} \sim x^{s+q} \qquad\qquad (6.177.a-c)$$

$$\frac{d}{dx}x^{s+p+q} = 0$$

from which is obtained

$$p+q-1 = \frac{q}{2}+p$$

$$2p+q-1 = s+q \qquad\qquad (6.178.a-c)$$

$$s+p+q = 0$$

or

$$p = -\frac{1}{3}, \quad q = 2, \quad s = -5/3. \qquad (6.179.a-c)$$

The dependence of variables are expected to be

$$\hat{u} = c_u x^{1/2}, \quad a = c_a x^2, \quad \Delta\hat{\rho} = \rho_a c_r x^{-5/3} \qquad (6.180.a-c)$$

and the area is defined by (6.175) so that

$$b \sim k_p x \qquad\qquad (6.181)$$

where k_p is the plume expansion rate and

$$\frac{db}{dx} = k_p. \qquad\qquad (6.182)$$

Substituting (6.180.a,b) in (6.174.a)

$$\frac{d}{dx}(\pi k_p^2 x^2 c_u x^{-1/3}) = 2\pi\alpha_e k_p x c_u x^{-1/3} \qquad (6.183)$$

or

$$\alpha_e = \frac{5}{6}k_p \qquad\qquad (6.184)$$

[For a jet the entrainment coefficient calculated with the same assumption of Gaussian velocity distribution would be

$$\alpha_{e_j} = \frac{1}{2} k_j \tag{6.185}$$

in comparison with plume]

Substituting (6.180.a–c) into (6.173.a–c) we find that

$$\text{as} \quad x \to 0, \quad q \to 0, \quad m \to 0, \quad p \to p_0 \tag{6.186}$$

analogous to the situation of pure plumes, i.e. the volume and momentum fluxes have to vanish at the origin, while the buoyancy has a finite value. For the pure plume, we will again find that the local plume densimetric Froude number is constant, having the following value in the case of circular plumes:

$$\text{Fr}_L = \frac{\hat{u}}{\left(\frac{\Delta\hat{\rho}}{\rho_a} g b\right)^{1/2}} = \left(\frac{5}{4}\right)^{1/2} \frac{\lambda}{\alpha_e^{1/2}} \tag{6.187}$$

6.2.4.2 Axisymmetric Buoyant Jet

The *buoyant circular jet* case can be solved by similar techniques used for 2-D jets, based on equations (6.176.a–c) with the initial conditions

$$a = a_0, \quad \hat{u} = u_0, \quad \Delta\hat{\rho} = \Delta\rho_0 \quad \text{at } x = 0, \tag{6.188}$$

Defining

$$X = 2\sqrt{\pi}\alpha_e \frac{x}{a_0^{1/2}} = 2\alpha_e \frac{x}{b_0}$$

$$A = \frac{a}{a_0} = \frac{b^2}{b_0^2}$$

$$P = \frac{\rho_a - \hat{\rho}}{\rho_a - \rho_0} \tag{6.189.a - d}$$

$$U = \text{Fr}\frac{\hat{u}}{u_0}$$

with source densimetric Froude number defined as

$$\text{Fr} = \frac{\alpha_e^{1/2}}{\lambda} \frac{u_0}{\sqrt{g \frac{\rho_a - \rho_0}{\rho_a} b_0}}, \tag{6.190}$$

Equations (6.176.a–c) become

$$\frac{d}{dX}AU = A^{1/2}U$$
$$\frac{d}{dX}AU^2 = AP \qquad (6.191.a-c)$$
$$\frac{d}{dX}AUP = 0$$

with initial conditions

$$A = 1, \quad P = 1, \quad U = \text{Fr}. \qquad (6.192.a-c)$$

The last equation (6.191.c) integrates to

$$AUP = \text{constanr} = [AUP]_{x=0} = \text{Fr} \qquad (6.193)$$

Defining

$$\frac{dQ}{dX} = M^{1/2}$$
$$\frac{dM}{dX} = \frac{\text{Fr}}{U} = \text{Fr}\frac{Q}{M}. \qquad (6.195.a,b)$$

After eliminating X,

$$\frac{dM}{dQ} = \text{Fr}\frac{Q}{M^{3/2}} \qquad (6.196)$$

which is then integrated to give

$$\frac{2}{5}M^{5/2} = \frac{1}{2}\text{Fr}Q^2 + C \qquad (6.197)$$

where C is found from (6.185.a–c)

$$C = \frac{2}{5} - \frac{1}{2}\text{Fr}. \qquad (6.198)$$

Substituting, (6.197) becomes

$$M^{5/2} = \frac{5}{4}\text{Fr}(Q^2 - 1) + 1 \qquad (6.199)$$

and X can be found from (6.195) as

$$
\begin{aligned}
X &= \int_0^\infty dX = \int_{\mathrm{Fr}}^Q M^{-1/2} dQ \\
  &= \int_{\mathrm{Fr}}^Q \frac{dQ}{\left(\frac{5}{4}\mathrm{Fr}\left(Q^2 - 1\right) + 1\right)^{1/5}}
\end{aligned}
\tag{6.200}
$$

which has to be numerically integrated.

References

1. Schlichting H (1968) Boundary layer Theory. Mc Graw Will, New York
2. Rajaratnam N (1976) Turbulent jets. Elsevier, Amsterdam
3. Abromovich GN (1963) The theory of turbulent jets. The MIT Press, Cambridge
4. Anwer HO (1969) Experiment on an effluent discharging from a slot into stationary or slow moving fluid of greater Density. J Hydraulic Res 7:411–431

Exercises

7

7.1 Chapter 1, Exercise 1

This reading exercise aims to investigate if adiabatic (or isentropic) conditions exist in the deepest parts of the earth's ocean, based on the assumption that external factors such as motion and diffusion processes have little influence on deep water properties. It is not really a well established fact if isentropic conditions can ever exist in the deep ocean, although we expect them to be rare occurrences in nature, for entropy is not necessarily conserved in equilibrium thermodynamics.

With these concerns and expectations, one is tempted to investigate seawater properties in the deepest pit of the world ocean, specifically at Challenger Deep of Mariana Trench in western Pacific Ocean, where a maximum depth of 11,000 m is found. Yet one finds that there has not been many measurement campaigns able to meet technological challenges for accurate measurements going the full depth of the pit. We only list few references for the benefit of the reader.

Isentropic conditions would imply a basic state in which potential temperature and density gradients must vanish. If the fluid is assumed to be in a basic state of rest, then only vertical gradients should be expected

$$\frac{d\theta}{dz} = 0 \quad \text{and} \quad \frac{d\Delta}{dz} = 0.$$

In particular, if there is no heating agent other than seawater compression, vanishing of potential temperature gradients $d\theta/dz = 0$ would imply non-adjusted, absolute temperature to increase with depth (in degrees Kelvin $°K$)

$$\frac{dT}{dz} = -\frac{g\alpha_T}{c_p}T$$

© Springer Nature Switzerland AG 2021   267
E. Özsoy, *Geophysical Fluid Dynamics II*, Springer Textbooks in Earth Sciences,
Geography and Environment,
https://doi.org/10.1007/978-3-030-74934-7_7

as a result of compression of sea water. Similarly the corresponding density gradient at rest would be given as

$$\frac{d\rho}{dz} = -\frac{g}{c_s{}^2}\rho$$

Observations

One could test if some of these expectations are confirmed by measurements in the ultimate depths of Mariana Trench. It could also be expected that for the range of pressure, temperature and salinity in the deep measurements, there may be some changes in the physical constants. Of these, the expansion coefficient seems to be reduced to half the value used in Sect. 1.3.3 (Lillibridge 1989), $\alpha_T = 0.75 \times 10^{-4}\,°\mathrm{C}^{-1}$, so that a renewed estimate of the temperature gradient is obtained as

$$\frac{dT}{dz} = -\frac{g\alpha_T}{c_p}T = -0.2 \times 10^{-6}\,\mathrm{m}^{-1} \times 282\,°\mathrm{C} \simeq -0.6 \times \times 10^{-4}\,°\mathrm{Cm}^{-1}$$

while actually Fig. 4 in Taira (1989) shows 0.1 °C increase in temperature in the depth range of 6000–11000 m, yielding a an estimate for the gradient

$$\frac{dT}{dz} = -\frac{0.08\,°\mathrm{C}}{5000\,\mathrm{m}} \sim -0.16 \times \times 10^{-4}\,°\mathrm{Cm}^{-1}.$$

Correspondingly, the density change is calculated based on a coefficient estimated as $g/c_s{}^2 = 3.8 \times 10^{-6}\,\mathrm{m}^{-1}$ and average density of $\rho = 1.062 \times 10^{-3}\,\mathrm{kg/m}^3$ to give

$$\frac{d\rho}{dz} = -\frac{g}{c_s{}^2}\rho = -4 \times 10^{-9}\,\mathrm{kg\ m}^{-4}$$

These first comparisons with the measurements did not seem to confirm isentropy and had quite variable estimates of gradients. In addition, measurements obtained by Taira (1989) also seemed to indicate a gradient of $-0.9 \times 10^{-6}\,°\mathrm{Cm}^{-1}$, with increasing potential temperature with depth, which seemed to be in denial of isentropy, possibly indicating some inconsistency in measurements, because the gradient would have been positive $d\theta/dz \geq 0$ at best in the absence of isentropy.

Further measurements at Mariana Trench by Van Haren (2017) showed that potential temperature gradient was close to zero and positive in the vertical direction ($d\theta/dz \geq 0$), as expected.

In fact, it is perhaps not a valid question to ask if isentropy could exist at all in the strict sense anywhere in the ocean, since we know even the deepest parts of the ocean can not be at absolute rest, without interference by viscous or turbulent motions, unless we go to absolute zero temperature. Besides, entropy has a tendency to increase even in the remotest parts of the universe.

References

Lillibridge, J. L., 1989. Computing the Seawater Expansion Coefficients Directly from the 1980 Equation of State, Journal of Atmospheric and Oceanic Technology, Vol. 6, p. 59–66.

Taira K., Yanagimoto, D. and Kitagawa, S., 2005. Deep CTD Casts in the Challenger Deep, Mariana Trench, Journal of Oceanography, Vol. 61, pp. 447–454.

van Haren, H., Berndt, C., Klaucke, I., 2017. Ocean mixing in deep-sea trenches: New insights from the Challenger Deep, Mariana Trench, Deep-Sea Research Part I, Vol. 129, pp. 1–9.

7.2   Chapter 3, Exercise 1

Equations (3.131) and (3.132) in Chap. 3 state conservation of relative vorticity q along the basic zonal flow $U(y.z)$

$$\left(\frac{\partial}{\partial t} + U(y.z)\frac{\partial}{\partial x} \right) q + R\frac{\partial \psi}{\partial x} = 0$$

where

$$q = \nabla^2 \psi + \frac{f_0^2}{\rho_r} \frac{\partial}{\partial z} \left(\frac{\rho_r}{N^2} \frac{\partial \psi}{\partial z} \right)$$

$$R(y, z) = \hat{\beta} - \frac{\partial^2 U}{\partial y^2} - \frac{f_0^2}{\rho_r} \frac{\partial}{\partial z} \left(\frac{\rho_r}{N^2} \frac{\partial U}{\partial z} \right).$$

with boundary conditions applied at the sides, surface and bottom:

$$\frac{\partial \psi}{\partial x} = 0 \quad \text{on} \quad y = 0, Y$$

$$\left(\frac{\partial}{\partial t} + U(y, z)\frac{\partial}{\partial x} \right) \frac{\partial \psi}{\partial z} - \left(\frac{\partial U}{\partial z} \right) \frac{\partial \psi}{\partial x} = 0 \quad \text{on} \quad z = 0, H$$

We also assume solutions that are periodic in x for m repeating periods of $(0, X)$

$$\psi(x, y, z, t) = \psi(x + mX, y, z, t) = \sum_{n=0}^{\infty} A_n(y, z, t)e^{ik_n(x+mX)}$$

where $k_n = 2\pi n/X$. The periodic solution could evolve from some initial condition, and since the problem is linear one could consider only a single component of the series which behaves as

$$\psi(x, y, z, t) \sim \psi_0 e^{ikx}$$

where k is the wave number and ψ_0 is the amplitude for this single wave.

We note that integration with respect to zonal direction x of the periodic solution vanishes

$$\int_0^X \frac{\partial}{\partial x}() \, dx = 0$$

and also the x-derivative of the solution is

$$\frac{\partial \psi}{\partial x} = -ik\psi = 0$$

which implies that the following can be applied as extended side boundary conditions used in integration of the equations:

$$\psi = 0, \quad \frac{\partial \psi}{\partial x} = 0, \quad \frac{\partial \psi}{\partial t} = 0 \quad \text{on} \quad y = 0, Y \quad \text{for all} \quad t.$$

Let us now define operators L_1 and L_2 as

$$L_1 = \frac{\partial}{\partial t} + U(y, z) \frac{\partial}{\partial x}$$

$$L_2 = \nabla^2 + \frac{1}{\rho_r} \frac{\partial}{\partial z} \frac{\rho_r}{N^2} \frac{\partial}{\partial z}.$$

We can now form the product of basic state density and stream-function $\rho_r \phi$ with the vorticity equation and integrate across the domain $(0, H)$, $(0, Y)$, $(0, X)$ represented by triple integrals in the following. We first set

$$I = I_1 + I_2 = \underbrace{\iiint \rho_r \psi L_1 L_2 \psi \, dxdydz}_{I_1} + \underbrace{\iiint \rho_r R \psi \frac{\partial \psi}{\partial x} \, dxdydz}_{I_2} = 0$$

where I_1 and I_2 represent the first and second terms, with the first term broken into parts $I_1 = I_3 + I_4$ as follows

$$I_1 = \underbrace{\iiint L_1(\rho_r \psi L_2 \psi) \, dxdydz}_{I_3} - \underbrace{\iiint (L_1 \rho_r \psi)(L_2 \psi) \, dxdydz}_{I_4} = I_3 + I_4$$

We then integrate the second term to obtain

$$I_2 = \iiint \rho_r R \psi \frac{\partial \psi}{\partial x} \, dxdydz = \iint \rho_r R \left(\int_0^X \frac{1}{2} \frac{\partial \psi^2}{\partial x} dx \right)^0 dydz = 0$$

which vanishes since the motion is periodic in x, noting that $R(y, z)$ is only a function of y and z. Next, we evaluate $I_3 = I_{31} + I_{32} + I_{33}$, broken into three sub-integrals

$$I_3 = \iiint \frac{\partial}{\partial t}(\rho_r \psi L_2 \psi)\, dxdydz + \cancelto{0}{\iint U \frac{\partial}{\partial x}(\rho_r \psi L_2 \psi)\, dxdydz}$$

$$= \frac{\partial}{\partial t} \iiint \rho_r \psi \left[\frac{\partial^2 \psi}{\partial x^2} + \frac{\partial^2 \psi}{\partial y^2} + \frac{1}{\rho_r}\frac{\partial}{\partial z}\frac{\rho_r}{N^2}\frac{\partial \psi}{\partial z}\right] dxdydz$$

$$= I_{31} + I_{32} + I_{33}$$

Integration by parts of each term yields

$$I_{31} = \frac{\partial}{\partial t} \iiint \rho_r \psi \left(\frac{\partial^2 \psi}{\partial x^2}\right) dxdydz$$

$$= \frac{\partial}{\partial t} \iiint \rho_r \frac{\partial}{\partial x}\cancelto{0}{\left(\psi\frac{\partial \psi}{\partial x}\right)} dxdydz - \frac{\partial}{\partial t} \iiint \rho_r \left(\frac{\partial \psi}{\partial x}\right)^2 dxdydz$$

$$= -\frac{\partial}{\partial t} \iiint \rho_r \left(\frac{\partial \psi}{\partial x}\right)^2 dxdydz$$

where the first part vanishes due to periodic conditions. Similarly, we can obtain

$$I_{32} = \frac{\partial}{\partial t} \iiint \rho_r \psi \left(\frac{\partial^2 \psi}{\partial y^2}\right) dxdydz$$

$$= \frac{\partial}{\partial t} \iiint \rho_r \frac{\partial}{\partial y}\left(\psi\frac{\partial \psi}{\partial y}\right) dxdydz - \frac{\partial}{\partial t} \iiint \rho_r \left(\frac{\partial \psi}{\partial y}\right)^2 dxdydz$$

$$= \frac{\partial}{\partial t} \iint \rho_r \cancelto{0}{\left[\psi\frac{\partial \psi}{\partial y}\right]_0^Y} dxdz - \frac{\partial}{\partial t} \iiint \rho_r \left(\frac{\partial \psi}{\partial y}\right)^2 dxdydz$$

$$= -\frac{\partial}{\partial t} \iiint \rho_r \left(\frac{\partial \psi}{\partial y}\right)^2 dxdydz$$

using side boundary conditions and evaluate the third term by integration by parts

$$I_{33} = \frac{\partial}{\partial t} \iiint \rho_r \psi \frac{f_0^2}{\rho_r}\frac{\partial}{\partial z}\frac{\rho_r}{N^2}\frac{\partial \psi}{\partial z}\, dxdydz$$

$$= \frac{\partial}{\partial t} \iiint \frac{\partial}{\partial z}\psi\frac{\rho_r f_0^2}{N^2}\frac{\partial \psi}{\partial z}\, dxdydz - \frac{\partial}{\partial t} \iiint \frac{\rho_r f_0^2}{N^2}\left(\frac{\partial \psi}{\partial z}\right)^2 dxdydz$$

$$\iint \left[f_0^2\frac{\rho_r}{N^2}\frac{\partial \psi}{\partial t}\frac{\partial \psi}{\partial z}\right]_0^H dxdy - \frac{\partial}{\partial t} \iiint \frac{\rho_r f_0^2}{N^2}\left(\frac{\partial \psi}{\partial z}\right)^2 dxdydz.$$

Combining $I_3 = I_{31} + I_{32} + I_{33}$ gives

$$I_3 = -\frac{\partial}{\partial t} \iiint \rho_r \left[\left(\frac{\partial \psi}{\partial x}\right)^2 + \left(\frac{\partial \psi}{\partial y}\right)^2 + \frac{f_0^2}{N^2} \left(\frac{\partial \psi}{\partial z}\right)^2 \right] dxdydz$$

$$+ \iint \left[\frac{\rho_r f_0^2}{N^2} \frac{\partial}{\partial t} \psi \frac{\partial \psi}{\partial z} \right]_0^H dxdy.$$

Next we let

$$I_4 = -\iiint (L_1 \rho_r \psi)(L_2 \psi)\, dxdydz$$

$$= -\iiint \left(\frac{\partial \rho_r \psi}{\partial t}\right)(L_2 \psi)\, dxdydz - \iiint U \left(\frac{\partial \rho_r \psi}{\partial x}\right) L_2 \psi\, dxdydz$$

$$= I_5 + I_6$$

broken into terms as such $I_4 = -I_5 + I_6$ which are then further split into parts as

$$I_5 = -(I_{51} + I_{52} + I_{53}).$$

The first term is evaluated as

$$I_{51} = \iiint \frac{\partial \rho_r \psi}{\partial t} \frac{\partial^2 \psi}{\partial x^2}\, dxdydz$$

$$= \iiint \frac{\partial}{\partial x} \left(\frac{\partial \rho_r \psi}{\partial t} \frac{\partial \psi}{\partial x}\right) dxdydz - \iiint \frac{\partial}{\partial x} \left(\frac{\partial \rho_r \psi}{\partial t}\right) \frac{\partial \psi}{\partial x}\, dxdydz$$

$$= -\iiint \rho_r \left(\frac{\partial}{\partial t} \frac{\partial \psi}{\partial x}\right)\left(\frac{\partial \psi}{\partial x}\right) dxdydz$$

$$= -\frac{\partial}{\partial t} \iiint \rho_r \frac{1}{2}\left(\frac{\partial \psi}{\partial x}\right)^2 dxdydz$$

in which the first term vanishes because of periodic conditions. We follow with

$$
\begin{aligned}
I_{52} &= \iiint \frac{\partial \rho_r \psi}{\partial t} \frac{\partial^2 \psi}{\partial y^2}\, dx dy dz \\
&= \iiint \rho_r \frac{\partial}{\partial y} \left(\frac{\partial \psi}{\partial t} \frac{\partial \psi}{\partial y} \right) dx dy dz - \frac{\partial}{\partial t} \iiint \frac{1}{2} \rho_r \left(\frac{\partial \psi}{\partial y} \right)^2 dx dy dz \\
&= \iint \rho_r \left[\frac{\partial \psi}{\partial t} \frac{\partial \psi}{\partial y} \right]_0^Y dx dz - \frac{\partial}{\partial t} \iiint \frac{1}{2} \rho_r \left(\frac{\partial \psi}{\partial y} \right)^2 dx dy dz \\
&= -\frac{\partial}{\partial t} \iiint \frac{1}{2} \rho_r \left(\frac{\partial \psi}{\partial y} \right)^2 dx dy dz
\end{aligned}
$$

where the time dependent term in the integral is evaluated to zero by virtue of the fact that

$$
\psi = 0 \ \text{ on } \ y - 0, Y
$$

consequently implies that

$$
\frac{\partial \psi}{\partial t} = 0 \ \text{ on } \ y == 0, Y
$$

at all time as noted in the beginning of the exercise. Similarly we integrate to obtain

$$
\begin{aligned}
I_{53} &= \iiint \rho_r \frac{\partial \psi}{\partial t} \frac{f_0^2}{\rho_r} \frac{\partial}{\partial z} \frac{\rho_r}{N^2} \frac{\partial \psi}{\partial z}\, dx dy dz \\
&= \iiint \frac{\partial}{\partial z} \left(\frac{\rho_r f_0^2}{N^2} \frac{\partial \psi}{\partial t} \frac{\partial \psi}{\partial z} \right) dx dy dz - \frac{\partial}{\partial t} \iiint \frac{1}{2} \frac{\rho_r f_0^2}{N^2} \left(\frac{\partial \psi}{\partial z} \right)^2 dx dy dz \\
&= \iint \left[\frac{\rho_r f_0^2}{N^2} \frac{\partial \psi}{\partial t} \frac{\partial \psi}{\partial z} \right]_0^H dx dy - \frac{\partial}{\partial t} \iiint \frac{1}{2} \frac{\rho_r f_0^2}{N^2} \left(\frac{\partial \psi}{\partial z} \right)^2 dx dy dz
\end{aligned}
$$

By combining the above we obtain

$$
\begin{aligned}
I_5 &= \frac{\partial}{\partial t} \iiint \frac{1}{2} \rho_r \left[\left(\frac{\partial \psi}{\partial x} \right)^2 + \left(\frac{\partial \psi}{\partial y} \right)^2 + \frac{f_0^2}{N^2} \left(\frac{\partial \psi}{\partial z} \right)^2 \right] dx dy dz \\
&\quad - \iint \left[\frac{\rho_r f_0^2}{N^2} \frac{\partial \psi}{\partial t} \frac{\partial \psi}{\partial z} \right]_0^H dx dy
\end{aligned}
$$

We then split I_6 as

$$I_6 = -\iiint U \frac{\partial \rho_r \psi}{\partial x} L_2(\psi)\, dx dy dz = -(I_{61} + I_{62} + I_{63})$$

to obtain the terms

$$I_{61} = \iiint \rho_r U \frac{\partial \psi}{\partial x} \frac{\partial^2 \psi}{\partial x^2}\, dx dy dz = \cancel{\iiint \rho_r U \frac{1}{2} \frac{\partial}{\partial x} \left(\frac{\partial \psi}{\partial x}\right)^2 dx dy dz}^{\,0} = 0$$

vanishing as a result of periodic conditions in x. Integrating I_{62} by parts

$$I_{62} = \iiint \rho_r U \frac{\partial \psi}{\partial x} \frac{\partial^2 \psi}{\partial y^2}\, dx dy dz$$

$$= \iiint \rho_r \frac{\partial}{\partial y} \left(U \frac{\partial \psi}{\partial x} \frac{\partial \psi}{\partial y}\right) dx dy dz - \iiint \rho_r \frac{\partial}{\partial y} \left(U \frac{\partial \psi}{\partial x}\right) \left(\frac{\partial \psi}{\partial y}\right) dx dy dz$$

$$= \cancel{\iint \left[\rho_r U \frac{\partial \psi}{\partial x} \frac{\partial \psi}{\partial y}\right]_0^Y dx dz}^{\,0} - \iiint \rho_r \frac{\partial U}{\partial y} \frac{\partial \psi}{\partial x} \frac{\partial \psi}{\partial y}\, dx dy dz$$

$$\quad - \iiint \rho_r U \left(\frac{\partial}{\partial y} \frac{\partial \psi}{\partial x}\right) \left(\frac{\partial \psi}{\partial y}\right) dx dy dz$$

$$= -\iiint \rho_r \frac{\partial U}{\partial y} \frac{\partial \psi}{\partial x} \frac{\partial \psi}{\partial y}\, dx dy dz - \cancel{\iiint \rho_r U \frac{1}{2} \frac{\partial}{\partial x} \frac{\partial}{\partial x} \left(\frac{\partial \psi}{\partial y}\right)^2 dx dy dz}^{\,0}$$

$$= -\iiint \rho_r \frac{\partial \psi}{\partial x} \frac{\partial \psi}{\partial y} \frac{\partial U}{\partial y}\, dx dy dz$$

where it is recognized that the extended side boundary conditions

$$\psi = 0, \quad \frac{\partial \psi}{\partial x} = 0, \quad \frac{\partial \psi}{\partial t} = 0 \quad \text{on} \quad y = 0, Y \quad \text{for all} \quad t$$

lead to cancellation of the above terms. Then we have for the last term we have

$$I_{63} = \iiint \rho_r U \frac{\partial \psi}{\partial x} \frac{f_0^2}{\rho_r} \frac{\partial}{\partial z} \frac{\rho_r}{N^2} \frac{\partial \psi}{\partial z} dxdydz$$

$$= \iiint \frac{\partial}{\partial z} \left(\frac{\rho_r f_0^2}{N^2} U \frac{\partial \psi}{\partial x} \frac{\partial \psi}{\partial z} \right) dxdydz - \iiint \frac{\rho_r f_0^2}{N^2} \frac{\partial}{\partial z} \left(U \frac{\partial \psi}{\partial x} \right) \frac{\partial \psi}{\partial z} dxdydz$$

$$= \iint \left[\frac{\rho_r f_0^2}{N^2} U \frac{\partial \psi}{\partial x} \frac{\partial \psi}{\partial z} \right]_0^H dxdz - \iiint \frac{\rho_r f_0^2}{N^2} \frac{\partial U}{\partial z} \frac{\partial \psi}{\partial x} \frac{\partial \psi}{\partial z} dxdydz$$

$$- \iiint \frac{\rho_r f_0^2}{N^2} U \left(\frac{\partial}{\partial z} \frac{\partial \psi}{\partial x} \right) \left(\frac{\partial \psi}{\partial z} \right) dxdydz$$

$$= \iint \left[\frac{\rho_r f_0^2}{N^2} U \frac{\partial \psi}{\partial x} \frac{\partial \psi}{\partial z} \right]_0^H dxdz - \iiint \frac{\rho_r f_0^2}{N^2} \frac{\partial U}{\partial z} \frac{\partial \psi}{\partial x} \frac{\partial \psi}{\partial z} dxdydz$$

$$- \cancel{\iiint \frac{\rho_r f_0^2}{N^2} U \frac{1}{2} \frac{\partial}{\partial x} \left(\frac{\partial \psi}{\partial z} \right)^{\cancel{2}\,0}} dxdydz$$

$$= \iint \left[\frac{\rho_r f_0^2}{N^2} U \frac{\partial \psi}{\partial x} \frac{\partial \psi}{\partial z} \right]_0^H dxdy - \iiint \frac{\rho_r f_0^2}{N^2} \frac{\partial U}{\partial z} \frac{\partial \psi}{\partial x} \frac{\partial \psi}{\partial z} dxdydz$$

Then these terms are combined as

$$I_6 = + \iiint \rho_r \left[\left[\frac{\partial \psi}{\partial x} \frac{\partial \psi}{\partial y} \frac{\partial U}{\partial y} + \frac{f_0^2}{N^2} \frac{\partial \psi}{\partial x} \frac{\partial \psi}{\partial z} \frac{\partial U}{\partial z} \right] dxdydz \right]$$

$$- \iint \left[\frac{\rho_r f_0^2}{N^2} U \frac{\partial \psi}{\partial x} \frac{\partial \psi}{\partial z} \right]_0^H dxdy.$$

In summary, we have

$$I = I_1 + I_2$$
$$= (I_3 + I_4) + I_2$$
$$= I_3 + (I_5 + I_6) + I_2$$
$$= -[I_{31} + I_{31} + I_{31}] - [I_{51} + I_{52} + I_{53}] - [I_{61} + I_{62} + I_{63}] + I_2$$

Combining these terms together we have:

$$\frac{\partial}{\partial t} \iiint \frac{1}{2}\rho_r \left[\left(\frac{\partial \psi}{\partial x}\right)^2 + \left(\frac{\partial \psi}{\partial y}\right)^2 + \frac{f^2}{N^2}\left(\frac{\partial \psi}{\partial z}\right)^2 \right] dxdydz$$

$$= \iiint \rho_r \left[\frac{\partial \psi}{\partial x}\frac{\partial \psi}{\partial y}\frac{\partial U}{\partial y} + \frac{f_0^2}{N^2}\frac{\partial \psi}{\partial x}\frac{\partial \psi}{\partial z}\frac{\partial U}{\partial z} \right] dxdydz$$

$$+ \iint \left[\frac{\rho_r f_0^2}{N^2}\left(\frac{\partial}{\partial t}\psi\frac{\partial \psi}{\partial z} - \frac{\partial \psi}{\partial t}\frac{\partial \psi}{\partial z} - \frac{\partial \psi}{\partial t}\frac{\partial \psi}{\partial z} - U\frac{\partial \psi}{\partial x}\frac{\partial \psi}{\partial z} \right) \right]_0^H dxdy$$

We can now see that the last term in the above equation would vanish. To show this, we can multiply the top and bottom boundary conditions by ψ:

$$\psi\frac{\partial}{\partial t}\frac{\partial \psi}{\partial z} + U\psi\frac{\partial}{\partial x}\frac{\partial \psi}{\partial z} - \left(\frac{\partial U}{\partial z}\right)\psi\frac{\partial \psi}{\partial x} = 0 \quad \text{on} \quad z = 0, H$$

which then can be expanded as

$$\frac{\partial}{\partial t}\left(\psi\frac{\partial \psi}{\partial z}\right) - \frac{\partial \psi}{\partial t}\frac{\partial \psi}{\partial z} + U\frac{\partial}{\partial x}\psi\frac{\partial \psi}{\partial z} - U\frac{\partial \psi}{\partial x}\frac{\partial \psi}{\partial z} - \frac{\partial U}{\partial z}\psi\frac{\partial \psi}{\partial x} = 0 \quad \text{on} \quad z = 0, H$$

Then we can integrate these equations for one period, $\int_0^X (\ ) dx$ to obtain

$$\int \left[\frac{\partial}{\partial t}\psi\frac{\partial \psi}{\partial z} - \frac{\partial \psi}{\partial t}\frac{\partial \psi}{\partial z} - U\frac{\partial \psi}{\partial x}\frac{\partial \psi}{\partial z} \right]_0^H dx = 0$$

and therefore show that the last term is zero. With this cancellation, we thus obtain the equation for conservation of total energy:

$$\frac{\partial}{\partial t} \iiint \frac{1}{2}\rho_r \left[\overline{\psi_x^2} + \overline{\psi_y^2} + \frac{f_0^2}{N^2}\overline{\psi_z^2} \right] dxdydz$$

$$= \iiint \rho_r \left[\overline{\psi_x\psi_y}U_y + \frac{f_0^2}{N^2}\overline{\psi_x\psi_z}U_z \right] dxdydz$$

which is equivalent to the total energy Eq. (3.152) derived in Sect. 3.3.6,

$$\frac{\partial}{\partial t} \int_0^H \int_0^Y \frac{1}{2}\rho_r \left[\overline{\psi_x^2} + \overline{\psi_y^2} + \frac{f_0^2}{N^2}\overline{\psi_z^2} \right] dydz$$

$$= \int_0^H \int_0^Y \rho_r \left[\overline{\psi_x\psi_y}U_y + \frac{f_0^2}{N^2}\overline{\psi_x\psi_z}U_z \right] dxdydz$$

where Eqs. (3.153) and (3.154) provide the corresponding relations to eddy variables

$$\psi_x = +v'$$
$$\psi_y = -u'$$
$$\psi_z = \frac{g}{f_0} \frac{\theta'}{\theta_r}.$$

7.3   Chapter 3, Exercise 2

Following early works on atmospheric dynamics [1], an analytical solution is presented for a simplified problem of baroclinic perturbations superposed on uniform zonal flow of linear variation with height, typical of tropospheric winds

$$U(z) = U_0 \left(\frac{z}{H}\right)$$

where planetary and zonal flow vorticity components contributing to quasi-geostrophic relative vorticity q have simply been ignored (by setting $R = 0$), as in the above reference to the Eady problem.

It is further assumed that $\rho_r \approx \rho_0 = $ constant, yielding a stratification parameter $N^2 = N_0^2 = $ constant, leading to a simpler vorticity equation

$$\left(\frac{\partial}{\partial t} + U_0 \frac{z}{H} \frac{\partial}{\partial x}\right) \left[\nabla^2 \psi + \frac{f_0^2}{N_0^2} \frac{\partial^2 \psi}{\partial z^2}\right] = 0$$

with boundary conditions

$$\frac{\partial \psi}{\partial x} = 0 \quad \text{on} \quad y = 0, Y$$
$$\frac{\partial}{\partial t} \frac{\partial \psi}{\partial z} - \frac{U_0}{H} \frac{\partial \psi}{\partial x} = 0 \quad \text{on} \quad z = 0$$
$$\left(\frac{\partial}{\partial t} + U_0 \frac{\partial}{\partial x}\right) \frac{\partial \psi}{\partial z} - \frac{U_0}{H} \frac{\partial \psi}{\partial x} = 0 \quad \text{on} \quad z = H.$$

To this set of equations we can propose a wave solution

$$\frac{p}{\rho_0 f_0} = \psi = \text{Re}\{\hat{F}(z)e^{ik(x-ct)} \sin \ell y\}$$

where $\hat{F}(z)$ is a dimensional amplitude function. We further define a non-dimensional amplitude $F(\zeta) = \hat{F}(z/H)$

$$\Psi = \text{Re}\{F(z)e^{i(\kappa\xi - C\tau)} \sin \pi y/Y\}$$

where

$$\ell = \frac{\pi}{Y}$$

represents a wave number of the sinusoidal solution $\sin \ell y = \sin(\pi y/Y)$ satisfying side boundary conditions (3.148.a), $v = \partial \psi/\partial x = 0$ at $y = 0, Y$.

Substituting the proposed solution into the vorticity equation yields

$$\left(-ikc + ikU_0 \frac{z}{H}\right)\left(-(k^2 + \ell^2)\hat{F}(z) + \frac{f_0^2}{N_0^2}\frac{d^2\hat{F}(z)}{dz^2}\right) = 0.$$

Excluding the multiplicative term in the first bracket gives

$$\frac{d^2\hat{F}}{dz^2} - \frac{N_0^2}{f_0^2}(k^2 + \ell^2)\hat{F} = 0$$

with top and bottom boundary conditions

$$-ikc\frac{d\hat{F}}{dz} - ik\frac{U_0}{H}\hat{F} \quad \text{on } z = 0$$

$$(-ikc + ikU_0)\frac{d\hat{F}}{dz} - ik\frac{U_0}{H}\hat{F} \quad \text{on } z = H$$

which can be readily solved to determine the amplitude $\hat{F}(z)$.

Non-dimensional vertical coordinate ζ and wave-number κ are defined as

$$\zeta = \frac{z}{H} \quad \text{and} \quad \kappa^2 = \left(\frac{N_0 H}{f_0}\right)^2 (k^2 + \ell^2)$$

where *Rossby radius of deformation* L_d is identified as

$$L_d = \frac{N_0 H}{f_0}$$

and dimensional wave-number as

$$\tilde{k}^2 = k^2 + \ell^2$$

allowing to redefine

$$\kappa = L_d \tilde{k}.$$

Also defining

$$C = \frac{c}{U_0},$$

the equations in normalized coordinates become

$$\frac{d^2 F}{d\zeta^2} - \kappa^2 F = 0$$

with boundary conditions

$$C \frac{dF}{d\zeta} - F \quad \text{on} \quad \zeta = 0$$

$$(1 - C) \frac{dF}{d\zeta} + F \quad \text{on} \quad \zeta = 1.$$

The solution

$$F(\zeta) = A \cosh \kappa\zeta + B \sinh \kappa\zeta$$

with coefficients A and B is obtained from boundary conditions

$$C\kappa B + A = 0$$
$$\kappa(1 - C)(A \sinh \kappa + B \cosh \kappa) - 1(A \cosh \kappa + B \sinh \kappa) = 0.$$

Use is made of the first equation, to obtain

$$A = -C\kappa B$$

letting an arbitrary amplitude value $B = -1$ determine the vertical amplitude function

$$F(\zeta) = C\kappa \cosh \kappa\zeta - \sinh \kappa\zeta.$$

Eliminating the two unknowns A, B between the above equations on boundary conditions gives

$$C\kappa[\kappa(1 - C) \sinh \kappa - \cosh \kappa] - [\kappa(1 - C) \cosh \kappa - \sinh \kappa] = 0$$

and by expanding terms

$$C^2 \kappa^2 \sinh \kappa - C\kappa[\kappa \sinh \kappa - \cosh \kappa + \cosh \kappa] + [\kappa \cosh \kappa - \sinh \kappa] = 0$$

and re-organizing,

$$C^2 - C + \frac{1}{\kappa^2} \left[\frac{\kappa \cosh \kappa - \sinh \kappa}{\sinh \kappa} \right] = 0.$$

The above equation is the *dispersion relation* for the quasi-geostrophic waves under the present restricted model, relating the wave speed $C = c/U_0$ to mean flow velocity U_0 and wave-number of horizontal waves $\kappa^2 = (N_0 H/f_0)^2(k^2 + \ell^2) = L_d^2 \tilde{k}^2$.

Re-defining wave speed c relative to the mean value $U_0/2$ of the velocity profile linearly varied with depth

$$C = \frac{1}{2} + \hat{C},$$

the dispersion relation is updated by substituting the above changes,

$$\frac{1}{4} + \hat{C} + \hat{C}^2 - \left[\frac{1}{2} + \hat{C}\right] + \frac{1}{\kappa^2}\left[\frac{\kappa \cosh \kappa - \sinh \kappa}{\sinh \kappa}\right] = 0$$

yielding

$$\hat{C}^2 = \frac{1}{\kappa^2}\left[\frac{\kappa^2}{4} - \frac{\kappa \cosh \kappa - \sinh \kappa}{\sinh \kappa}\right] = \frac{1}{\kappa^2}\left[\frac{\kappa^2}{4} - \kappa \coth \kappa + 1\right].$$

The next step is needed to show

$$
\begin{aligned}
\left[\frac{\kappa}{2} - \tanh\frac{\kappa}{2}\right]\left[\frac{\kappa}{2} - \coth\frac{\kappa}{2}\right] &= \frac{\kappa^2}{4} - \frac{\kappa}{2}\left[\coth\frac{\kappa}{2} + \tanh\frac{\kappa}{2}\right] + 1 \\
&= \frac{\kappa^2}{4} - \frac{\kappa}{2}\left[\frac{e^{\kappa/2} + e^{-\kappa/2}}{e^{\kappa/2} - e^{-\kappa/2}} + \frac{e^{\kappa/2} - e^{-\kappa/2}}{e^{\kappa/2} - e^{-\kappa/2}}\right] + 1 \\
&= \frac{\kappa^2}{4} - \frac{\kappa}{2}\left[\frac{(e^{\kappa/2} + e^{-\kappa/2})^2 + (e^{\kappa/2} + e^{-\kappa/2})^2}{e^{\kappa} - e^{-\kappa}}\right] + 1 \\
&= \frac{\kappa^2}{4} - \frac{\kappa}{2}\left[\frac{(e^{\kappa} + e^{-\kappa} + 2) + (e^{\kappa} + e^{-\kappa} - 2)}{e^{\kappa} - e^{-\kappa}}\right] + 1 \\
&= \frac{\kappa^2}{4} - \kappa \coth \kappa + 1
\end{aligned}
$$

so that the dispersion relation becomes

$$\hat{C}^2 = \frac{1}{\kappa^2}\left[\frac{\kappa^2}{4} - \kappa \coth \kappa + 1\right] = \left(\frac{1}{\kappa}\right)^2 \left[\frac{\kappa}{2} - \tanh\frac{\kappa}{2}\right]\left[\frac{\kappa}{2} - \coth\frac{\kappa}{2}\right].$$

The terms $\kappa/2$, $\tanh(\kappa/2)$ and $\coth(\kappa/2)$ are plotted as functions of κ in Fig. 7.1 to show intersections of these functions at points $(0, 0)$ and $(2.4, 1.2)$ marked by the black dots in the same figure.

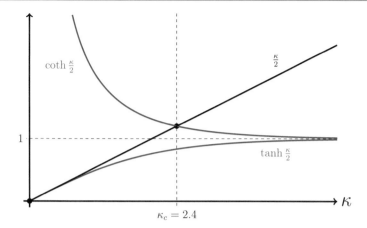

Fig. 7.1 Terms in dispersion relation

If we now return to the total phase velocity $C = \hat{C} + 1/2$, we will obtain two roots of the dispersion equation corresponding to different modes of wave propagation:

$$C^{\pm} = \frac{1}{2} \pm \frac{1}{\kappa}\sqrt{\left[\frac{\kappa}{2} - \tanh\frac{\kappa}{2}\right]\left[\frac{\kappa}{2} - \coth\frac{\kappa}{2}\right]}.$$

The behavior of individual terms in the dispersion equation can be understood by considering intersections of individual functions plotted in Fig. 7.1. It can be observed that the first term $\kappa/2 - \tanh(\kappa/2)$ is always positive, while the multiplicative term $\kappa/2 - \coth(\kappa/2)$ changes sign at

$$\kappa_c = 2.399 \simeq 2.4$$

(approximated above) the product being negative for smaller values $\kappa < \kappa_c$, and positive for $\kappa > \kappa_c$. This implies that the wave velocity is a complex number

$$C = C_r + iC_i$$

at least in the region $\kappa < \kappa_c$, resulting from the negative square root. Wave speeds in the corresponding ranges of κ are therefore given as

region $\kappa < \kappa_c$

$$C_r = \frac{1}{2}$$

$$C_i^+ = +\frac{1}{\kappa}\sqrt{\left[\frac{\kappa}{2} - \tanh\frac{\kappa}{2}\right]\left[\coth\frac{\kappa}{2} - \frac{\kappa}{2}\right]}$$

$$C_i^- = -\frac{1}{\kappa}\sqrt{\left[\frac{\kappa}{2} - \tanh\frac{\kappa}{2}\right]\left[\coth\frac{\kappa}{2} - \frac{\kappa}{2}\right]}$$

Fig. 7.2 Dispersion diagram for the Eady problem of baroclinic waves on a uniform zonal flow

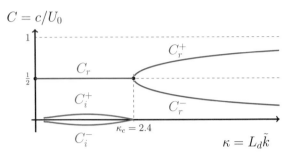

region $\kappa > \kappa_c$

$$C_r = C_r^+ = \frac{1}{2} + \frac{1}{\kappa}\sqrt{\left[\frac{\kappa}{2} - \tanh\frac{\kappa}{2}\right]\left[\frac{\kappa}{2} - \coth\frac{\kappa}{2}\right]}$$

$$C_r = C_r^- = \frac{1}{2} - \frac{1}{\kappa}\sqrt{\left[\frac{\kappa}{2} - \tanh\frac{\kappa}{2}\right]\left[\frac{\kappa}{2} - \coth\frac{\kappa}{2}\right]}$$

$$C_i = 0$$

In the region $\kappa < \kappa_c$ (long wavelengths), the phase velocity has real and imaginary parts, C_r and C_i respectively shown in blue and red colors in Fig. 7.2. In the region $\kappa > \kappa_c$ (short wavelengths) the wave speed does not have an imaginary part, instead the solution now has two real branches C_r^+ and C_r^-. The positive branch with faster wave speed approaching mean current velocity $C = c/U_0 \rightarrow 1$, while the negative branch has slower wave speeds approaching $C \rightarrow 0$, as $\kappa \rightarrow \infty$.

The fact that phase velocity has an imaginary part implies temporal changes other than simple wave motion. Long waves, with wave-numbers $\kappa < \kappa_c$ imply exponential (unstable) growth represented by $\alpha = \kappa C_i^+ > 0$, while short waves with $\kappa > \kappa_c$ imply exponential (stable) decay with $\alpha = \kappa C_i^- < 0$, by inserting the corresponding components in the non-dimensional stream-function Ψ, which we will temporarily represent as

$$\Psi \sim \text{Re}\{F(z)e^{\kappa C_i t} e^{i(\kappa \cdot \mathbf{x} - \kappa C_r t)}\}\sin(\pi y/Y) = \text{Re}\{F(z)e^{\alpha t} e^{i(\kappa \cdot \mathbf{x} - \Omega t)}\}\sin(\pi y/Y).$$

The product κC has real and imaginary components $\alpha = \kappa C_i$ and $\Omega = \kappa C_r$, representing growth rate and frequency of the waves. Figure 7.3 illustrates the corresponding wave dispersion relation, where α and Ω are plotted against wave-number κ.

It can be observed that long waves in the range $\kappa < \kappa_c$ are non-dispersive, with constant wave speed $\Omega/\kappa = C_r = 1/2$. Then, for short waves $\kappa > \kappa_c$, the negative branch approaches near constant frequency $\Omega = \kappa C_r^- \rightarrow \kappa/2$, while the positive branch approaches a non-dispersive regime with increased wave speed approaching $\Omega/\kappa \rightarrow C_r^+ \simeq 1$.

Fig. 7.3 Dispersion diagram for wave frequency and growth rate versus horizontal wave-number

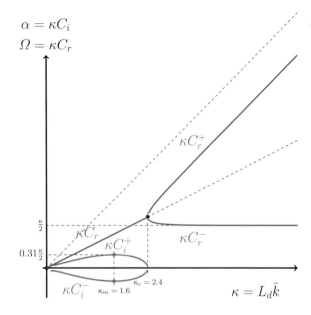

Stable Disturbances

Wave speed components in the wave-numbers range $\kappa > \kappa_c = 2.4$ are specified with the real numbers C_r^+ and C_r^-. However, it is noted that in the range $\kappa C > 1$, the two branches of the dispersion diagram approach asymptotic limits

$$\kappa C_r^+ \to \kappa$$
$$\kappa C_r^- \to 1.$$

Therefore, by slightly reorganizing the functional form of $F(\zeta)$,

$$F(\zeta) = \underbrace{\cosh \kappa\zeta}_{>1}(\underbrace{\kappa C}_{>1} - \underbrace{\tanh \kappa\zeta}_{<1}) > 0$$

it can be confirmed that the vertical profile must always have positive sign at all heights $F(\zeta) > 0$ in the range of wave-numbers $\kappa > \kappa_c$.

As indicated above, there will be two real case solutions $F^+(\zeta)$ and $F^-(\zeta)$ corresponding to the two branches in the dispersion diagram. Amplitude profiles corresponding to the two branches are displayed in Fig. 7.4. It can be observed that the solution $F^+(\zeta)$ has a minimum at ζ^+ in the lower part, reaching greater amplitude in the upper part, while the solution $F^-(\zeta)$ has much smaller amplitude, reaching a minimum at ζ^- in the upper part, continuing with a mild slope to the top.

An important characteristic of short waves $\kappa > \kappa_c$ is that they are *phase locked* with respect to depth, i.e. without phase lag and no change of sign in amplitude with respect to depth. All wave motions have the same phase of oscillation, although with varying amplitude, as shown in Fig. 7.5.

Fig. 7.4 Profiles of $F_+(\zeta)$ (red) with minimum at $\zeta^- = 0.231$, $F_-(\zeta)$ (blue) with minimum at $\zeta^+ = 0.693$, for wave number $\kappa = 3.0$

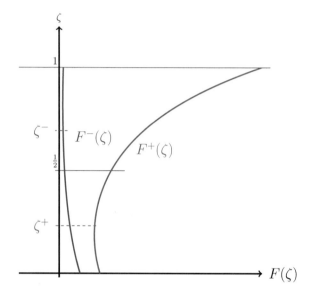

Fig. 7.5 Temperature distribution corresponding to (upper figure) amplitude $F^+(\zeta)$, (lower figure) amplitude $F^-(\zeta)$, depth ζ and zonal coordinate $\kappa\xi$ for two periods -2π to 2π, at wave number $\kappa = \kappa_c = 1.6$

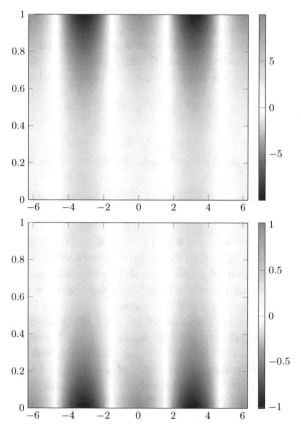

An interesting property noted both in Figs. 7.4 and 7.5 is that the solution $F^+(\zeta)$ has minimum at ζ^+ in the lower half although intensified at the top, while $F^-(\zeta)$ has minimum at ζ^- in the upper half, intensifying near the bottom.

The vertical positions ζ^+ and ζ^- can be obtained by differentiation with respect to depth

$$\frac{dF}{d\zeta} = \kappa[C\kappa \sinh \kappa\zeta - \cosh \kappa\zeta] = \kappa \underbrace{\sinh \kappa\zeta}_{>0}[C\kappa - \coth \kappa\zeta] = 0$$

and utilizing the facts that $\sinh \kappa\zeta > 0$ and both $C\kappa > 1$, $\coth \kappa\zeta > 1$ we can obtain

$$\zeta^\pm = \frac{1}{\kappa} \coth^{-1}(C^\pm\kappa)$$

where the value of inverse hyperbolic cotangent function $\coth^{-1}(x)$ is given by

$$\coth^{-1}(x) = \ln \frac{x+1}{x-1} \quad \text{for } x > 1$$

in the given range of interest, $\kappa > \kappa_c = 2.4$ and $C^\pm\kappa > 1$. For instance, the limit at $\kappa_c = 2.4$ is obtained as

$$\zeta^\pm = 1/2$$

and therefore the minima are found in the range

$$0 < \zeta^+ < \frac{1}{2} \quad \text{and} \quad \frac{1}{2} < \zeta^- < 1.$$

Detailed structure of key variables pressure, temperature, density and vertical velocity are displayed in Fig. 7.6, showing their phase relationships to each other along the bottom, top and in the vertical. Dependency of these variables are obtained in the following.

Amplitude of pressure (stream-function) is shown to be in phase at all heights for each of the modes $F^\pm(\zeta)$.

Similarly, density and temperature distributions are obtained for stable motions, by noting that

$$\frac{\rho}{\rho_0} = -\frac{T}{T_0}$$

and using hydro-static equation to obtain density perturbation

$$\rho = -\frac{1}{g}\frac{\partial p}{\partial z} = -\frac{\rho_0 f_0}{gH}\frac{\partial \Psi}{\partial \zeta} \cos[k(x-ct)]\sin(\pi y/Y)$$

$$= -\frac{\rho_0 f_0}{gH}\frac{\partial F}{\partial \zeta} \cos(\kappa\xi - C\tau)\sin(\pi y/Y)$$

Fig. 7.6 Line graphs of
temperature (red), pressure
(blue), vertical gradient of
vertical velocity (green) at
the bottom and top, as a
function of zonal distance,
and vertical velocity as a
function of depth, all in
relative magnitude, for cases
corresponding to (upper
figure) positive amplitude
function F^+, (lower figure)
negative amplitude function
F^-, at critical value of wave
number $\kappa = \kappa_c = 1.6$

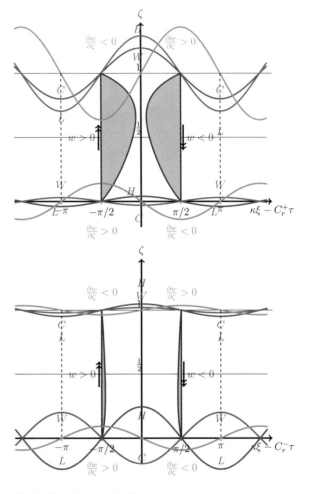

where use is made of the amplitude function defined as

$$\frac{dF}{d\zeta} = \kappa[C\kappa \sinh \kappa\zeta - \cosh \kappa\zeta]$$

which means that the density changes sign over depth at each of the minima $\zeta^\pm$ so
that density and temperature are out of phase with pressure, such that

for $\zeta < \zeta^\pm$

$p > 0 \quad (H) \rightarrow \rho > 0 \text{ and } T < 0 \quad (C)$

$p < 0 \quad (L) \rightarrow \rho < 0 \text{ and } T > 0 \quad (W)$

for $\zeta > \zeta^\pm$

$p > 0 \quad (H) \rightarrow \rho < 0 \text{ and } T > 0 \quad (W)$

$p < 0 \quad (L) \rightarrow \rho > 0 \text{ and } T < 0 \quad (C)$

which is true for either case of the modes $F^{\pm}(\zeta)$, although the amplitude shape differs according to the fast and slow modes of the short waves as indicated.

Finally, we can seek the relationship between vertical velocity and divergence (hence vorticity) with respect to changes in pressure. We recall that the vertical velocity is directly tied with relative vorticity

$$f_0 \frac{\partial w}{\partial z} = \{\frac{\partial}{\partial t} + U(z)\frac{\partial w}{\partial z}\}\nabla^2\psi = -(k^2 + \ell^2)\{\frac{\partial}{\partial t} + U(z)\frac{\partial}{\partial z}\}\psi$$

$$= -(\tilde{k}^2)(ik)(C - U(z))\hat{F}(z)\sin(kx - ct)\sin(\pi y/Y)$$

Integrating to obtain vertical velocity w gives

$$w = \left[\int_0^\zeta (C - \zeta)\hat{F}(\zeta)d\zeta\right]\sin(\kappa\xi - C\tau)\sin(\pi y/Y)$$

To integrate, we make use of

$$F(\zeta) = C\kappa\cosh\kappa\zeta - \sinh\kappa\zeta$$

$$\int \zeta\cosh\kappa\zeta = \frac{\zeta}{\kappa}\sinh\kappa\zeta - \frac{1}{\kappa^2}\cosh\kappa\zeta$$

$$\int \cosh\kappa\zeta = \frac{1}{\kappa}\sinh\kappa\zeta$$

$$\int_0^\zeta (C - \zeta)\hat{F}(\zeta)d\zeta = \frac{C}{\kappa}[C\kappa\sinh\kappa\zeta - \cosh\kappa\zeta]$$

$$- C\kappa\left[\frac{\zeta}{\kappa}\sinh\kappa\zeta - \frac{1}{\kappa^2}\cosh\kappa\zeta\right]$$

$$+ \left[\frac{\zeta}{\kappa}\cosh\kappa\zeta - \frac{1}{\kappa^2}\sinh\kappa\zeta\right]$$

$$= \left[-\frac{C\!\!\!/}{\!\!\!/\kappa} + \frac{C\!\!\!/}{\!\!\!/\kappa} + \frac{\zeta}{\kappa}\right]\cosh\kappa\zeta + \left[C^2 - C\zeta - \frac{1}{\kappa^2}\right]\sinh\kappa\zeta$$

$$= \frac{\zeta}{\kappa}\cosh\kappa\zeta + \left[C(C - \zeta) - \frac{1}{\kappa^2}\right]\sinh\kappa\zeta$$

we can expand the square brackets in the last expression as

$$
\begin{aligned}
\left[C(C - \zeta) - \frac{1}{\kappa^2} \right] &= C(C - 1) + C(1 - \zeta) - \frac{1}{\kappa^2} \\
&= \left(\frac{1}{2} + \frac{1}{\kappa}\sqrt{\frac{\kappa}{2}} - \tanh\frac{\kappa}{2} \right)\left(-\frac{1}{2} + \frac{1}{\kappa}\sqrt{\frac{\kappa}{2}} - \coth\frac{\kappa}{2} \right) + C(1 - \zeta) - \frac{1}{\kappa^2} \\
&= \frac{1}{\kappa^2}\left[\frac{\kappa}{2} - \tanh\frac{\kappa}{2} \right]\left[\frac{\kappa}{2} - \coth\frac{\kappa}{2} \right] - \frac{1}{4} + C(1 - \zeta) - \frac{1}{\kappa^2} \\
&= \frac{1}{\kappa^2}\left[\frac{\kappa\!\!\!/^2}{4} - \kappa\coth\kappa + \!\!\!/1 \right] - \frac{1\!\!\!/}{4} + C(1 - \zeta) - \frac{1\!\!\!/}{\kappa^2\!\!\!/} \\
&= -\frac{1}{\kappa}\coth\kappa + C(1 - \zeta)
\end{aligned}
$$

so that

$$
w = \frac{\zeta}{\kappa}\cosh\kappa\zeta + \left[-\frac{1}{\kappa}\coth\kappa + (1 - \zeta)\left(\frac{1}{2} + \frac{1}{\kappa}\sqrt{\frac{\kappa}{2}} - \tanh\frac{\kappa}{2} \right) \right]
$$

Unstable Disturbances

Unstable disturbances occur in the long wave regime $\kappa < \kappa_c$. The real part C_r corresponds to propagating waves, while the imaginary part $\kappa C_i^+ > 0$ corresponds to a growing component and a decaying one with $\kappa C_i^- < 0$, respectively shown by blue and red lines in the dispersion diagram of Fig. 7.3.

The wave form is described as

$$
\Psi = \mathrm{Re}\{F(z)e^{kC_i t}e^{ik(x - C_r t)}\}\sin\ell y,
$$

with amplitude $F(\zeta)$ as a function of the normalized vertical coordinate $\zeta = z/H$, wave speed $C = c/U_0$, which in turn is a function of wave-number κ. The complex amplitude has an absolute value $|F(\zeta)|$ and phase angle $\theta(\zeta)$ that can be specified as

$$
F(\zeta) = |F(\zeta)|e^{i\theta(\zeta)}.
$$

The amplitude $|F(\zeta)|$ and phase $\theta(\zeta)$ of the wave solution can be displayed, with distinct variations between the upper and lower atmosphere, as shown in Fig. 7.7a, b.

For unstable waves we take the positive root of C

$$
C = C_r + iC_i^+ = \frac{1}{2} + i\frac{1}{\kappa}\sqrt{\left[\frac{\kappa}{2} - \tanh\frac{\kappa}{2} \right]\left[\coth\frac{\kappa}{2} - \frac{\kappa}{2} \right]}.
$$

Fig. 7.7 Profiles of (upper figure) $|F(\zeta)|$ for $\kappa = 0.6$ (brown), $\kappa = 1.8$ (blue) and $\kappa = \kappa_c = 2.4$ (red), (lower figure) $\theta(\zeta)$ for $\kappa = 2.4$ (red), $\kappa = 2.0$ (blue), $\kappa = \kappa_m = 1.6$ (brown) and $\kappa = 0.1$ (violet)

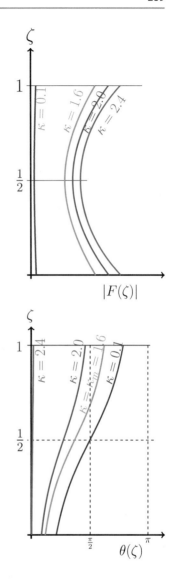

The complex amplitude is given as

$$F(\zeta) = C\kappa \cosh \kappa\zeta - \sinh \kappa\zeta = C_r \kappa \cosh \kappa\zeta - \sinh \kappa\zeta + i C_i^+ \kappa \cosh \kappa\zeta$$

with the magnitude and direction (phase angle) obtained as

$$|F(\zeta)|^2 = (C_r \kappa \cosh \kappa\zeta - \sinh \kappa\zeta)^2 - (C_i^+ \kappa \cosh \kappa\zeta)^2$$

$$= (\frac{\kappa}{2} \cosh \kappa\zeta - \sinh \kappa\zeta)^2 - \frac{1}{\kappa^2}\left[\frac{\kappa}{2} - \tanh \frac{\kappa}{2}\right]\left[\coth \frac{\kappa}{2} - \frac{\kappa}{2}\right]\kappa^2 \cosh^2 \kappa\zeta$$

$$= \frac{\kappa^2}{4} \cosh^2 \kappa\zeta - \kappa \cosh \kappa\zeta \sinh \kappa\zeta + \sinh^2 \kappa\zeta$$

$$\qquad - \frac{1}{\kappa^2}\left[\frac{\kappa^2}{4} - \kappa \coth \kappa + 1\right]\kappa^2 \cosh^2 \kappa\zeta$$

$$= -\kappa \cosh \kappa\zeta \sinh \kappa\zeta + \sinh^2 \kappa\zeta + [\kappa \coth \kappa - 1]\cosh^2 \kappa\zeta$$

$$= \kappa \cosh \kappa\zeta [\cosh \kappa\zeta \coth \kappa - \sinh \kappa\zeta] - 1$$

$$= \kappa \cosh \kappa\zeta \left[\frac{\cosh \kappa\zeta \cosh \kappa - \sinh \kappa\zeta \sinh \kappa}{\sinh \kappa}\right] - 1$$

$$= \left[\frac{\kappa \cosh \kappa\zeta \cosh \kappa(\zeta - 1)}{\sinh \kappa}\right] - 1$$

where use have been made of the hyperbolic identities

$$\cosh^2 \kappa\zeta - \sinh^2 \kappa\zeta = 1$$

$$\cosh \kappa\zeta \cosh \kappa - \sinh \kappa\zeta \sinh \kappa = \cosh(\kappa\zeta - \kappa).$$

We recognize that the absolute value of the amplitude

$$|F(\zeta)| = \sqrt{\left(\frac{\kappa \cosh \kappa\zeta \cosh \kappa(\zeta - 1)}{\sinh \kappa}\right) - 1} \quad > 0$$

is always positive and symmetrical about the mid-depth, as displayed in upper Fig. 7.7.

The phase is obtained as

$$\theta(\zeta) = \arctan\left(\frac{C_i^+ \kappa \cosh \kappa\zeta}{\frac{\kappa}{2} \cosh \kappa\zeta - \sinh \kappa\zeta}\right) = \arctan\left(\frac{C_i^+ \kappa}{\frac{\kappa}{2} - \tanh \kappa\zeta}\right)$$

$$= \arctan\left(\frac{\sqrt{[\frac{\kappa}{2} - \tanh \frac{\kappa}{2}][\coth \frac{\kappa}{2} - \frac{\kappa}{2}]}}{\frac{\kappa}{2} - \tanh \kappa\zeta}\right).$$

Applying the real part of C_r for long wave-numbers $\kappa < \kappa_c$ selected profiles are plotted in Fig. 7.7. It can be observed that long waves for selected values of $\kappa = 0.1, 1.6, 2.0, 2.4$ evolve from zero to a cusp shape with maximum values at $\kappa = \kappa_c = 2.4$.

The phase relationships for pressure (blue) and temperature (red) fields are plotted in Fig. 7.8, for selected values of wave number $\kappa = 2.0$ and $\kappa = 2.0$, by displaying of

Fig. 7.8 Wave pattern of zero points for two periods along the zonal coordinate $\kappa\xi$ at fixed time for (upper figure) $\kappa = 2.0$ and (lower figure) $\kappa = 1.6$

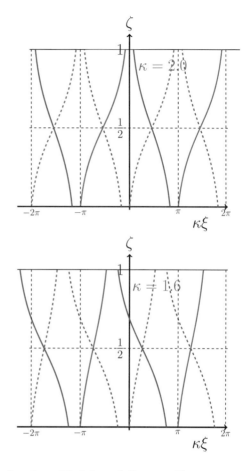

the zero-points of these variables as a function of height and distance. Temperature and pressure are observed often to be out of phase for growing disturbances. For smaller wave numbers (longer waves) the phase difference between top and bottom become greater as indicated by Figs. 7.7 and 7.8.

Pressure patterns for unstable waves are displayed for the most stable wave number $\kappa = \kappa_m = 1.6$ and for long waves $\kappa = 0.2$ in Fig. 7.9, showing phase differences between top and bottom.

Density field for the most stable wave number $\kappa = \kappa_m = 1.6$ is displayed in Fig. 7.10.

The time-dependent components either grow as $e^{\kappa C_i^+ t}$ or decay as $e^{-\kappa C_i^- t}$. The maxima of the imaginary parts κC_i^+ and κC_i^- occur approximately at

$$\kappa = \kappa_m \simeq 1.6$$

in the range $0 < \kappa < \kappa_c$, where $\kappa_c = 2.4$. Maximal value for the non-dimensional wave growth rate α_m occurs at $\kappa = \kappa_m \simeq 1.6$, at which we evaluate

Fig. 7.9 Wave pressure ψ amplitude as a function of depth ζ and zonal coordinate $\kappa\xi$ for two periods -2π to 2π with (upper figure) $\kappa = \kappa_m = 1.6$ and (lower figure) $\kappa = 0.2$

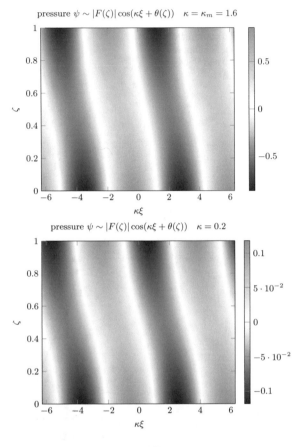

pressure $\psi \sim |F(\zeta)| \cos(\kappa\xi + \theta(\zeta))$ $\kappa = \kappa_m = 1.6$

pressure $\psi \sim |F(\zeta)| \cos(\kappa\xi + \theta(\zeta))$ $\kappa = 0.2$

Fig. 7.10 Density amplitude as a function of depth ζ zonal coordinate $\kappa\xi$ for two periods -2π to 2π for $\kappa = \kappa_m = 1.6$

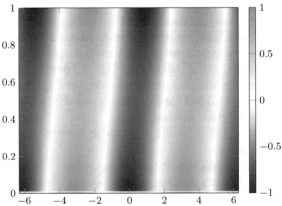

$$\alpha_m = 0.31 \frac{\kappa_m}{2} = 0.31 \frac{1.6}{2} = 0.25.$$

Defining

$$L_d \equiv \frac{N_0 H}{f_0}$$

and remembering that

$$\kappa = L_d \tilde{k} \leq 2.4$$

yields the dimensional wave-numbers range for unstable waves

$$L_d \leq \frac{2.4}{\tilde{k}}$$

where $\tilde{k}^2 = (k^2 + \ell^2)$ and $\ell = \pi/Y$ is a constant related to the channel width Y, and the wave-number $k = 2\pi/\lambda$ in x-direction is related to the wave-length λ.

We recognize *Rossby radius of deformation* $L_d \equiv N_0 H / f_0$ defined in the earlier sections, so that

$$L_d \leq \frac{2.4}{\sqrt{(\frac{2\pi}{\lambda})^2 + (\frac{\pi}{Y})^2}}.$$

Since the horizontal wave-number can take any value $k \geq 0$, the more restrictive criteria must be

$$L_d \leq 2.4 \frac{Y}{\pi} \quad \text{or} \quad Y \geq 1.3 L_d$$

which leads to conclude that the zonal channel width must be greater than the Rossby radius scale.

However, considering the fact that zonal channel width is too restrictive, we let $Y \to \infty$, or $\ell \to 0$, and the wave-length based criteria becomes

$$L_d \leq 2.4 \frac{\lambda}{2\pi} \quad \text{or} \quad \lambda \geq 2.6 L_d.$$

We can test this criteria with appropriate values (i) for the atmosphere, letting $N_0 = 10^{-2} \, \text{s}^{-1}$, $f_0 = 10^{-4} \, \text{s}^{-1}$ and $H = 10 \, \text{km}$ to yield $L_d = 1000 \, \text{km}$ and $\lambda > 2600 \, \text{km}$, (ii) for the ocean, letting same values of N_0, f_0 and $H = 100 - 1000 \, \text{m}$ to yield $L_d = 10 - 100 \, \text{km}$ leading to $\lambda > 260 \, \text{km}$ which appear at the upper limits of the meso-scale motions both for the atmosphere and ocean.

Next we would like to determine the growth rate of the most unstable wave, in the range $\kappa < \kappa_c$, reviewed above.

We have seen the maximum growth rate occurs at dimensionless wave-number $\kappa_m = 1.6$ and that the greatest growth rate occurs at $\alpha_m = 0.25$.

Since $\ell = \pi/Y$ is a constant, we can define an effective dimensional wave-number $\tilde{k}$ and where it has the maximum as

$$\tilde{k}_m = \sqrt{k_m^2 + \ell^2} \equiv \frac{2\pi}{\lambda_m}$$

But then, we can relate this maximum growth rate to the corresponding dimensional wave-number k_m

$$\kappa_m = L_d \tilde{k}_m = L_d \frac{2\pi}{\lambda_m} = 1.6$$

yielding the wave-length of the fastest growing wave

$$\lambda_m = 3.9 L_d.$$

Providing the maximum dimensionless growth rate defined above we can manipulate

$$\alpha_m = 0.25 = \kappa_m C_i^+ = L_d \tilde{k}_m \frac{c_{im}^+}{U_0}$$

to obtain a dimensional growth rate

$$\tilde{k}_m c_{im}^+ = 0.25 \frac{U_0}{L_d}.$$

We can assume a typical top velocity of $U_0 = 20$ m/s for the upper atmosphere and $U_0 = 1$ m/s for the upper ocean, while still estimating the Rossby radii as $L_d = 1000$ km and $L_d = 100 - 1000 m$ for the respective media. Therefore we can estimate the dimensional growth rate respectively for the atmosphere as $\tilde{k}_m c_{im}^+ = 0.4 d^{-1}$ and for the ocean $\tilde{k}_m c_{im}^+ = (0.025 - 0.25) d^{-1}$. The typical growth (e-folding) time scales

$$t_m = (\tilde{k}_m c_{im}^+)^{-1} = 4.0 \frac{L_d}{U_0}$$

7.4   Chapter 4, Exercise 1

The internal wave motion of a fluid with constant stratification N between vertical boundaries at $z = 0$ and $z = H$ is described by (4.44)–(4.46) governing the amplitude

$$\frac{d^2 \hat{w}}{dz^2} + \left(\frac{N^2}{\omega^2} - 1 \right) k^2 \hat{w} = 0,$$

Fig. 7.11 Dispersion curves relating frequency ω to wave number k for modes $n = 1, 2, 3 \ldots$

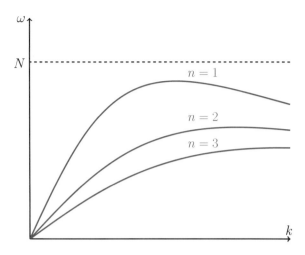

and boundary conditions

$$\hat{w}(0) = \hat{w}(H) = 0.$$

The solution is obtained as

$$\hat{w} = A \sin\left(\frac{n\pi z}{H}\right), \quad \text{for } n = 1, 2, 3 \ldots$$

where n is an integer and $m = n\pi/H$ is the vertical wave number yielding

$$-\left(\frac{n\pi}{H}\right)^2 + \left(\frac{N^2}{\omega^2} - 1\right)k^2 = 0.$$

The dispersion relation is identical to (4.51), only m taking integer values of n. and a relation between ω and k can be determined for these integer values of n. The dispersion relation describes a relation between ω and k as plotted in Fig. 7.11.

The vertical velocity is by Eq. (4.44)

$$w = Re\left\{\hat{w}(z) e^{i(kx - \omega t)}\right\} \sim A \sin mz \cos(kx - \omega t).$$

Horizontal component of velocity can be obtained from the continuity equation

$$\frac{\partial u}{\partial x} + \frac{\partial w}{\partial z} = 0$$

as

$$u = A\frac{m}{k} \cos mz \sin(kx - \omega t) \tag{4.79}$$

Fig. 7.12 Reflected internal
wave pattern in bounded
space

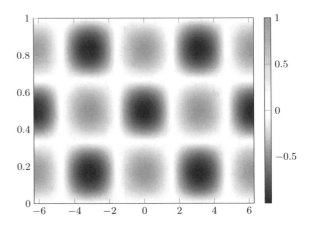

An example of the motion pattern at a fixed time for mode $n = 3$ is displayed in
Fig. 7.12.

The motion is divided into compartments and each compartment (cell) includes a
vortex with alternating sense of rotation. With increasing time, whole cellular pattern
of motion propagates to the right. since the cells are identical, the pattern may be
extended to an infinite domain, representing standing wave solutions in the vertical
direction only.

7.5   Chapter 4, Exercise 2

Internal gravity waves in a uniformly stratified medium with constant N have been
studied earlier. In the case of variable stratification $N = N(z)$, solutions can be
obtained by solving the ordinary differential equation (4.46) with a variable coefficient, depending strongly on the specified form of the stratification parameter $N(z)$.

A simple example is displayed in Fig. 7.13 for a fluid with sharp changes of
density typical of summer stratification near the "thermocline" or "pycnocline" in
the upper ocean.

Fig. 7.13 Profiles of density
and stability frequency near
a thermocline

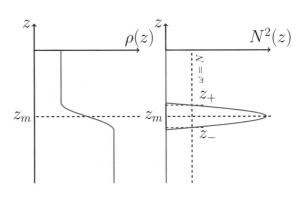

As the sharpest gradient of density occurs near the interface ($z = z_m$), the stability frequency N reaches a maximum N_m. Periodic wave solutions $w = \hat{w}(z)e^{i(kx-\omega t)}$ are expected with fixed frequency ω, according to (4.44) and (4.46). The limiting depths for such motions will be at $z = z_+$ and $z = z_-$ close to depth $z = z_m$ of maximum N, for any given frequency $\omega = N$. Oscillatory solutions in z will only be possible in the depth interval $z_- < z < z_+$ where $N > \omega$. Outside this region the solutions would be exponentially decaying.

Concentrating near the interface and assuming an unlimited fluid domain outside this zone, a reasonable model of stratification near the interface would be

$$N^2(z) = N_m^2 \left[1 - (z - z_m)^2/a^2\right]$$

which specifies a parabolic variation. Letting

$$\xi = \left(\frac{N_m k}{a\omega}\right)^{1/2}(z - z_m)$$

as a coordinate transformation, Eq. (4.46) takes the following form:

$$\frac{d^2\hat{w}}{d\xi^2} + \left[\lambda - \xi^2\right]\hat{w} = 0$$

where

$$\lambda = \left(N_m^2 - \omega^2\right)\left(\frac{ak}{N_m\omega}\right).$$

Boundary conditions imply decaying solutions outside the pycnocline region

$$\hat{w} \to 0 \text{ as } \xi \to \pm\infty.$$

It is noted that he problem specification is similar to the equation for a harmonic oscillator in quantum mechanics theory. To investigate asymptotic behavior of the solution, a functional specification is applied

$$\hat{w}(\xi) = e^{f(\xi)},$$

and the transformed equation becomes

$$f'' + \left(f'\right)^2 + \lambda - \xi^2,$$

where primes denote differentiation with respect to z.

As $\xi \to \pm\infty$ the solution must be decaying, dropping some of the terms allows simplification to

$$\left(f'\right)^2 - \xi^2 \cong 0 \text{ as } \xi \to \pm\infty.$$

Fig. 7.14 Internal wave modes $n = 1, 2, 3$ at an interface layer

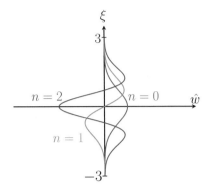

with the approximated solution

$$f \simeq \pm\xi^2/2 \ \text{ as } \ \xi \to \pm\infty.$$

Here it is noted only the negative sign is acceptable for solution to be decaying. This asymptotic behavior suggests solutions of the form

$$\hat{w}(\xi) = H(\xi)e^{-\xi^2/2}.$$

Substitution gives

$$H_{\xi\xi} - 2\xi H_\xi + (\lambda - 1) H = 0$$

which is known as the *Hermite's equation*.

In agreement with boundary conditions to (4.90) we must require that $H(\xi)$ can not grow faster than $\sim e^{\xi^2/2}$. A *Frobenius Series* solution for this equation

$$H(\xi) = \sum_{m=0}^{\infty} a_m \xi^m$$

leads to a *polynomial* only if λ should be an odd integer

$$\lambda = \lambda_n = 2n + 1, n = 0, 1, 2 \cdots$$

which directly yields a dispersion relation:

$$\frac{\omega}{k} = \frac{a}{N_m (2n + 1)} \left(N_m^2 - \omega^2 \right).$$

Solutions are given as

$$\hat{w} (\xi) = A H_n (\xi) e^{-\xi^2/2}$$

where $H_n(\xi)$ represents a Hermite polynomial of order n, given by the generating function

$$H_n(\xi) = (-1)^n e^{\xi^2} \frac{d^n}{d\xi^n} \left(e^{-\xi^2} \right)$$

for which the first few are listed below

$$H_0(\xi) = 1$$
$$H_1(\xi) = 2\xi$$
$$H_2(\xi) = 4\xi^2 - 2$$
$$H_3(\xi) = 8\xi^3 - 12\xi$$
$$H_4(\xi) = 16\xi^4 - 48\xi^2 + 12$$
$$\text{etc.}$$

From the solutions sketched in Fig. 7.14, it can be observed that internal modes are trapped roughly around the region where $N > \omega$, near the maximum of the density gradient of the interface.

7.6 Chapter 5, Exercise 1

First of all, review Eq. (5.144) for a container with circular cross-section with radius $r(z)$, area $A(z)$ and circumference $C(z)$ varying with height,

$$\frac{d^2 T^I}{dz^2} + \left(\frac{1}{A(z)} \frac{dA(z)}{dz} - \frac{M_0}{kA(z)} \right) \frac{dT^I}{dz} - \frac{s_0}{\cos\gamma} \frac{C(z)}{A(z)} (T^I - \hat{T}) = 0.$$

where $\tilde{T} = (T^I - \hat{T})$ is the temperature difference between the fluid interior T^I and the external environment $\hat{T}$.

The area and circumference is zero at the bottom, reaching values

$$A_1 = \pi r_1^2, \qquad C_1 = 2\pi r_1 \qquad \text{at} \qquad z = H.$$

Defining a normalized coordinates, non-dimensional value of the volume flux at the bottom and the difference in temperature

$$\xi = z/H, \qquad \mu = \frac{M_0 H}{kA_1}, \qquad \tilde{T} = (T^I - \hat{T})$$

and specifying the geometry in normalized coordinates

$$\tilde{r}(\xi) = \frac{r(z)}{r_1}, \qquad \tilde{A}(\xi) = \frac{A(z)}{A_1} = (a\xi)^{2q}, \qquad \tilde{C}(\xi) = \frac{C(z)}{C_1} = (a\xi)^q$$

the non-dimensional form of the governing equation is obtained.

$$\frac{d^2\tilde{T}}{d\xi^2} + \left(\frac{1}{\tilde{A}(\xi)}\frac{d\tilde{A}(\xi)}{d\xi} - \frac{\mu}{\tilde{A}(\xi)}\right)\frac{d\tilde{T}}{d\xi} - \frac{s_0}{\cos\gamma}\frac{\tilde{C}(\xi)}{\tilde{A}(\xi)}\tilde{T} = 0$$

7.7   Chapter 5, Exercise 2

A typical basin with the simplest geometry could be a case with monotonously increasing radius with height and 'parabolic' form of any order q

$$r(z) \sim z^q$$

which however would only make sense for a closed basin of concave form if $q < 1$. Selecting the case $q > 1$ would make the boundaries diverge from a central axis with increasing height. As such, adapting a profile that starts with a bottom at $z = 0$ and expanding as in the above expression until a height H is chosen. Specifying the geometry in normalized coordinates

$$r(\xi) = (a\xi)^q, \qquad \tilde{A}(\xi) = \frac{A(\xi)}{A_1} = (a\xi)^{2q}, \qquad \tilde{C}(\xi) = \frac{C(\xi)}{C_1} = (a\xi)^q$$

the non-dimensional form of the governing equation is obtained.

$$\frac{d^2\tilde{T}}{d\xi^2} + \left(\frac{2q}{a\xi} - \frac{\mu}{(a\xi)^{2q}}\right)\frac{d\tilde{T}}{d\xi} - \frac{s_0}{\cos\gamma}\frac{1}{(a\xi)^q}\tilde{T} = 0.$$

This second order ODE with variable coefficients however does not look very promising for analytical solution, but likely to be solved for a selected value of q, otherwise by making use of Frobenius series or by numerical methods

Here we can consider the parabolic container with $a = 1$ and $q = 1/2$ in the above form

$$r(\xi) = (\xi)^{1/2}$$

which yields the relatively simpler thermal stratification equation

$$\xi\frac{d^2\tilde{T}}{d\xi^2} + (1 - \mu)\frac{d\tilde{T}}{d\xi} - \frac{s_0}{\cos\gamma}\xi^{1/2}\tilde{T} = 0$$

which can be solved with boundary conditions

$$T = T_0 \ \text{ at } \ \xi = 0, \qquad T = T_1 \ \text{ at } \ \xi = 1.$$

This again involves some difficulty for direct analytical solution because of the third term representing heat flux through the boundary. Neglecting last term, as done for an insulated container in the main text, a simpler form will be considered:

$$\xi \frac{d^2\tilde{T}}{d\xi^2} + (1 - \mu)\frac{d\tilde{T}}{d\xi} = 0.$$

This equation is reformed as

$$(\xi\tilde{T})'' - \mu\tilde{T}' = 0$$

where primes denote differentiation with respect to ξ. Defining a new variable

$$S = \xi\tilde{T}'$$

the order of the equation is reduced to give

$$S' - \frac{\mu}{\xi}S = 0$$

and multiplying by an integrating factor,

$$I(\xi) = e^{-\mu\int \frac{1}{\xi}d\xi} = e^{-\mu \ln \xi} = \xi^{-\mu}$$

the equation is further reduced

$$\xi^{-\mu}S' - \mu\xi^{-\mu-1}S = (\xi^{-\mu}S)' = 0$$

which is integrated to give

$$\xi^{-\mu}S = c_0.$$

Finally replacing $S = \xi T'$

$$T' = c_0\xi^{\mu-1}$$

integrated to give the solution

$$\tilde{T} = (T_1 - T_0)\xi^{\mu} + T_0$$

which satisfies the boundary conditions $\tilde{T} = T_0$ at $\xi = 0$ and $\tilde{T} = T_1$ at $\xi = 1$.

7.8   Chapter 5, Exercise 3

Historical uses of pottery ceramics in cooking and keeping cooled drinks are well known. Earthenware pots have made use of applied heat together with evaporation heat loss. Traditional methods of water cooling by evaporation have been widely used in the Mediterranean and other arid regions of the world such as central Asia, Africa and Americas, throughout history, before the age of refrigeration [2].

An ingenious device utilizing these principles since the beginning of civilization is the common earthenware water jug, where evaporation through the porous surface has been used for cooling. The revered "botijo" from Spain and household "testi" in Turkey and "desti" in Iran (Fig. 7.15) are typical examples of water jugs utilizing this effective cooling mechanism. Recently, traditional cooling water jugs are being rediscovered, with revived interest on scientific, climatic and anthropogenic implications. As a tool of drinking water adaptation in arid climates, the cooling jug is also enjoyed as part of local culture, its use not necessarily limited to water, but also of wine.[1,2]

For common use, the jug is filled with water, let rest in a corner while the contents are cooled. Reports have declared cooling on the order of $\sim 10\,°C$ in about an hour. After some period of resting, the jug is turned sideways for cooled water drawn out into a cup for drinking, or through an outlet without touching one's mouth as in the case of festive events of western Mediterranean.

The evaporative cooling mechanism however has not been fully investigated by experimental methods and fluid dynamical theory, exposing roles of container geometry, materials and the external environment. Investigations accounting for evaporation losses as well as other heat flow components have so far only been considered for uniform temperature in a small container varying as a function of time [3], without considering the development of interior stratification, effects of thermal diffusion and boundary layers.

Going back to ancient civilizations, unglazed earthenware water jugs can be found in many parts of the old world. Examples of ancient artifacts of Hittite style in Fig. 7.16 dating back to 2nd millenium BC demonstrate beautiful designs of aesthetic and technical perfection. We must recognize however, that these small containers were not only used to cool water but also as "libation" instruments presenting other liquids (water, wine, milk, blood!) to deities.

Interestingly, the ancient containers of Hittite period exemplified in Fig. 7.16 seem to have an increasing cross section in the lower part, used for storage of the cooled liquid, narrowing towards a mouth region in the upper part. This seems to be a perfect design for cooling container. Upon turning the jug sideways, the cooled liquid stored in the lower part is possibly emptied through a beautiful beaked mouth, where hydraulic control allows smooth pouring of the liquid, offered for drinking.

---

[1] https://www.atlasobscura.com/articles/botijo-jug.
[2] https://www.notechmagazine.com/2012/04/botijos.html.

Fig. 7.15 Earthenware shop producing water jugs "testi" at Avanos, Cappadocia, Turkey, and Spanish melon seller accompanied by "botijo" painted by William Knight Keeling (1807–1886)

The above review of cooling jugs motivates detailed analyses of thermal stratification proposed below. The release of latent heat by evaporation at the wet surface of a container made of porous material such as terracotta is expected to influence momentum and heat transfer in the container. Evolution of the thermal stratification in the container is represented by Eq. (5.106.a,b) of the interior, with boundary conditions representing heat loss by evaporated mass flux e. In short, Eqs. (5.108) and (5.109.a,b), specifying surface heat flux are modified by adding the evaporation component

Fig. 7.16 Examples of terracotta jugs of the Hittite style from Maşat Höyük (Tokat Museum, Turkey), Alacahöyük (Çorum Museum, Turkey), second millenium B.C., demonstrating the beaked mouth, the linear/concave/convex trends of the container profile in the upper and lower parts

$$\tilde{q} = \left\{ -k \frac{\partial T^I}{\partial \eta} \right\}_{\eta=0} = - \left\{ \frac{k_w}{d}(T^I - \hat{T}) + \frac{\lambda}{c_p} e \right\}_{\eta=0}$$

where k and k_w are respectively the heat diffusivity of fluid and the container wall, d the wall thickness, λ the latent heat of evaporation, c_p the specific heat of water and e the rate of evaporation at the wet outer surface of the container. Latent heat of evaporation is typically given as $\lambda = 583$ kcal/kg, while the specific heat $c_p = 1$ kcal/(kg °C) for water. The evaporation speed e in units of m/s is the rate of evaporation, yielding a net volume flux in m^3/s when multiplied with the surface area. Typically, surface evaporation is physically tied to external environment through simple mass transfer relationship

$$e = \rho k'(h_s - h)$$

where k' is a mass transfer coefficient that represents parameterized external effects such as wind velocity, surface types and temperature, h humidity of air and h_s its saturation value.

For a container with radial symmetry, making use of the earlier approximations in (5.96.a), (5.108) and (5.109.a,b), and making note of Eqs. (5.107), (5.111) and (5.116), the boundary layer flux is re-evaluated as

$$M_B = \oint_C m_B dl = k \frac{dA(z)}{dz} - \frac{C(z)}{\frac{\partial T^I}{\partial x} \cos \gamma} \left[k s_0 (T^I - \hat{T}) + \frac{\lambda}{c_p} e \right]$$

where $A(z)$ is the area and $C(z)$ is the circumference of container as a function of depth z, which in the present case has circular geometry.

Because of mass ejection by evaporation, a time dependent continuity equation is needed reflecting changes in the level of free surface, in consequence to the decrease in total volume of fluid in the container. However in the present problem, evaporation as well as time dependent temperature and level changes at the free surface will be ignored, precedence given to the large latent heat loss at the outer surface alone. The steady state integrated continuity equation at any level z is assumed to be valid at any time, updated by evaporation flux contribution in M_B

$$M_B + w^I A(z) = M_0$$

yielding internal vertical velocity w^I

$$w^I = \frac{M_0}{A} - \frac{k}{A}\frac{dA}{dz} + \frac{C}{A\frac{\partial T^I}{\partial x}\cos\gamma}\left[ks_0(T^I - \hat{T}) + \frac{\lambda}{c_p}e\right]$$

Substituting these in (5.102), we obtain

$$\frac{\partial T^I}{\partial t} + \left(\frac{M_0}{A} - \frac{k}{A}\frac{dA}{dz}\right)\frac{\partial T^I}{\partial z} + \frac{C}{A\cos\gamma}\left[ks_0(T^I - \hat{T}) + \frac{\lambda}{c_p}e\right] = k\frac{\partial^2 T^I}{\partial z^2}$$

By letting

$$E = \frac{\lambda}{kc_p}e$$

$$\frac{A(z)}{k}\frac{\partial T^I}{\partial t} + \frac{\partial}{\partial z}\left(A(z)\frac{\partial T^I}{\partial z}\right) - \frac{M_0}{k}\frac{\partial T^I}{\partial z} - \frac{C(z)}{\cos\gamma}\left[s_0(T^I - \hat{T}) + E\right] = 0$$

the above equation determines the interior temperature distribution for given mass flux M_0 diffusion coefficient k, ambient temperature $\hat{T}$, wall thickness parameter s_0 and geometrical parameters $A(z)$ and $\cos\gamma$ of the container. Note that the equation automatically satisfies the boundary conditions on the inclined walls of the container.

For a container with radial symmetry with parabolic shape of order q specified in dimensionless coordinates

$$\frac{r(z)}{r_0} = \left(\frac{z}{H}\right)^q = \xi^q \quad \text{where} \quad \xi = \frac{z}{H}$$

the variable coefficients in the governing equation in the vertical are evaluated as

$$A(z) = \pi r(z)^2 = \pi r_0^2 \xi^{2q},$$
$$C(z) = 2\pi r(z) = 2\pi r_0 \xi^q,$$

$$\frac{1}{\cos\gamma} = \frac{\sqrt{dr^2 + dz^2}}{dz} = \sqrt{1 + \left(\frac{dr}{dz}\right)^2} = \sqrt{1 + \left(\frac{r_0}{H}\right)^2 q^2 \xi^{2q-2}}$$

and replacing these in the equation gives

$$\frac{\pi r_0^2 \xi^{2q}}{k} \frac{\partial T^I}{\partial t} + \frac{1}{H^2} \frac{\partial}{\partial \xi}\left(\pi r_0^2 \xi^{2q} \frac{\partial T^I}{\partial \xi}\right) - \frac{1}{H} \frac{M_0}{k} \frac{\partial T^I}{\partial \xi}$$
$$- 2\pi r_0 \xi^q \sqrt{1 + \left(\frac{r_0}{H}\right)^2 q^2 \xi^{2q-2}}\left[s_0(T^I - \hat{T}) + E\right] = 0$$

The last term represents sidewall heating $s_0(T^I - \hat{T})$, where the temperature differential is denoted as $T^* = T^I - \hat{T}$ with reference to temperature $\hat{T}$ of the container wall.

We have only derived equations describing the evolution of stratification in containers of relatively simple shape. The solutions to such complicated equations with variable coefficients either requires the use of series expansions or brute force numerical methods, which we leave for the reader to seek.

7.9   Chapter 5, Exercise 4

Certain simplified cases based on Exercise 3 could be found by taking integer or fractional orders q of parabolic shape functions, yet analytical solutions can not be easily obtained. In this exercise, the simplified case of a conically shaped container in Fig. 7.17 with linear increase in radius $q = 1$ will be addressed. Taking the simplest route, we can only propose solutions for the lower part of the container, for water level starting from a total depth of H at time $t = 0$ and decreasing for $z < H$ afterwards following the specified terms

$$\frac{r_0^2 \xi^2}{k} \frac{\partial T^*}{\partial t} + \frac{r_0^2}{H^2} \frac{\partial}{\partial \xi}\left(\xi^2 \frac{\partial T^*}{\partial \xi}\right) - \frac{M_0}{\pi k H} \frac{\partial T^*}{\partial \xi}$$
$$- 2r_0 \sqrt{1 + \left(\frac{r_0}{H}\right)^2} \xi\left[s_0 T^* + E\right] = 0.$$

Fig. 7.17 The geometry of a container with a conic lower part

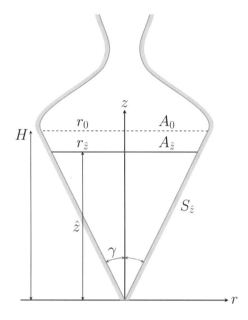

With further definition of constants

$$\sigma = \frac{H^2}{k}, \qquad \mu = \frac{M_0 H}{k \pi r_0^2} = \frac{M_0 H}{k A_0},$$

$$h = 2H\sqrt{1 + \left(\frac{H}{r_0}\right)^2} = 2H\sqrt{1 + \frac{1}{\tan^2 \gamma}} = \frac{2}{\sin \gamma} H$$

and dividing by the constant r_0^2/H^2 gives

$$\sigma \xi^2 \frac{\partial T^*}{\partial t} + \frac{\partial}{\partial \xi}\left(\xi^2 \frac{dT^*}{d\xi}\right) - \mu \frac{\partial T^*}{\partial \xi} - h s_0 T^* \xi - h E \xi = 0.$$

In the simple solution presented here, the thermal heating by the environment will be totally ignored by letting $s_0 = 0$, with evaporation effects dominating the other heat flow components that could be included. Since there is no through the system, volume flux component is set to zero $M_0 = 0$, or $\mu = 0$, reducing the equations to a much simpler but limited system

$$\sigma \xi^2 \frac{\partial T^*}{\partial t} + \frac{\partial}{\partial \xi}\left(\xi^2 \frac{\partial T^*}{\partial \xi}\right) - \epsilon \xi = 0$$

where we define

$$\epsilon = hE.$$

Seeking a solution by separation of variables

$$T^*(\xi, t) = P(t)Q(\xi)$$

is substituted in the equation

$$\sigma\xi^2 \frac{\partial P}{\partial t} Q + P\left(\xi^2 Q'\right)' - \epsilon\xi = 0$$

where the primes denote differentiation with respect to ξ. Dividing by $\xi^2 P(t)Q(\xi)$ with separable parts equated with a common function of time $\alpha^2(t)$,

$$\sigma \frac{1}{P}\frac{dP}{dt} = -\frac{1}{\xi^2 Q}\left[\left(\xi^2 Q'\right)' - \frac{\epsilon}{P(t)}\xi\right] = -\alpha^2(t)$$

so that the time dependent part produces an ordinary differential equation (ODE)

$$\frac{1}{P}\frac{dP}{dt} = \frac{d\ln P}{dt} = -\frac{\alpha^2(t)}{\sigma}$$

the solution to which is obtained as

$$P(t) = e^{-\int \frac{\alpha^2(t)}{\sigma}dt}.$$

The remaining part is accepted as an ODE, but with time dependent coefficient which is unknown at present

$$\left(\xi^2 Q'\right)' - \frac{\epsilon}{P(t)}\xi = \alpha^2 \xi^2 Q.$$

This technique deviates a from the standard method of separation of variables, with time dependent constant remaining in the vertical equation. We modify the first term by noting

$$\xi Q' = (\xi Q)' - Q$$

which then is multiplied by ξ

$$\xi^2 Q' = \xi(\xi Q)' - \xi Q$$

is used to reduce the equation to

$$\left[\xi(\xi Q)' - \xi Q\right]' - \frac{\epsilon}{P(t)}\xi = \alpha^2(t)\xi^2 Q.$$

Inserting a new variable further reduces terms

$$S = \xi Q,$$

$$\left(\xi^2 Q'\right)' = \left[\xi(\xi Q)' - \xi Q\right]' = \left(\xi S' - S\right)' = \xi S'' + S' - S'$$

yielding the final form of the vertical equation

$$S'' - \alpha^2(t)S = \frac{\epsilon}{P(t)}$$

where homogeneous and particular solution components are added together to obtain

$$S = S(\xi, t) = S_h(\xi, t) + S_p(t)$$

where

$$S_h(\xi, t) = A(t)e^{+\alpha(t)\xi} + B(t)e^{-\alpha(t)\xi}$$

$$S_p(t) = -\frac{\epsilon}{\alpha^2(t)P(t)}$$

combined to give the total solution

$$T^* = P(t)Q(\xi, t) = P(t)\frac{S(\xi, t)}{\xi}$$

$$= P(t)\frac{1}{\xi}\left(Ae^{+\alpha(t)\xi} + Be^{-\alpha(t)\xi} - \frac{\epsilon}{\alpha^2(t)P(t)}\right)$$

$$= P(t)\frac{1}{\xi}\left(Ae^{+\alpha(t)\xi} + Be^{-\alpha(t)\xi}\right) - \frac{\epsilon}{\alpha^2(t)\xi}$$

where A and B are constants. In essence $\alpha^2(t)$ is a non-dimensional geometric factor multiplying the basic diffusion time scale k/H^2. It is also noted that the time dependence of the solution represented by the separation coefficient $\alpha^2(t)$ and $P(t)$ have to be unified between the homogeneous and particular solutions:

$$\alpha^2(t) = \frac{1}{P(t)} = e^{+\int \frac{\alpha^2(t)}{\sigma}dt}.$$

Logarithmic differentiation of both sides give

$$\frac{d \ln \alpha^2}{dt} = \frac{1}{\alpha^2}\frac{d\alpha^2}{dt} = \frac{\alpha^2}{\sigma},$$

$$\frac{d\alpha^2}{dt} = \frac{\alpha^4}{\sigma}.$$

Letting $\phi = \alpha^2$ and integrating in time

$$\int \frac{1}{\phi^2} \frac{d\phi}{dt} dt = \frac{1}{\sigma} \int dt$$

results in

$$-\frac{1}{\phi} = \frac{1}{\sigma} t + C,$$

so that by letting the constant $C = 0$ and $\phi(t) = \alpha^2(t)$ it is found that

$$\alpha^2(t) = -\frac{\sigma}{t}, \qquad P(t) = -\frac{t}{\sigma}$$

and defining

$$\tau = \frac{t}{\sigma}$$

the solution becomes

$$T^* = -\frac{1}{\sigma} \frac{t}{\xi} \left(A \sin \xi \sqrt{\frac{\sigma}{t}} + B \cos \xi \sqrt{\frac{\sigma}{t}} + \epsilon \right)$$

$$= -\frac{\tau}{\xi} \left(A \sin \frac{\xi}{\sqrt{\tau}} + B \cos \frac{\xi}{\sqrt{\tau}} + \epsilon \right).$$

Checking limits as $\xi \to 0$

$$\lim_{\xi \to 0} \frac{\tau}{\xi} \sin \left(\frac{\xi}{\sqrt{\tau}} \right) = \lim_{\xi \to 0} \frac{\sin(\xi/\sqrt{\tau})}{\xi/\tau} = \lim_{\xi \to 0} \frac{\tau^{-1/2} \cos(\xi/\sqrt{\tau})}{\tau^{-1}} \to \sqrt{\tau}$$

$$\lim_{\xi \to 0} \frac{\tau}{\xi} \left[\cos \left(\frac{\xi}{\sqrt{\tau}} \right) - 1 \right] = \lim_{\xi \to 0} \left[\frac{\cos(\xi/\sqrt{\tau}) - 1}{\xi/\tau} \right] = \lim_{\xi \to 0} \frac{\tau^{-1/2} \sin(\xi/\sqrt{\tau})}{\tau^{-1}} \to 0$$

Limits at initial time $\tau \to 0$ are confirmed as

$$\lim_{\tau \to 0} \frac{\tau}{\xi} \sin \left(\frac{\xi}{\sqrt{\tau}} \right) \to 0$$

$$\lim_{\tau \to 0} \frac{\tau}{\xi} \cos \left(\frac{\xi}{\sqrt{\tau}} \right) \to 0$$

so that the initial condition for the temperature evolution in the container start with a uniform profile

$$T^* = T^I - \hat{T} = 0$$

with of interior temperature equal to that of the exterior $T^I = \hat{T}$.

Considering asymptotic behavior as $\tau \to \infty$, the following limits are noted

$$\lim_{\tau \to \infty} \frac{\tau}{\xi} \sin\left(\frac{\xi}{\sqrt{\tau}}\right) = \lim_{\tau \to \infty} \frac{\sin(\xi/\sqrt{\tau})}{\xi/\tau} = \lim_{\tau \to \infty} \frac{-\frac{1}{2}\tau^{-3/2}\cos(\xi/\sqrt{\tau})}{-\tau^{-2}}$$

$$= \lim_{\tau \to \infty} \frac{\frac{1}{2}\cos(1/\sqrt{\tau})}{1/\sqrt{\tau}} \to \infty$$

$$\lim_{\tau \to \infty} \frac{\tau}{\xi} \cos\left(\frac{\xi}{\sqrt{\tau}}\right) = \lim_{\tau \to \infty} \frac{\cos(\xi/\sqrt{\tau})}{\xi/\tau} = \lim_{\tau \to \infty} \frac{\frac{1}{2}\tau^{-3/2}\sin(\xi/\sqrt{\tau})}{-\tau^{-2}}$$

$$= \lim_{\tau \to \infty} \frac{-\frac{1}{2}\sin(\xi/\sqrt{\tau})}{1/\sqrt{\tau}} = \lim_{\tau \to \infty} \frac{1}{2}\frac{-\frac{1}{2}\xi\tau^{-3/2}\cos(\xi/\sqrt{\tau})}{-\frac{1}{2}\tau^{-3/2}} \to -\frac{\xi}{2}.$$

The constants A and B need to be determined by boundary conditions. However, it is noted that the sinusoidal terms multiplied by $1/\xi$ become unbounded at $\xi = 0$, the container bottom. For this purpose the above limits have to be considered for a realistic solution to be constructed.

To allow bounded solutions, the only choice is to select $B = -\epsilon$ as the coefficient of cosine term, leading to a finite value which actually should take care of the bottom boundary condition at vertex of the conical container. Constant A can be left free to satisfy boundary condition applied at the free surface of the water, moving down as a result of the total evaporation losses.

The surface boundary condition for evaporation heat transfer (neglecting contact heating by conduction) is

$$\frac{dT^*}{dz} = -\frac{\lambda}{kc_p}e_s \quad \text{at} \quad z = \hat{z}$$

where e_s is the evaporation at free surface level $z = \hat{z}$, with a rate of evaporation much smaller than that at the wet outer surface of the container $e_s << e$, because of restricted air exchange at the narrow mouth of container, which in normal use is kept closed. Therefore it would be convenient to assume that $e_s \simeq 0$, leading to a normal boundary condition at the free surface.

Note that this is a nonlinear boundary condition applied at the surface $z = \hat{z}(t)$ which decreases with time as a result of evaporation. Although the above nonlinear boundary condition could in principle be applied to determine constant A, it is inappropriate to be specified on a moving surface in a linear solution. Further caution is also necessary in applying it to the restricted analytical solution of the simplified problem ignoring many other heat flow processes. We present this simplified, linear separation of variables solution by noting these deficiencies. We leave to resolve these questions for future analyses, temporarily requiring $A = -\epsilon$, which seems to be a natural but restricted choice accounting only for evaporation component of heat transfer. With these considerations, the trial solution becomes

$$T^* = \epsilon \frac{\tau}{\xi}\left[-\sin\frac{\xi}{\sqrt{\tau}} + \left(\cos\frac{\xi}{\sqrt{\tau}} - 1\right)\right]$$

Fig. 7.18 Profiles of functions $F_1(\xi) = -\frac{1}{\chi}\sin\chi$ (red) and $F_2(\xi) = \frac{1}{\chi}[\cos\chi - 1]$ (blue), (upper figure) as a function of transformed coordinate χ in the range $0 - 4\pi$, (lower figure) enlarged view of the same functions near the bottom of conical container in the boundary region $0 \leq \chi \leq 0.002$, showing convergence of the functions to zero as $\chi \to 0$ (within computational accuracy)

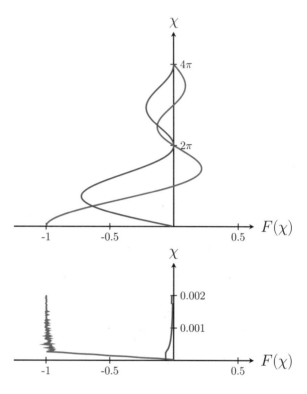

so arranged as to avoid singularity at the vertex $\xi = 0$ of the cone. Looking back, the functional forms (analyzed below) with natural choices of constants are not capable to satisfy nonlinear surface conditions on a moving boundary, the solution within the valid range $0 < \xi < \hat{\xi}$, with $\hat{\xi}(\tau)$ being time dependent free surface height. This is the main weakness of the trial solution noted earlier.

Alternatively, defining combined coordinates $\chi = \xi/\sqrt{\tau}$

$$F(\chi) \equiv \frac{T^*}{\epsilon\sqrt{\tau}} = \frac{1}{\chi}[-\sin\chi + (\cos\chi - 1)]$$

allows one to plot part of the solution in terms of the combined natural coordinate χ as shown in Fig. 7.18. Firstly, we can recognize $\chi = \xi/\sqrt{\tau}$ as a natural coordinate for the diffusion operator, limited in the vertical range $0 < \xi < \hat{\xi}$ and diffusion time $\sqrt{\tau}$ linked together except for a separate amplitude factor $\sqrt{\tau}$. At initial time $\tau = 0$, the function $F(\chi) = 0$ but the range is $\chi \to \infty$. This means that for initial period of time the coordinate range of χ is enlarged when the solution is oscillatory with respect to χ, and as time is increased, the range is shortened. In fact, the first part of solution $\frac{1}{\chi}\sin\chi$ tends to a limit increasing with $\sqrt{\tau}$. But even before that happens, the decreased volume due to evaporation limits the solution, with the free surface height $\hat{\xi}$ decreasing with time as will be shown below.

Fig. 7.19 (upper figure) Solution profiles in the full depth range ($0 \leq \xi \leq 1$) of functions $f_1(\xi) = -\frac{\tau}{\xi} \sin\left(\frac{\xi}{\sqrt{\tau}}\right)$ (red) and $f_2(\xi) = \frac{\tau}{\xi}[\cos\left(\frac{\xi}{\sqrt{\tau}}\right) - 1]$ (blue) for values of time $\tau = 0.001, 0.01, 0.1, 1$, (lower figure) in the total (dashed) and valid depth range ($0 \leq \xi \leq \hat{\xi}$) as the free surface is lowered to level $\hat{\xi}$ (bold, brown) of total solution $f_1(\xi) + f_2(\xi)$ for values of time $\tau = 0.001, 0.01, 0.1, 0.25$

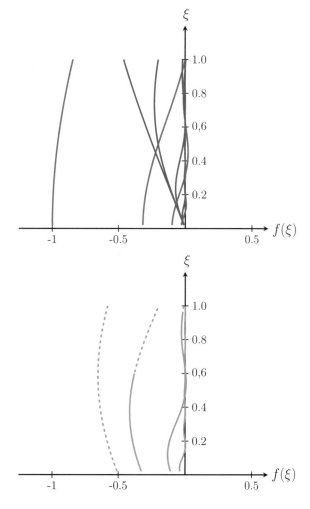

The components of this solution are plotted as profiles depending on vertical coordinate ξ, for selected values of time τ, excluding the multiplicative constant ϵ, which is the evaporation loss parameter. The first term, a sinusoidal function (plotted in red in Fig. 7.19) has growing values in the whole depth with time, including a finite value at the bottom $\xi = 0$, with a growth rate $\sim \sqrt{\tau}$ as shown by the above limits as $\tau \to \infty$. This growth however will be terminated when all water in the container will be finished when $\xi = 0$ is reached at time $\tau = \tau_e$. The second term $(\cos(a\xi) - 1)/\xi$ is selected so as to have a finite limit of zero at $\xi = 0$ (plotted in red in Fig. 7.19). The solution starts from an initial value of zero everywhere, evolving towards a linear profile of $-\xi/2$ in the vertical, as predicted by the limit $\tau \to \infty$.

Volume conservation determines water level change by

$$\frac{dV_{\hat{z}}}{dt} = -e_s A_{\hat{z}} - e S_{\hat{z}}$$

where e_s and e respectively are the evaporation rates at free surface and container walls, $A(\hat{z})$ and $S(\hat{z})$ being the area of respective surfaces. The cone geometry is described as

$$\frac{r_0}{H} = \frac{r_{\hat{z}}}{\hat{z}} = \tan\gamma$$

$$V_{\hat{z}} = \int_0^{\hat{z}} A(z)dz = \int_0^{\hat{z}} \pi r^2(z)dz = \pi \left(\frac{r_0}{H}\right)^2 \int_0^{\hat{z}} z^2 dz$$

$$= \frac{1}{3}\pi \left(\frac{r_0}{H}\right)^2 \hat{z}^3 = \frac{1}{3}\pi(\tan\gamma)^2 \, \hat{z}^3$$

$$A_{\hat{z}} = \pi r_{\hat{z}}^2 = \pi \left(\frac{r_0}{H}\right)^2 \hat{z}^2 = \pi(\tan\gamma)^2 \hat{z}^2$$

$$S_{\hat{z}} = \pi r_{\hat{z}}\sqrt{\hat{z}^2 + r_{\hat{z}}^2} = \pi \hat{z}^2 \frac{r_{\hat{z}}}{\hat{z}}\sqrt{1 + \frac{r_{\hat{z}}^2}{\hat{z}^2}}$$

$$= \pi \hat{z}^2 \frac{r_0}{H}\sqrt{1 + \frac{r_0^2}{H^2}} = \pi \hat{z}^2 \tan\gamma\sqrt{1 + \tan^2\gamma}.$$

Making use of the above, rate of change of volume is

$$\frac{1}{3}\pi \left(\frac{r_0}{H}\right)^2 \frac{d\hat{z}^3}{dt} = -e\pi \left(\frac{r_0}{H}\right)^2 \hat{z}^2 - e\pi \left(\frac{r_0}{H}\right)\sqrt{1 + \frac{r_0^2}{H^2}\hat{z}^2}$$

and with cancellations of constants, the rate of change in free surface elevation is obtained as

$$\frac{d\hat{z}}{dt} = -\left(e_s + \frac{1}{\sin\gamma}e\right) = -\tilde{e}$$

where $\tilde{e}$ is the total evaporation rate, this integrates to

$$\hat{z} = H - \tilde{e}\,t,$$

with the time of complete water loss in the container

$$t_e = \frac{H}{\tilde{e}}.$$

The rate of change of level in dimensionless form is

$$\hat{\xi} = 1 - \hat{s}\tau = 1 - \frac{H\tilde{e}}{k}\tau = 1 - \frac{H^2}{kt_e}\tau$$

with

$$\hat{s} = \frac{H^2}{kt_e} = \frac{\sigma}{t_e} = \frac{t_d}{t_e}$$

expressing the ratio of diffusion time $t_d = \sigma$ (as defined earlier) to the emptying time t_e. Diffusion time is typically much larger than the emptying time, $t_d/t_e \gg 1$. As a result of water loss, depth dependent solutions are therefore vary from the bottom to free surface height $0 < \xi \leq \hat{\xi}$.

Now is the time to plug in some numbers to see what happens. The emptying of a container of height $H = 0.5\,\text{m}$ can be estimated on the order of days, (taken as $t_e = 5d = 5 \times 86400\,\text{s}$ for this example), to give $\tilde{e} = H/t_e \simeq 1.1 \times 10^{-6}\,\text{m/s}$. Taking the heat diffusivity of water as $k = 0.15 \times 10^{-6}\,\text{m}^2/\text{s}$ gives the change in surface elevation as

$$\hat{\xi} = 1 - \hat{s}\tau, \quad \text{where} \quad \hat{s} = \frac{H\tilde{e}}{k} = \frac{H^2}{kt_e} \simeq 4$$

so that the non-dimensional time for emptying is about $\tau_e = 0.25$.

Assuming a conic container with $\gamma = 45°$ would be equivalent to evaporation rate e divided by a factor varying in the range $\sqrt{2}$ to $1 + \sqrt{2}$ depending on if e_s is either vanishing or not.

Calculating the amplitude, we find an unrealistically large number,

$$\epsilon = \left(\frac{H\tilde{e}}{k}\right)\left(\frac{2\lambda}{c_p}\right) = 4 \times 2 \times 583/1 = 2322$$

which however is estimated from assumed magnitudes of physical parameters for a fluid at rest with given molecular heat diffusivity, setting arbitrary evaporation rate and geometry. These results also remind us of the insufficiency in analyses based on linearized problem, with aberration of evaporation as a single factor in thermal evolution. We leave further investigations to more complete analyses to be performed, and hope that the concepts used in the present problem can lead to better understanding of real world cases.

7.10   Chapter 5, Exercise 3

Simple development of thermal stratification in containers have been exemplified in Exercises 1–4. This reading exercise aims to update investigations of stratified basins at geophysical scales of earth geography. Only a few key references are provided in the following, as starters to understand and solve real world problems on the evolution of stratified basins.

References

1. Eady E (1949) Long waves and cyclone waves. Tellus 1:33–52
2. Martinez de Azagra A, Del Río J (2015) World map of potential areas for the use of water cooling pitchers (botijos). J Maps 11(2):240–244
3. Zubizarreta JI, Pinto G (1995) An ancient method for cooling water explained by means of mass and heat transfer. Chem Eng Educ 29:96–99

Recommended Video Resources

Laboratory experiments and physical models have been effective means of fluid mechanics investigation, with great value in deeper understanding of fluid behavior and in connecting theory with observations. At present age, however, computational developments have enabled experiments to be carried out through computer models, often replacing former capacity and demand for physical testing facilities. Yet, laboratory demonstration and visualization of fluid mechanics concepts have always had great value in teaching, allowing a better grasp of the meaning and significance of theoretical concepts through visual experience. Therefore fluid mechanics videos and films are valued as great teaching resources helping to expand knowledge on fluid properties and behavior in different realms.

ASCHER SHAPIRO FLUID MECHANICS FILMS

One of the best teaching resources is the *National Committee for Fluid Mechanics Films* series (also known as MIT fluid mechanics films) started by Ascher Shapiro in 1961.

Laboratory demonstrations available in video formats:
http://web.mit.edu/hml/ncfmf.html

also found under *NSF Fluid Mechanics Series*
https://www.youtube.com/playlist?list=PL0EC6527BE871ABA3

The NCFMF Book of Film Notes, National Committee for Fluid Mechanics, MIT Press, 251p. - out of print, contents accessible at:
http://web.mit.edu/hml/notes.html

OTHER FLUID MECHANICS FILMS AND VISUALIZATIONS

Gallery of Fluid Motion
https://gfm.aps.org/

The Colorful Fluid Mixing Gallery
http://www.bakker.org/

© Springer Nature Switzerland AG 2021
E. Özsoy, *Geophysical Fluid Dynamics II*, Springer Textbooks in Earth Sciences, Geography and Environment,
https://doi.org/10.1007/978-3-030-74934-7

efluids resources:
http://www.efluids.com

ITSC Fluids Movie Archive (old)
http://www.maths.manchester.ac.uk/~mheil/MATTHIAS/Fluid-Animations/Fluid-Animations.html

GFD VISUALIZATIONS AND LABORATORY DEMONSTRATIONS

GFDL Data Visualizations
http://www.gfdl.noaa.gov/visualization

GFD Experiments - MIT
http://paoc.mit.edu/labweb/experiments.htm

GFD U. Leeds
http://www1.maths.leeds.ac.uk/~sdg/teaching/gafd/

Physical Oceanography Lab Demos
http://www.phys.ocean.dal.ca/programs/doubdiff/labdemos.html

Physical Oceanography Demo Movies
http://www.po.gso.uri.edu/demos/

ATMOSPHERE-OCEAN CLIMATE

Atmosphere and Ocean in a Laboratory
http://dennou-k.gaia.h.kyoto-u.ac.jp/library/gfd_exp/index.htm

NASA scientific Visualization Studio
https://svs.gsfc.nasa.gov/

Science on a Sphere
http://www.pmel.noaa.gov/co2/story/Science+On+a+Sphere

WHOI in motion
http://www.whoi.edu/VideoGallery/research.html

Ocean Portal - Planet Ocean
http://ocean.si.edu/planet-ocean

Top documentaries
http://topdocumentaryfilms.com/planet-ocean/

Ocean Motion
http://oceanmotion.org/

LEARNING ZONE

UCAR Learning Zone
http://scied.ucar.edu/resources

Michael E McIntyre - louder buzz of bees
http://www.atm.damtp.cam.ac.uk/people/mem/

Fluid Dynamics Lecture Notes of M. E. McIntyre
http://www.atm.damtp.cam.ac.uk/people/mem/FLUIDS-IB/

iFluids - MIT Program for Fluid Mechanics
http://web.mit.edu/fluids-modules/www/

Harvard Colloquia on Fluid Mechanics
https://video.seas.harvard.edu/category/Colloquia%3Efluid+mechanics/13151451

Recommended Text Books

1. Abarbanel DI, Young WR (1987) General circulation of the ocean. Springer
2. Aubrey DG, Friedrichs CT (eds) (1996) Buoyancy effects on coastal and estuarine dynamics, Coastal and Estuarine Studies. American Geophysical Union
3. Badin G, Crisciani F (2018) Variational formulation of fluid and geophysical fluid dynamics: mechanics. springer, symmetries and conservation laws
4. Bardos C, Fursikov A (2008) Instability models connected with fluid flows I. Springer
5. Bardos C, Fursikov A (2008) Instability models connected with fluid flows II. Springer
6. Batchelor GK (1970) An introduction to fluid dynamics. Cambridge University Press, Cambridge
7. Bühler O (2009) Waves and mean flows. Cambridge University Press, Cambridge
8. Cavallini F, Crisciani F (2013) Quasi-geostrophic theory of oceans and atmosphere: topics in the dynamics and thermodynamics of the fluid Earth. Springer
9. Chandrasekhar S (1961) Hydrodynamic and hydromagnetic stability. Oxford University Press, Oxford
10. Chemin J-Y, Gallagher I, Grenier E (2006) Mathematical geophysics: an introduction to rotating fluids and the Navier-Stokes equations. Clarendon Press, Oxford
11. Childress S (2009) An introduction to theoretical fluid mechanics, courant institute of mathematical sciences
12. Clercx HJH, Van Heijst GF (2018) Mixing and dispersion in flows dominated by rotation and buoyancy. Springer
13. Cullen M (2006) A mathematical theory of large-scale atmosphere / ocean flow. Imperial College Press
14. Cushman-Roisin B (1994) Introduction to geophysical fluid dynamics. Prentice-Hall
15. Cushman-Roisin B, Beckers J-M (2009) Introduction to geophysical fluid dynamics: physical and numerical aspects. Academic Press
16. Csanady GT (1982) Circulation in the coastal ocean. Springer
17. Dijkstra HA (2008) Nonlinear physical oceanography: a dynamical systems approach to the large scale ocean circulation and El Niño. Springer
18. Dijkstra HA (2008) Dynamical oceanography. Springer
19. Durran D (1999) Numerical methods for wave equations in geophysical fluid dynamics. Springer
20. Egbers C, Pfister G (eds) (2000) Physics of rotating fluids. Springer

© Springer Nature Switzerland AG 2021

E. Özsoy, *Geophysical Fluid Dynamics II*, Springer Textbooks in Earth Sciences, Geography and Environment,

https://doi.org/10.1007/978-3-030-74934-7

21. Friedlander S (1980) An introduction to the mathematical theory of geophysical fluid dynamics. North-Holland
22. Gertenbach JD (2001) Workbook on aspects of dynamical meteorology: a self discovery mathematical journey for inquisitive minds, 1st edn. Author's copyright
23. Ghil M, Childress S (1987) Topics in geophysical fluid dynamics: atmospheric dynamics, dynamo theory, and climate dynamics. Springer
24. Gill AE (1982) Atmosphere-ocean dynamics. Academic Press
25. Greenspan HP (1980) The theory of rotating fluids. Cambridge University Press, Cambridge
26. Grimshaw R (2003) Environmental stratified flows. Kluwer Academic
27. Holton JR (2004) An introduction to dynamic meteorology. Academic Press
28. Hoskins BJ, James IN (2014) Fluid dynamics of the midlatitude atmosphere. Blackwell, Wiley -
29. Huang RX (2010) Ocean circulation: wind-driven and thermohaline processes. Cambridge University Press, Cambridge
30. Hutter K, Wang Y (2016) Fluid and thermodynamics volume 1: basic fluid mechanics. Springer
31. Hutter K, Wang Y (2016) Fluid and thermodynamics volume 2: advanced fluid mechanics and thermodynamic fundamentals. Springer
32. Hutter K (2012) Nonlinear internal waves in lakes. Springer
33. Ibragimov NK, Ibragimov RN (2011) Applications of lie group analysis in geophysical fluid dynamics. World Scientific
34. Kantha LH, Clayson CA (2000) Small scale processes in geophysical fluid flows. Academic Press
35. Klinger B, Haine TWN (2019) Ocean circulation in three dimensions. Cambridge University Press, Cambridge
36. Landau LD, Lifshitz EM (1959) Fluid mechanics. Pergamon Press
37. LeBlond PH, Mysak LA (1981) Waves in the ocean. Elsevier
38. Lin Y-L (2007) Mesoscale dynamics. Cambridge University Press, Cambridge
39. Loper DE (2017) Geophysical waves and flows. Cambridge University Press, Cambridge
40. Majda A, Introduction to PDEs and waves for the atmosphere and ocean, courant institute of mathematical sciences
41. Majda AJ, Wang X (2006) Non-linear dynamics and statistical theories for basic geophysical flows. Cambridge University Press, Cambridge
42. Malanotte-Rizzoli P, Robinson AR (1994) Ocean processes in climate dynamics: global and mediterranean examples. Kluwer
43. Marshall J, Plumb RA (2008) atmosphere, ocean, and climate dynamics: an introductory text. Elsevier Academic Press
44. Massel SR (2015) Internal gravity waves in shallow seas. Springer
45. McWilliams JC (2010) Fundamentals of geophysical fluid dynamics. University of California
46. Monin AS (1990) Theoretical geophysical fluid dynamics. Kluwer Academic Publishers
47. Mooers CNK (1986) Baroclinic processes on continental shelves. American Geophysical Union
48. Morozov EG (2018) Oceanic internal tides: observations, analysis and modeling. Springer
49. Müller P, von Storch H (2004) Computer modelling in atmospheric and oceanic sciences: building knowledge. Springer
50. Müller P (2006) The equations of oceanic motions. Cambridge University Press, Cambridge
51. Olbers D, Willebrand J, Eden C (2012) Ocean dynamics. Springer
52. Paldor N (2015) Shallow water waves on the rotating earth. Springer
53. Palmén E, Newton CW (1969) Atmospheric circulation systems. Academic Press
54. Panchev S (1985) Dynamic meteorology. Springer
55. Pedlosky J (1987) Geophysical fluid dynamics. Springer
56. Pedlosky J (1998) Ocean circulation theory. Springer
57. Pedlosky J (2003) Waves in the ocean and atmosphere, introduction to wave dynamics. Springer
58. Phillips OM (1977) The dynamics of the upper ocean. Cambridge University Press, Cambridge
59. Pratt LJ, Whitehead JA (2007) Rotating hydraulics: nonlinear topographic effects in the ocean and atmosphere. Springer

60. Radko T (2013) Double diffusive convection. Cambridge University Press, Cambridge
61. Randall DA (2005) General circulation of the atmosphere. Colorado State University
62. Rajaratnam N (1976) Turbulent jets. Elsevier
63. Reid WH (1971) Mathematical problems in the geophysical sciences: geophysical fluid dynamics. American Mathematical Society
64. Salmon R (1998) Lectures on geophysical fluid dynamics. Oxford University Press, Oxford
65. Samelson RM, Wiggins S (2006) Lagrangian transport in geophysical jets and waves: the dynamical systems approach
66. Samelson RM (2011) The theory of large scale ocean circulation. Cambridge University Press, Cambridge
67. Sarkisyan AS, Sündermann JE (2009) Modelling ocean climate variability. Springer
68. Schlichting H (1968) Boundary layer theory. Mc Graw-Hill
69. Schlichting H, Gersten K (2017) Boundary layer theory, 8th edn. Springer
70. Shames IH (1962) Mechanics of fluids. Mc Graw-Hill
71. Shevchuk IV (2016) Modelling of convective heat and mass transfer in rotating flows. Springer
72. Sokolovskiy MA, Verron J (2014) Dynamics of vortex structures in a stratified rotating fluid. Springer
73. Stern ME (1975) Ocean circulation physics. Academic Press, New York
74. Stocker T, Hutter K (1987) Topographic waves in channels and lakes on the f-plane. Springer
75. Sutherland BR (2010) Internal gravity waves. Cambridge University Press, Cambridge
76. Turner JS (1973) Buoyancy effects in fluids. Cambridge University Press, Cambridge
77. Vallis GK (2017) Atmospheric and oceanic fluid dynamics: fundamentals of large-scale circulation. Cambridge University Press, Cambridge
78. Vanyo JP (1993) Rotating fluids in engineering and science. Butterworth-Heinemann
79. Velasco Fuentes OU, Sheinbaum J, Ochoa J (eds) (2003) Nonlinear processes in geophysical fluid dynamics: a tribute to the scientific work of Pedro Ripa. Kluwer Academic
80. Von Schwind JJ (1981) Geophysical fluid dynamics for oceanographers. Prentice-Hall
81. Vlasenko V, Stashchuk N, Hutter K (2005) Baroclinic tides, theoretical modeling and observational evidence. Cambridge University Press, Cambridge
82. White FM (1974) Viscous fluid flow. Mc Graw-Hill
83. Whiteman CD (2000) Mountain meteorology fundamentals and applications. Oxford University Press, Oxford
84. Yang H (1991) Wave packets and their bifurcations in geophysical fluid dynamics. Springer
85. Yih C-S (1965) Dynamics of non-homogeneous fluids. McMillan
86. Yih C-S (1969) Fluid mechanics. Mc Graw-Hill
87. Zeitlin V (2018) Geophysical fluid dynamics: understanding (almost) everything with rotating shallow water models. Oxford University Press, Oxford
88. Munk WH (1966) Abyssal recipes. Deep-Sea Res 13:707–730
89. Woods AW (1991) Boundary-driven mixing. J Fluid Mech 226:625–654
90. Garrett C (1991) Marginal mixing theories. Atmosphere-Ocean 29(2):313–339
91. Garrett C, MacCready P, Rhines P (1993) Boundary mixing and arrested Ekman layers: Rotating stratified flow near a sloping boundary. Ann Rev Fluid Mech 25:291–323
92. Munk W, Wunsch C (1998) Abyssal recipes II: energetics of tidal and wind mixing. Deep-Sea Res 45:1977–2010
93. Garrett C, Laurent LSt, (2002) Aspects of deep ocean mixing. J Oceanography 58:11–24
94. Müller P, Garrett C (2004) Near-boundary processes and their parameterization. Oceanography 17(1):107–117
95. Miller MD, Yang X, Tziperman E (2020) Reconciling the observed mid-depth exponential ocean stratification with weak interior mixing and Southern Ocean dynamics via boundary-intensified mixing. Eur Phys J Plus 135:375

nted in the United States
Baker & Taylor Publisher Services

Pr
by